W9-CBT-973

RIGHT HAND, LEFT HAND

Library & Media Ctr.
Carroll Community College
1601 Washington Rd.
Westminster, MD 21157

WITHDRAWN

RIGHT HAND, LEFT HAND
The Origins of Asymmetry in Brains, Bodies,
Atoms and Cultures

Chris McManus

Harvard University Press
Cambridge, Massachusetts
2002

First published in Great Britain in 2002 by
Weidenfeld & Nicolson, Ltd.

© 2002 by Chris McManus

All rights reserved. No part of this publication may be
reproduced, stored in a retrieval system, or transmitted
in any form or by any means, electronic, mechanical,
photocopying, recording, or otherwise, without the prior
permission of the copyright owner.

The moral right of Chris McManus to be identified as
the author of this work has been asserted in accordance
with the Copyright, Designs and Patents Act of 1988.

Typeset by Selwood Systems, Midsomer Norton

Printed in the United States of America

Library of Congress Cataloging-in-Publication Data

McManus, I. C.
Right hand, left hand : the origins of asymmetry in brains,
bodies, atoms, and cultures / Chris McManus.
p. cm.
Includes bibliographical references and index.
ISBN 0-674-00953-3 (alk. paper)
1. Laterally. 2. Cerebral dominance. 3. Left- and right- handedness. I. Title

QP385.5 .M38 2002
152.3'35—dc21   2002017275

For Christine, Franziska and Anna

CONTENTS

FIGURES

TABLES

PREFACE

I can hardly count the number of times when, browsing around in second-hand bookshops, my eye has been caught by a book entitled *Left Hand, Right Hand!* Disappointment, however, always followed. Instead of what I had hoped to find, a book about why our right and left hands are so different, I had actually found the first volume of the autobiography of the English writer, Sir Osbert Sitwell. It said little about what I was looking for. With my present title, I hope I may satisfy those few others who have perhaps also looked for that seemingly elusive book about our two hands and sides – a book that explores why the world, indeed the universe, so ubiquitously but so seemingly irrationally seems full of asymmetries.

Sitwell called his autobiography *Left Hand, Right Hand!* 'because, according to the palmists, the lines of the left hand are incised inalterably at birth, while those of the right hand are modified by our actions and environment, and the life we lead'. It is an intriguing statement, with its mixture of a very modern emphasis on the twin roles of genes and environment in shaping our lives, and the reference to palmistry, which most scientists would immediately reject, and which has components of the universal symbolisms attached to left and right. Sitwell, himself, is certainly not so gullible as to see the idea as anything other than a loose metaphor, and although he acknowledges that 'I believe all men, including myself, to be superstitious', he does reject 'the childish boundaries of chiromancy'.

From our present perspective, we do not need seriously to ask whether such a claim has any empirical truth in it, but it can nevertheless act as a starting point for asking how our hands actually do differ. We can also see it as part of that vast repertoire of symbolism associated with right and left, which permeates so much of our daily lives, from the political left and right wing, through dexterity and 'cack-handedness', to 'left-handed marriages'. If we do have such symbolisms, *why* do we have them? Is it perhaps because our hands are so asymmetrical, or because our hearts are asymmetric, or even because the universe in which we live is asymmetric? These are wonderfully deep questions, which will take us into many and varied aspects of social life and anthropology, and into the workings of the brain and the subtleties of human language. They will require us to look at our asymmetric bodies, with their asymmetric heart, liver, stomach, kidneys and even testicles; down into the amino acids and sugars, the building blocks of our body, which are also asymmetric; and, finally, into the asymmetries of physics itself, and hence of the universe.

In his introduction to *Left Hand, Right Hand!* Sitwell muses that, 'Already I am nearing fifty and the grey hairs are beginning to show, I have reached the watershed and can see the stream which I must follow downhill towards the limitless ocean, cool and featureless. It is time to begin.' My own grey hairs have long been apparent, and I have been studying and researching, some would say obsessing, about handedness and lateralisation since my first paper on the subject in 1972. The interest has never gone away – the questions, in fact, becoming ever more interesting – and the interest of others has continued similarly to grow. Half a dozen times a year, I am phoned by radio and TV producers who have suddenly noticed that ten per cent of their audience is left-handed and who therefore thought it would be an exceedingly novel idea to do a programme on left-handedness. When I do those interviews, the same old questions are asked and much interest is shown, but there is little time to provide proper answers. I hope here that I have provided broader answers for those audiences, who will realise that the field is far too interesting to be crammed into ten minutes. In recent years, there have been all sorts of advances, in everything from molecular asymmetries, through anatomy and developmental biology, to neuroscience, psychology, anthropology, sociology, and even cosmology. It is time, therefore, to take stock. Even if I am uncomfortably aware that I am not an expert in even a fraction of the requisite areas, I can claim to be an enthusiast.

If I have perhaps climbed high enough into the mountains to see over many terrains, I hope I have still not yet reached that col, that saddle point, that watershed, which Sitwell described from where one can only gently descend to the distant ocean. The terrain to be covered by this book is varied; sometimes rocky and thorny, and occasionally sufficiently dense to make impossible any distant perspective. A brief map of the territories to be travelled may therefore be helpful. Chapter 1 begins with the nineteenth-century case history of a patient, John Reid, who had his heart on the right side of his body. As was realised by Dr Thomas Watson

– a now somewhat neglected figure whose reputation deserves rehabilitation – explaining why the heart is usually found on the left is extremely difficult. Watson realised that it is even more difficult to account for why John Reid should also have been right-handed rather than left-handed. Two other great nineteenth-century discoveries – Pasteur's demonstration that the molecules in living organisms are asymmetric, and Dax and Broca's finding that language is usually in the left half of the brain – set the scene for the rest of the book. Chapter 2 stands back and surveys the near-universal human interest in right and left, and the left–right symbolism that seems to be found in all cultures, underpinning many of the phenomena encountered in this book. Chapter 3 looks at Kant's problem concerning the philosophical difficulties of describing left and right, and chapter 4 considers how the words 'left' and 'right' have evolved in various languages, exploring, in particular, why so many people have problems in using these terms consistently and without confusion. Chapter 5 returns to Watson's problem of why the heart should be on the left, setting this in the broader biological perspective of why bodies are usually symmetric and how the gross asymmetry of the heart could have evolved. Chapter 6 digs more deeply into the sub-microscopic details of biology, looking at the asymmetry of molecules, in particular the amino acids, and considers how such asymmetries could have arisen either from sub-atomic physics or in the cold wastes of interstellar space. Chapter 7 returns to a more mundane level, looking at right- and left-handedness in their everyday sense, and describing a genetic model that accounts for how handedness is inherited and runs in families. Chapter 8 looks at the asymmetry of the brain, considering both the typical language-related processes that occur in the left hemisphere, and the more holistic functions of the right hemisphere, as well as the ways in which the two sides work together to carry out complex psychological processes. Chapter 9 concludes the section on handedness by asking about historical and cross-cultural differences in the rate of right- and left-handedness. It looks at archaeological evidence for right-handedness in early hominids, and its absence in apes and other animals, and then asks what the underlying difference is between the left and right hemispheres of the brain. Chapter 10 is about how social interactions are important in determining such lateralised behaviour as the direction of writing (from left to right in the case of English), or driving on a particular side (such as the left in Britain), or the relative advantage of left-handers in the ritualised fighting of sport. Chapter 11 considers the social processes that influence what it is like to be in the left-handed minority of a society in which most people are right-handed, and the effects on language, perception and stigmatisation. Chapters 12 and 13 are diversions that nonetheless have their deeper aspects; those in chapter 12 looking at some of the many errors that haunt much thinking about lateralisation and the many fictions that have resulted from these concerning left and right, whereas chapter 13 is unashamedly lightweight: a collection of handedness trifles, trivia and miscellanea, some fun, some amusing, but hopefully all instructive in their way. Chapter 14 returns to the more solid substance of the book but instead of looking at asymmetry it considers the thus far neglected concept that has lurked behind so much discussion of asymmetry, and that is so central to scientific theorising – symmetry. In a finale to the book, chapter 15 argues for the triumph of asymmetry over symmetry and proposes a single picture of asymmetries, from the sub-atomic, through the biochemical and the anatomical, to the neurological, the cultural and the social.

Hypernotes

Since I am an academic, I wrote footnotes and endnotes in profusion while preparing this book, not least so that I could defend and justify some tricky points to my ever critical colleagues. I did, though, remember the words of Toby Mundy suggesting that popular science books should be 'scholarly but not academic'. The notes allowed me to be an academic and also to wander off into some arcane backwaters. In the interests of space and the reader's patience, however, many of these have been stripped from the book, much as one takes down the scaffolding after building a house. Some of those deleted may be of interest to the occasional reader. In the notes at the end of the book I have therefore included the symbol ☞www☜ to indicate a more extensive note, a *hypernote*, which can be accessed on the Internet. The site to look for is www.righthandlefthand.com. This website also has additional material, and offers the opportunity to take part in experiments and research studies.

ACKNOWLEDGEMENTS

Many people have helped at various points in the writing of this book, and my thanks go to them all. None is responsible for the inevitable errors that are still present, in particular the inevitably embarrassing left–right confusions and reversals, but all have helped reduce their number. My first thanks must go to the Wellcome Trust, for awarding me the prize that stimulated the writing of the book, and in particular those who made that decision, including Sue Blackmore, Matt Ridley and the late lamented and left-handed Douglas Adams. At the Trust, I was fortunate in having much support from Laurence Smaje and Sarah Bronsdon. My special thanks go to my agent, Felicity Bryan, and to my editors at Weidenfeld & Nicolson; firstly Toby Mundy, for providing perspective and support, and then Peter Tallack, whose advice, comments and detailed criticisms were invaluable. I must also thank those at my publishers, and in particular Nicky Jeanes and Tom Wharton, who have helped to meet a tight production schedule. Writing the book would not have been possible without much support from University College London, particularly from Oliver Braddick and David Ingram, who provided the time and space I needed, and from the superb Library and its ever helpful staff.

This book has, in some sense, been in preparation for nearly three decades, and I must thank Nick Humphrey and Michael Morgan for fostering my first interests, and for their continued interest, enthusiasm and friendship ever since. For a decade and more, I was privileged to visit the University of Waterloo and work with the late Phil Bryden and the talented team of researchers that assembled around him, including Taha Amir, Arve Asbjornsen, Russ Boucher, Pam Bryden, Barbara Bulman-Fleming, Lorin Elias, Gina Grimshaw, Yukihide Ida, Manas Mandal, Todd Mondor, Maharaj Singh, Runa Steenhuis, and Dan Voyer. Whenever my interest in lateralisation waned, they reinvigorated me. If I have one great regret, it is that Phil and I could not sit and chat about the book.

Among those who have provided various forms of help, and particularly discussion and reference to obscure corners of the literature, I must thank Rosalind Arden, Peter Ayton, Oliver Braddick, Nigel Brown, Diane Cheung, John Cronin, Jules Davidoff, Sergio Della Sala, Finn Fordham, Stephen Gangestad, Peter Halligan, Lauren Harris, Jörg Hennig, Peter Hepper, Ben Heydecker, Liz Hornby, Kenneth Hugdahl, Bob Jacobs, Steve Jones, Stephen Lea, Richard Lee, Jim MacIntyre, Mark McCourt, John Marshall, Geoffrey Miller, Hannah Mitchison, Michael Morgan, R. Nagarajan, Mike Nicholls, Sonja Ofte, Richard Palmer, Michael Peters, Sandra Pizzarello, Andrew Pomiankowski, Marvin Powell, Nigel Sadler, Howard Taylor, Keith Tipton, Don Tucker, Luca Turin, Steve Upham, Giorgio Vallortigara, Sir Andrew Watson, Stephen Wilson, Chuck Wysocki and Ron Yeo. John Hewson deserves particular mention for setting up the book's website. I especially thank Christine Pleines, Jonathan Cooke, Dick Jefferies, Belinda Winder and Lewis Wolpert for their careful reading of parts of the manuscript, and Jonathan Cooke, in particular, for allowing me to marvel with my own eyes at the developing chick heart. Jo Parker and the staff of the Vestry House Museum were extremely helpful in allowing me to revisit their exhibition on left-handedness, and I especially thank Nigel Sadler, curator of the exhibition, for allowing me access to his unpublished notes, and to use his survey data from the exhibition.

There is hardly a book written in which the acknowledgements do not end with heartfelt thanks to the author's family for putting up with the ever more demanding cuckoo that has been spirited into their nest. I can now see why. Christine, Franziska and Anna deserve much more than thanks, and it is to them that this book is justly dedicated.

DR WATSON'S PROBLEM

John Reid was forty-eight in October 1835, when he died in the Middlesex Hospital in the centre of London. We know little about John Reid in life, except that he appears to have had no problems with his chest, no-one having bothered to examine it very thoroughly; which, as it happened, was unfortunate. Reid had been looked after by a dynamic young physician, Dr Thomas Watson, who wanted to know why his patient had died. A post-mortem was carried out, probably by Watson himself. No-one expected what was found – John Reid's heart was on the wrong side. Unlike the vast majority of people, who have their heart on the left, John Reid had his heart on the right side. Indeed, everything else was back to front as well. His liver was on the left rather than the right, and his stomach and spleen were on the right rather than the left. As Watson put it, the organs 'exacted the appearance which the same viscera...would present if seen reflected [in] a plane mirror' (see Figure 1.1). Watson described the condition then as heterotaxy, but the more usual name in modern scientific literature is *situs inversus*, meaning an inverted site or location for each of the organs.[1]

Dr Watson (Figure 1.2) had a keen, enquiring mind, and an obvious desire to reach the top of the medical profession, which he not only achieved, but managed while still remaining beloved of his peers; a difficult task in a competitive, self-critical profession like medicine. Born in 1792, he only started studying medicine at the age of twenty-seven, previously studying mathematics at Cambridge where he had been tenth wrangler. By 1835, he was a Censor of the College of Physicians. In 1842, his wife died of puerperal sepsis ('childbed fever'), and that same year he was the first to suggest that infection might be prevented by disinfecting the hands or by using what we now call disposable surgical gloves. In 1859, he was appointed Physician to the Queen, and in 1861 was one of the three doctors looking after Prince Albert when the prince died of typhoid fever.

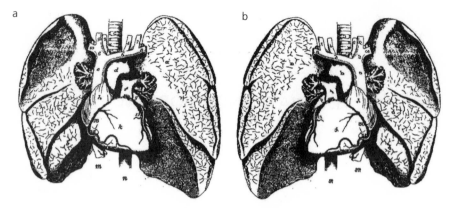

Figure 1.1 a, The normal layout of the organs of the chest (*situs solitus*) with the heart pointing towards the left, the arch of the aorta leaving the heart and looping over towards the left-hand side, the right lung having three lobes, and the left lung having two lobes. **b,** The layout of the organs of the chest in *situs inversus*, in which the pattern is a mirror-image of the case in *situs solitus* (in this case literally, the original engraving having been reversed or, as Watson put it, 'reflected in a plane mirror').

In 1862 he was elected President of the College of Physicians, a post he held for five years, and in 1866 he was created a baronet, having his portrait painted by George Richmond, the eminent society painter. Watson's entry in *Munk's Roll*, the historical record of the Fellows of the College of Physicians, describes him as 'the Nestor of English Physicians', a reference to the King of Pylos in Ancient Greece, who was the oldest and wisest of the chieftains who had gone to the siege of Troy. T. H. Huxley described him simply as 'the very type of a philosophical physician'.[2]

On 30 May 1836, Watson presented his findings on John Reid to an evening meeting of the College of Physicians. Sitting on the table was Reid's heart, which had been preserved in a bottle and was kept in the pathological museum of King's College, London. Later that year, Watson wrote a long account of Reid, along with a description of several dozen other cases, in an article for the *London Medical Gazette*. It was a *tour de force*, citing cases published in English, French and Latin, including a doggerel verse in French (written by Leibniz, the co-inventor with Newton of the calculus), and a quotation in Greek from Galen, the Roman physician who was second only to Hippocrates in the ancient medical world, and deemed an authority for a millennium and a half. Watson had canvassed opinions from several other eminent doctors in London, and had received a detailed history of a similar case from Sir Astley Cooper, one of the most distinguished and intellectual of London surgeons. Watson quoted verbatim from Cooper's description of this case, which concerned a woman called Susan Wright.[3]

We know somewhat more about Susan Wright than about John Reid. She was seventy-three when, on 19 March 1836, ten weeks before the meeting at the College of Physicians, she had died of an attack of diarrhoea. Her health previously had been generally good, and she was well known in the hospital, acting as an attendant on the wards, where it was commented on how adroit she was at dressing the patients' wounds and ulcers. We are told of her kindness and humanity, and that 'her habits were extremely temperate, although she had been on a few occasions observed in a state of inebriation'. She never married, 'and the appearances after death shewed that she had died a maiden, as the hymen still remained…'. The post-mortem was carried out by a Mr Braine, a surgeon of St James's Square, who opened the abdomen and found, to his surprise, that the intestines were completely reversed. The chest was then opened, found also to be completely reversed, and word was sent to Cooper, who examined the body the next day. As he then puts it, 'The viscera were removed with care…and sent to my house, where, having injected them, I had them dried and preserved.' These specimens were also on the table at the meeting of the College of Physicians.

In his article in the *London Medical Gazette*, Watson reflects on how one should describe these strange anatomical findings. He considers 'malformation' and 'monstrosity', and rejects each – there is no malformation, since there is nothing faulty or malformed about these organs, each being merely back to front; and likewise he rejects monstrosity since there is 'nothing constructed amiss, nothing wanting, nothing superabundant; therefore nothing monstrous. Every part and every organ is as perfect as in the ordinary fashion.' There is the rub, as Watson recognised. There is no obvious reason why the 'ordinary fashion' should be precisely that, ordinary, whereas the mirror image is extremely rare. Watson could see no difference in function between the two types: 'We are not sensible of any advantage, or comfort, from having the heart on the sinister and the liver on the dexter side; and therefore we cannot expect that any embarrassment or uneasiness should arise when that arrangement happens to be reversed.'

Watson struggled to make sense of the predilection of one side over the other when it would seem either way around would do equally well. He was aware that there is a biological dimension, and talked of how in most molluscs the spiral of the shell twists from left to right, while in a few individuals, or even whole species in some cases, the shells curl in the opposite direction. He played briefly with the idea that reversed viscera do not occur in animals, but was told by a student from King's College that Galen mentions it, and by a butcher that the phenomenon is sometimes found in sheep. This made him wonder if *situs inversus* in humans

a

b

Figure 1.2 Sir Thomas Watson **a,** An etching of 1854, by F. Hall after a drawing by Richmond. **b,** A photograph in 1867.

is perhaps actually not as rare as it seems, but merely difficult to diagnose in life.

To a modern medical eye the surprising thing is that most of these cases were not diagnosed in life, but only on opening the body at postmortem. However, we must remember that the French physician Laënnec had invented the stethoscope only sixteen years earlier, in 1819, and that its use was still difficult, leaving room for error. We know that Watson himself certainly didn't like using the stethoscope, describing it many years later as 'more of a hindrance than a help', and saying that although 'he could not do without it…he did without it as much as he could'. Watson was not a Dr Lydgate, the thrusting young physician in George Eliot's novel *Middlemarch*, who believed fervently in science in medical practice, and who, in 1829, just a decade after Laënnec's discovery, when he examined the ageing Dr Casaubon, 'not only used his stethoscope (which had still not become a matter of course in practice at that time), but sat quietly and watched him'. Even so, Watson surely had no need of a stethoscope to make the diagnosis. One of the first things that all medical students learn is to find by touch the 'apex beat', the thrusting tip of the heart, usually located at the level of the fifth rib, on the left side, on the vertical line running down through the middle of the collar bone, the clavicle. It is not hard to find, and I have certainly seen cases of *situs inversus* where even from the end of the bed the apex beat is visibly palpitating on the right-hand side. At the end of his paper, Watson had to admit that, even in life, 'by carefully repeated examinations, no reasonable doubt could remain'. Perhaps, like Eliot's Lydgate, Watson should just have sat quietly and watched John Reid.[4]

If the mere existence of these unusual individuals with their mirror-image anatomy was a problem for Watson, there was another more troublesome issue lurking, one to which Watson devoted a large amount of his discussion. Even if John Reid and Susan Wright were complete mirror images of the normal situation, that would be difficult to explain. Harder still was the fact that not everything had been reversed in these individuals. Humans have several asymmetries, including that of the hands, most people being right-handed. *And John Reid and Susan Wright were right-handed.* Astley Cooper is specific in his description of Susan Wright that 'all who had observed her agreed, that she gave the preference to the use of the right hand'. These two cases cannot be written off as statistical flukes, since Watson reviews all the cases of *situs inversus* that he could find in the medical literature, and, while there is no evidence that any were left-handed, several others were indubitably right-handed.[5]

Watson was justifiably surprised that John Reid and Susan Wright were right-handed. Most people assume that individuals with *situs inversus* will be left-handed. The novelist H. G. Wells was no exception. In his science fiction tale, *The Plattner Story*, Gottfried Plattner manages to get blown up by a mysterious green powder, and disappears. On his return,

> the right lobe of his liver is on the left side, the left on his right; while his lungs, too, are similarly contraposed. What is still more singular, unless Gottfried is a consummate actor we must believe that his right hand has become his left … Since this occurrence he has found the utmost difficulty in writing except from right to left across the paper with his left hand.

Watson's counter-intuitive conclusion, that individuals with *situs inversus* are usually right-handed, has been amply confirmed by modern studies. In the most comprehensive of these, published in 1950 as a spin-off from mass X-ray screening examining for tuberculosis, Johan Torgersen, a Norwegian physician, described how 122 people with *situs inversus* had been found amongst 998,862 people examined using miniature radiographs, about one in ten thousand of the population. Torgersen also knew about another seventy cases. Although he did not know the handedness in all cases, of the 160 in which he did, only eleven (6.9 per cent) were left-handed, a proportion similar to that found in people with their heart on the normal side, the left.[6]

Perhaps the fact that John Reid and Susan Wright were right-handed was not so much of a problem as it seemed at first sight. After all, right-handedness is only a behaviour, and many behaviours are learned. We would hardly be surprised to find that John Reid and Susan Wright wrote from left to right, since that was what everyone else in Britain had learned to do. Maybe handedness is also a cultural convention, most people writing and carrying out other skilled tasks with the right hand because of what Watson called 'the mere force of custom or prejudice'. Watson tried to argue such a case, but he eventually dismissed it, coming down on the side 'that the dextral tendency is implanted by nature' (or, to put it in more modern terms, it is biological, and hence probably genetic in origin). He does though hedge his bets a little, adding the caveat that the tendency 'cannot be very strong, since, in spite of the aid of training, it is so often overcome by slight external influences' – and here he is probably on weaker ground, overcoming innate handedness being surprisingly difficult.

Watson's basis for concluding that right-handedness was biological in origin lay in cross-cultural studies. If right-handedness were indeed an arbitrary cultural convention in an otherwise symmetric world, then half of all societies should be right-handed and half left-handed. Watson's description of the evidence is clear and forceful:

> The employment of the right hand in preference to the left hand is universal throughout all nations and countries. I believe no people or tribe of left-handed persons has ever been known to exist... Among the isolated tribes of North America which have the most recently become known to the civilised world, no exception to the general rule has been met with. Captain Back has informed me that the wandering families of the Esquimaux, whom he encountered in his several expeditions towards the North Pole, all threw their spears with the right hand, and grasped their bows with the left.[7]

There is no modern evidence to refute Watson, and plenty to support it.

But Watson now had himself on the horns of a dilemma; one that he was completely unable to resolve and, indeed, that continues to be one of the most fascinating questions for modern research. If right-handedness is biological in origin, then how can people with *situs inversus*, with everything else in their bodies reversed, not also have their handedness reversed?

The mystery, though, was to deepen, becoming, in the words used by Sir Winston Churchill to describe the Soviet Union, 'a riddle wrapped in a mystery inside an enigma'. The riddle and the enigma of lateralisation were still to come. The riddle involved simple molecules, and the enigma

Figure 1.3 Louis Pasteur in 1852, four years after he had described the two types of tartaric acid crystal.

was neurological, concerning that ultimate cultural achievement, language. Both were discovered in France, within a decade or two of Watson's presentation at the College of Physicians; the first by Louis Pasteur and the second by Dr Marc Dax. We will look at them only briefly here, but return to them in greater detail in later chapters.

Louis Pasteur (Figure 1.3) was perhaps the greatest of French scientists. On 22 May 1848, at the age of twenty-five, he presented a remarkable paper to the Paris Academy of Sciences. While looking at the mechanism by which wine goes sour, he had compared the natural tartaric acid produced from grapes with racemic acid, which is produced industrially. He found that, although chemically identical to tartaric acid, racemic acid differs in its effect on polarised light. When dissolved in water, the natural tartaric acid rotates the polarised light clockwise, whereas the synthetic racemic acid has no effect at all on the polarised light. Looking down the microscope, Pasteur saw that the natural tartaric acid consisted entirely of one type of crystal, whereas the tiny crystals of racemic acid were a mixture of two types, one the mirror-image of the other. He is said to have shouted, *'Tout est trouvé!'*, perhaps best translated as 'Eureka!' Pasteur then conducted a tedious but crucial experiment, using a microscope and dissecting needle to sort the crystals of racemic acid into their two types (see Figure 1.4). A solution of one type rotated polarised light clockwise, just like natural tartaric acid, whereas the other rotated it anticlockwise. A mixture of the two had no effect on light, each balancing out the other, as in racemic acid. An even more dramatic result was to come ten years later. Micro-organisms, Pasteur discovered, could survive and breed on the racemic acid which turned light clockwise, but could not metabolise the racemic acid that turned the light anticlockwise.[8]

a

b

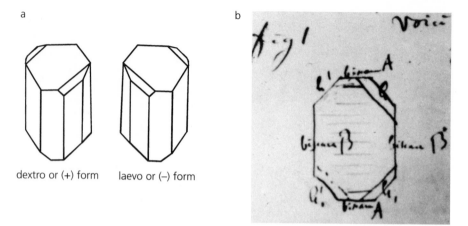

dextro or (+) form   laevo or (−) form

Figure 1.4 a, The crystals of racemic acid that Pasteur would have seen under his microscope. Natural tartaric acid from wine showed only the dextro or (+) form (left-hand figure). **b,** Pasteur's own drawing of one of the two types of crystal.

Pasteur's finding revolutionised biochemistry. Very many of the molecules that comprise the body occur in the laboratory in two forms, one of which is the mirror-image (the stereo-isomer) of the other. But although two forms *can* occur, in the human body only one form of each molecule occurs. For sugars it is what we now call the D- (or dextral) form (from the Latin for right), and for amino-acids it is the L- (or laevo) form (from the Latin for left), according to whether they rotate polarised light to the right or left. This total dominance of one type over the other is not unique to humans, but applies to almost every living organism found on our planet (with occasional intriguing exceptions that we will come back to in chapter 6).

What would have been the implications for Watson twelve years earlier if he had known about this? They would have been profound. Watson had assumed that the basic building blocks, the bricks from which the body is built, are symmetric, for there was no reason to believe otherwise. However, when a house, a machine, a body or whatever is built out of asymmetric bricks, then building a mirror-image is very much more difficult. Look at Figure 1.5. This is a simple spiral staircase built from stone steps that are symmetric. If the stones are merely turned over, then the staircase will turn in the opposite direction. One set of components will build either a right- or left-hand staircase. Now look at the staircase in Figure 1.6. Each stone step is asymmetric. When these steps are stacked on one another they make a right-handed staircase. Imagine, though, trying to use these same stones to build a left-handed staircase. It is simply not possible. The asymmetry of the components determines the asymmetry of the structures built with them.[9]

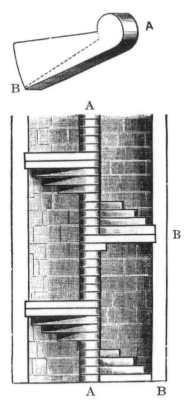

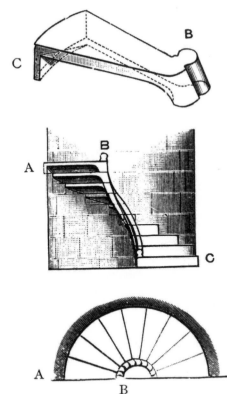

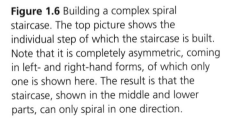

Figure 1.5 Building a simple spiral staircase. The top picture shows an individual step of which the staircase is built. It is symmetric, so that a staircase built from it (lower picture) can spiral in one direction (shown here), or in the opposite direction.

Figure 1.6 Building a complex spiral staircase. The top picture shows the individual step of which the staircase is built. Note that it is completely asymmetric, coming in left- and right-hand forms, of which only one is shown here. The result is that the staircase, shown in the middle and lower parts, can only spiral in one direction.

If the body is composed entirely of one set of stereo-isomeric molecules, does that explain why we almost all have our heart on the left? Maybe. However, if so, the greater difficulty remains in explaining how it is possible to build a body *at all* with the heart on the right, and also in determining whether we can any longer regard individuals with *situs inversus* as being mere mirror-images; individuals who just happen to be one way round rather than another. It is as if we had tried to print something to be looked at in a mirror but had used only normal letters such as 'a', 'b', 'c' and 'd', which are asymmetric. Here, then, is the first of Watson's additional problems – he'd claimed that '*a priori*, we can perceive no reason why the viscera should be disposed in the one way rather than in the other'. Well, now there was indeed a reason, a very strong reason, even if the route from molecule to viscera was extremely obscure.

If Pasteur's discovery was the riddle, then the enigma, which was to provide so much difficulty for Watson, had in fact manifested itself only a few weeks after his presentation at the College of Physicians, although neither he nor anyone else took note of it for a quarter of a century. Just as the experiments of that obscure monk, Gregor Mendel, were ignored for decades until the time was ripe for understanding them, so the work of Dr Marc Dax was equally ignored. Dax, who was born in 1770, spent most of his life, from 1800 until his death, practising medicine in Sommières, a small town twenty miles north-west of Montpellier in southern France. Dax is now remembered for a paper he presented in July 1836 at a medical conference in Montpellier, *Le Congrès Méridional*. Loosely translated, the paper's title was 'Damage to the left half of the brain associated with forgetting the signs of thought [that is, a loss of words]'. The paper was read aloud but never printed, and Dax himself died the next year, at the age of sixty-six, his ideas all but disappearing into oblivion. However, nearly thirty years later, in 1865, his son, Dr Gustave Dax, published the manuscript of his father's paper, the topic of language localisation having become one of the hottest in scientific Paris, as a result of the claim made by Dr Paul Broca that language was located in just one half of the seemingly symmetric human brain.[10]

Dax senior's 1836 paper begins with an incident from September 1800, when he saw a patient, a captain in the cavalry, who had been wounded in the head by a sabre blow and had difficulty in remembering words. Dax had been reading the work of Gall, the phrenologist, who believed that different mental functions were located in different parts of the brain. Dax therefore asked where the cavalry officer had been wounded, and was told the left parietal region (that is, on the side of the head). Gall had never suggested that mental functions could be located on only *one* side of the brain, and hence the whole story was mystifying. However, over the years Dax saw more and more patients and eventually conclud-ed that, despite it making little sense in Gall's typology, loss of language was associated with damage to the *left* half of the brain.

Dax junior published his father's manuscript because of an extremely acrimonious and lively debate taking place in the Académie de Médecine, the Société d'Anthropologie and the Société Anatomique in Paris. This started originally as a debate on the location of language in the brain, with Jean-Baptiste Bouillaud claiming that language was located in the frontal lobes, just above the orbit of the eye. In 1861, Paul Broca (Figure 1.7), a surgeon with wide-ranging interests in anatomy and anthropol-ogy, saw two patients both of whom had problems with language.

Figure 1.7 Dr Paul Broca, photographed in his fifties, a decade or so after he had described the brains of Tan and Lélong.

Remarkably, we have photographs of the brains of these two patients, both having been preserved in Paris. The first patient was called Leborgne, but had long been nicknamed *Tan* from the single spoken word that he could produce. He had been an epileptic since childhood, initially being paralysed in the right arm and then, later, in the right leg. He had been institutionalised at the Bicêtre Hospital for the previous twenty-one years. His death came suddenly at 11.00 am on 17 April 1861, from a neglected cellulitis and gangrene of the right leg. Twenty-four hours later, a post-mortem was carried out, the brain being removed and, within a few hours, displayed at the Société d'Anthropologie, after which it was preserved in alcohol. It can be seen in Figure 1.8, where, somewhat unusually, the brain has been preserved in a vertical position. If the page is rotated ninety degrees anti-clockwise the view is more conventional. The precise cause of Tan's illness is still unclear, but what is obvious, even from the relatively poor-quality photograph, is that a large area of damage is present in the left frontal lobe. That the damage was so localised was what principally interested Broca, particularly when, a little later, a second patient showed almost precisely the same damage.[11]

The name of the second patient was Lélong. In the spring of 1860, as an eighty-three-year-old man, he collapsed with a stroke, after which his daughter reported he was unable to speak. Eighteen months later, on 27 October 1861, he fell over and fractured the neck of the left femur. In the era before hip-replacement surgery, a broken hip was effectively a death sentence, and twelve days later Lélong died, under the care of Broca, to whose surgical ward he had been admitted. The post-mortem

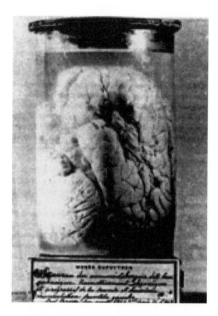

Figure 1.8 The brain of Leborgne (Tan), preserved in the Musée Dupuytren. The brain is stored vertically, with the frontal lobes at the top and the cerebellum visible in the lower left-hand corner. Only the left cerebral hemisphere can be seen in this photograph. The large, dark, horseshoe-shaped area in the middle, about a third of the way down the image, is the damage in Broca's area.

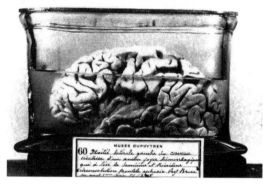

Figure 1.9 The brain of Lélong, preserved in the Musée Dupuytren. The left cerebral hemisphere can be seen. Unlike the brain of Tan in the figure above, this is stored in a more conventional position, with the frontal lobes to the left and the cerebellum not visible (it would be beneath the cerebral hemispheres on the right side of the picture). There is a large area of damage visible just above the left-hand half of the label.

examination of the brain was very clear – there was an area of damage in almost exactly the same area as that seen in Leborgne (Figure 1.9).

Broca's principal interest at this time was that the areas of damage were localised within the frontal lobes (in the place now known as Broca's area). In April 1863, when he had seen eight patients, all with damage in the left hemisphere, he commented that it was 'remarkable how in all these patients the lesion was on the left side. I do not dare to draw any conclusion from this and am waiting for new data.' Later in the same year, he would describe no less than twenty-five further patients with what he called aphemia (but soon became known as aphasia); that is, a loss of language. All the patients had left-sided brain damage. These patients were all diagnosed when alive, the side of brain damage being known because each had a hemiplegia, or paralysis, of the right side of their body. One of the strange features of the nervous system is that the right side of the brain controls the left side of the body and vice versa, the fibres from brain to body 'crossing over' in the brain stem. This meant that Broca's patients with a right-sided paralysis had damage to the left half of their brain.

For Broca the implications were clear:

> From the physiological point of view this is a most serious matter … [I]f it were shown that one particular and perfectly well determined faculty … can be affected only by a lesion in the left hemisphere, it would necessarily follow that the two halves of the brain do not have the same attributes – quite a revolution in the physiology of nervous centres. I must say that I could not easily resign myself to accept such a subversive consequence.

Broca was absolutely right – it would indeed be subversive and was, in fact, 'quite a revolution'. How could two seemingly identical masses of grey matter of the brain be so different? One, the left, was responsible for language – that highest faculty of the human mind and the jewel in the crown of civilised life – and the other, almost identical in form, permitted just a few monosyllabic grunts of the sort made by Tan and Lélong. Yet, so it is. The basic finding has not been disputed since the time of Broca, everyday experience of any practising physician or neurologist repeatedly confirming it.[12]

One of those practising physicians was Thomas Watson himself. In 1871, at the age of seventy-nine, he published the fifth edition of his *Lectures on the Principles and Practice of Physic*, one of the most successful medical textbooks in Victorian England, the first edition of which had been published nearly forty years earlier in 1843, based on his lectures at King's College, London. Watson was honest enough about the ease of replicating the basic finding: 'On looking back on brief notes, kept through many years, I find frequent evidences of the conjunction of some form of aphasia with right hemiplegia.' Watson had even published such cases; for instance, in the first edition of his textbook. Despite such cases, Watson was still sceptical about Broca's theory that language resided in specific parts of the brain: 'I cannot accept – I put no faith in – the theory upheld by M. Broca.' He cited an opinion which he attributes to Dr John Hughlings Jackson, the great London neurologist, then at the height of his powers, that 'the faculty of language resides nowhere in the brain, because it resides everywhere'. The problem with that, though, is clear – if language is everywhere, how can it be that damage to the left cerebral hemisphere results in aphasia, whereas damage to the right hemisphere typically does not? Watson struggled for some sort of theoretical model, much as he had done thirty-five years earlier, and he returned to an old interest – handedness.[13]

Since it is clear the brain itself is visibly symmetric, then the highly asymmetric association of aphasia with right hemiplegia seemed unlikely to come from the brain itself. Instead, Watson pointed out that our two eyes, lungs and kidneys all behave from birth as a pair, each responsible for half of normal functioning. The two halves of the brain, he argued, would also do the same were it not for 'one notorious exception. We all grow up right-handed.' According to Watson, the result is that the brain is educated for language on one side only, the left side. Language lateralisation on the left therefore follows on from, and is directly caused by, right-handedness. Watson moderated this statement a few sentences later, noting that 'the right side may indeed sometimes appropriate [language], just as some few persons are left-handed'. Here, at least, is some sort of explanation of why not everyone has left-sided language. But the theoretical mess was about to get deeper, for Watson was still trying to make sense of why we are mostly right-handed (which he again claimed to be universal – 'common … to all nations and races').[14]

Now Watson came up with an entirely different theory; one that is utterly but interestingly flawed. He returned to the anatomy of the body and looked at the blood supply to the brain. The arteries in the chest leading from the heart are as asymmetric as the heart itself, which means that the carotid arteries providing the main blood supply to the cerebral hemispheres are also asymmetric (Figure 1.1).

> Owing to the well-known arrangement of the arteries which rise from the aortic arch, the left hemisphere receives by the carotid a more direct and therefore freer supply of blood than does the right. Probably from this cause … the convolutions of the anterior lobe of the left hemisphere [that is, Broca's area] are developed at an earlier period than those of the opposite side … In this same fact we find a probable reason for most men being right-handed.

On first impressions it looks like a good theory, but a few moments of reflection reveal that it cannot possibly be correct. Watson's own observations of 1836 undermine the argument, for if we are right-handed and have left-sided language because of the asymmetry of the blood supply to the brain, then individuals with *situs inversus* should mostly be *left-handed*. Watson himself, however, had emphatically shown that they are not.[15]

Sir Thomas Watson has helped us map out the ground we are going to explore. The problems he highlighted are as acute today as when he first raised them, and there are others related to these that are also equally problematic. There are, then, exciting times ahead. Definite progress has been made in recent years, much of it in the past two decades. Since the early 1960s, there have been huge amounts of research into handedness and the difference between the cerebral hemispheres. At a more anatomical level, biologists in the 1980s, after years of almost totally ignoring the question, realised that there are interesting and profound problems in understanding how and why vertebrates have their heart on the left side. Again, great progress has been made. In the process, there has also been much study of the relationship between anatomical asymmetries and biochemical asymmetries – the big question being whether they are directly related in any causal sense. The answer is probably yes, as we shall see.

In 1991, the Ciba Foundation realised that real advances were being made in understanding why biological organisms were asymmetric, and it organised a meeting entitled 'Biological Asymmetry and Handedness'. For three days, twenty-nine scientists – from physicists, chemists and biochemists, through anatomists, development biologists and palaeontologists to psychologists and neuroscientists – discussed and debated these issues. All of those disciplines were needed, as it was increasingly realised that full understanding of any one involves all the others. The meeting was chaired by Lewis Wolpert, himself a left-hander, who, as an experimental scientist, couldn't resist starting the meeting by testing a pet hypothesis of his own: that left-handers would be over-represented among the participants. They were not – two participants were left-handed, representing seven per cent; a proportion comparable with the general population and with laterality researchers in general. At the end of one wonderfully varied but heavy discussion session, Lewis commented light-heartedly, 'From molecules to brains in one easy session!' It had indeed spanned that range, although 'easy' would probably not be most people's description. Here we will try to link those areas, looking for evidence from a very wide range of disciplines, although it will not always be easy. It should, however, take us into many of the nooks and crannies of the physical, biological, cognitive and social worlds.[16]

In studying handedness in all its forms, we will traverse scales from the very smallest to the very largest, from the sub-atomic to the cosmological. Although we will find that there seem to be asymmetries, or forms of handedness, at each of those levels, we will also find that much of their meaning and importance comes from a universal human desire to treat left and right as symbolically different. That, therefore, is where this book starts.

☞   ☜

DEATH AND THE RIGHT HAND

The death was as pointless as so many others in the First World War. It took place twenty miles east of where, a year later, 650,000 French and German soldiers would die in the 'mincing machine' of the battle for Verdun which A. J. P. Taylor described simply as 'the most senseless episode in a war not distinguished for sense anywhere'. At 2.50 on the afternoon of 13 April 1915, a spring day with perfect weather conditions, a French lieutenant leading the men under his command climbed out of his trench, walked ten metres towards the enemy and fell, mortally wounded. His two sub-lieutenants followed and, in their turn, fell almost immediately. They had covered no more then a dozen steps of the three hundred yards of open country separating them from the enemy lines, when, in clear view, they were mown down by the German machine guns. A few moments later twenty-two soldiers lay dead. One of the sub-lieutenants killed was Robert Hertz (Figure 2.1), a sociologist and anthropologist, who was just thirty-three years old when he died. His body, lying side by side with the other two officers, was recovered the following night. Hertz's last letter to his wife, Alice, showed his sense of foreboding prior to that day, finishing, *'un baiser grave et pieux – pour toujours'* (a solemn, devoted kiss – for ever).

The attack on the village of Marchéville that Hertz had been part of was unlikely ever to have been successful – the officers were well aware that they were walking to almost certain death. It was described later by Hertz's friend, colleague and posthumous editor, the sociologist Marcel Mauss, as *'l'attaque inutile'* (the useless attack). The military objective was undoubtedly important. The plain of the Woevre, which led directly to the German fortress town of Metz, was bisected by the River Meuse, and in September 1914 the German army had gained a rapid tactical advantage by establishing a bridgehead across the river at St Mihiel. The battle of the Woevre, in which Hertz took part, failed to regain St Mihiel, this

Robert Hertz
1881–1915

Figure 2.1 Robert Hertz.

only returning to French hands in September 1918, two months before the war's end, when it was attacked by General Pershing's First American Army.[1]

only returning to French hands in September 1918, two months before the war's end, when it was attacked by General Pershing's First American Army.[1]

Hertz was a student of the eminent sociologist Émile Durkheim, who died in 1917 at the age of fifty-nine having seen almost his entire group of young dynamic researchers die before him. In 1925, Durkheim's nephew, Marcel Mauss, whose most important contribution to anthropology was his work on the nature and symbolism of gift exchange, wrote an obituary for the entire group. It records a grim toll. 'Hertz, David, Bianconi, Reynier, Gelly…were all killed at the front'; Beuchat 'died for Science' in 1914, of hunger and cold, on an ethnographic expedition to Wrangell Island in the Gulf of Alaska; and Laffitte died 'of a long and cruel malady perhaps accelerated by his two wounds'. Finally, there was Durkheim's own son, André, a linguist, whose death 'was doubly felt, paternally and intellectually, and was one of the causes of his father's death [from a stroke]'. André Durkheim died in a Bulgarian hospital in December 1915 of wounds received while commanding a rearguard platoon during the rout and retreat from Serbia.[2]

The Durkheimian school was characterised by several features. Principally, its approach was 'functionalist', arguing that when one looks at a remote, different and seemingly primitive society which exhibits what might seem to be bizarre behaviours, then it is necessary, first, to be non-judgmental and, second, to try to assess the possible functional advantages of such behaviours and social organisation. To put it bluntly, people are not stupid, and if they persist in doing something it is more likely that they do it for a reason as yet not understood than for no reason at all. For example, in his book on gifts, Mauss describes how the 'potlatch' – the ceremonial and ritualised exchange of lavish gifts amongst the tribes of the American North West – can be understood as disposing of surplus produce while simultaneously maintaining the social structure through its three obligations, 'to give, to receive, to reciprocate'.[3]

The second innovation of Durkheim and his school was to suggest that primitive classifications of the world are inevitably somewhat limited, since individuals in the primitive world are not and have never been the neutral, rational, information-gathering scientists that we now aspire to be. As a result, they inevitably describe the world in terms of the one system that is readily apparent to them – the social world which they inhabit, with its relationships of family, tribe and so on.

The third important innovation of the Durkheimian school was methodological, emphasising the 'comparative method' – systematically comparing large numbers of societies and cultures rather than immersing itself in just one. Anthropologists became creatures of the archive and the library, scouring the field notes of earlier anthropologists who had learned the local language and documented habits and customs. The weakness of this method is that reading is never the same as doing – 'knowing that' and 'knowing how' can be worlds apart. Hertz himself realised the difference. He had steeped himself in Dayak culture, even learning the language, so that he felt he knew them as flesh and blood. In 1912, though, while visiting a remote corner of the Alps near Aosta to study the cult of the Roman legionary, Saint Besse, he observed, 'How much more alive than the work in the library, this direct contact with realities.'[4]

Hertz also anticipated a very modern attitude in anthropology: that one does not have to go to the far corners of the earth to find strange, remote cultures, with complex belief systems. Nowadays, anthropologists study almost any sub-group in modern society – from scientists in nuclear power stations, to shelf-stackers in supermarkets, to surgeons in operating theatres. Raymond Firth commented that 'One of the functions of anthropology is to ask questions about the obvious ... Why do we greet people with a shake of the right hand and why do we seat a guest of honour at our right side?' The answers to such seemingly obvious questions are far from obvious. The shaking of hands, for instance, is replete with meaning, as John Bulwer pointed out in the seventeenth century, 'To shake the given hand is an expression usual in friendship, peacefull love, benevolence, salutation, entertainment, and bidding welcome; reconciliation, congratulation, giving thanks, valediction, and wellwishing.' Nevertheless, the modern form of shaking hands seems to be an

invention of the eighteenth or nineteenth century. It might have originated in England, if Flaubert's *Madame Bovary* is anything to go by, when
Léon Dupuis says farewell to Emma by shaking hands, her comment
being, '*A l'anglaise donc*' – 'English fashion, then.'[5]

In 1907, Hertz published his first proper academic paper, on the collective representation of death in different societies. That essay and its companion piece on right–left symbolism were linked in the sparse and
dramatic title of their English translation, half a century later, *Death and
the Right Hand*. Hertz extensively researched his essay on death in several
libraries, including ten months spent consulting books and papers in the
British Museum. He started by emphasising that he is dealing primarily
with a social phenomenon that transcends the simple biological facts of
life and death; death not only ending the bodily life of a person but also
that person's social function.[6]

Hertz was particularly struck by the sheer variety of ways that different
cultures and societies respond to death. In particular, he looked at the so-
called 'mortuary rites' of Indonesia, and the phenomenon of second
burial, in which the final disposal of the body takes place weeks or even
months after death. Some of the mortuary rites are, quite frankly, revolt-
ing to a modern Western reader, although it must be admitted that our
practices with regard to our loved ones would appear equally bizarre to
those carrying out second burial. Typical examples described by Hertz are
the custom in Bali of keeping the body in the house for many weeks in a
special coffin that has been pierced at the bottom so that the liquids of
decomposition can be collected in a bowl and ceremonially emptied each
day; or the practice of the Dayak in Borneo of collecting the liquids and
mixing them with rice, which is then eaten as part of the mourning; or,
in many cases, of smearing the liquids over the body of a bereaved relative. The eventual, second burial may be associated with scraping the
remaining flesh from the bones, before cremating it and then pulverising
the bones to a powder, which is smeared over the body in a paste, as in
some South American tribes.

What was clear to Hertz in all this variety was that there had to be
some explanation of the practices. Resisting the temptation to find ratio-
nalist explanations for death practices and deep underlying commonalities across societies, Hertz cautions, 'We must beware of attributing the
various representations a generality…which they do not have.' Here we
do not need to look any further at what Hertz wrote about death, except
to note that he found himself forced to think very differently about right
and left and to acknowledge commonalities across cultures.

In 1909, two years after his essay on death, Hertz published the article
for which he is best remembered, and which describes the ideas that

pervade this book. Entitled *The Pre-eminence of the Right Hand*, it is concerned with the issue of left–right symbolism: how and why it is that left
and right so often have symbolic, metaphorical and ritual meanings, as
well as merely indicating locations in space. The article starts in a poetic,
rhetorical manner, reminiscent of Shakespeare's 'What a piece of work is
a man'. Regrettably, one can hardly imagine any editor accepting it
nowadays in a scholarly scientific journal.

> What resemblance more perfect than that between our two hands! And
> yet what a striking inequality there is!
> To the right hand go honours, flattering designations, prerogatives: it
> acts, orders, and *takes*. The left hand, on the contrary, is despised and
> reduced to the rôle of a humble auxiliary: by itself it can do nothing; it
> helps, it supports, it *holds*.
> The right hand is the symbol and model of all aristocracy, the left
> hand of all common people.
> What are the titles of nobility of the right hand? And whence comes
> the servitude of the left?

As in the essay on death, Hertz begins with biology, but again concludes
that the symbolism of right and left cannot be reduced merely to biological differences between the hands. These are complex social phenomena;
if they are social there will be a collective basis for them and if there is a
collective basis, then the sacred and profane will be a part of that story.
As Durkheim put it, 'Hertz showed that the causes of the primacy [of the
right hand] are essentially religious.'[7]

The key difference between Hertz's essay on death and his essay on the
right hand is that while he is happy to emphasise cultural differences in
the handling of death, albeit finding a deep similarity of purpose and
function beneath the surface, he has far greater difficulty in doing so for
right–left differentiation. Although right–left symbolism is in many ways
arbitrary, Hertz has to confront the fact that in almost all societies it is
the *right* hand that is pre-eminent. Something must cause that uniformity. Just as Sir Thomas Watson recognised that if handedness were a
purely social phenomenon, then half of all societies should be right-
handed and half should be left-handed, so the very same problem applies
to left–right symbolism. If it is arbitrary, then why is it almost always the
same way round? Right is almost always seen as good and left as bad,
rather than vice versa.

Before thinking about why right and left have symbolic meanings, and
what it means for something to be symbolic, we need to have a quick
'Cook's Tour' of symbolism in a range of different contexts. We will start,

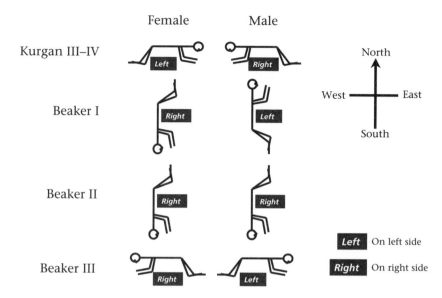

Figure 2.2 Schematic representations of the positions for burial of males and females in the Kurgan III–IV and Beaker I, II and III cultures.

as Hertz would have liked us to, by looking at right–left symbolism in death.

One of the earliest pieces of left–right symbolism found by archaeologists is the burial patterns of early proto-Indo-European peoples, the Kurgans, who probably came from the area between the Don, Volga and Ural rivers in Russia and Kazakhstan, and who dominated Europe in the fourth millennium BC. Great care was taken over the burial of the dead, which was carried out in highly stylised ways that tell us much about thought patterns and social life. Hertz would have recognised many of the rituals, including the frequently practised sinking of a shaft so that the dead could be provided with food and water. The dead were laid in a semi-flexed, almost fetal, position, perhaps indicating a belief in the possibility of rebirth. When a body is buried semi-flexed it can be laid out in various patterns on either its right or its left side. For the Kurgans and Indo-Europeans, those patterns were far from random. The four main types are shown schematically in Figure 2.2, the compass indicating the direction in which the body was laid.

Among three of the four groups, male and female bodies were buried differently. There seems little doubt that we are dealing with some form of symbolism here. It would be difficult to come up with any practical reason derived from the exigencies of daily life, or from the physics of the world, that produces this sort of consistency within cultures and yet such differences between them. What symbolic system, then, is being followed

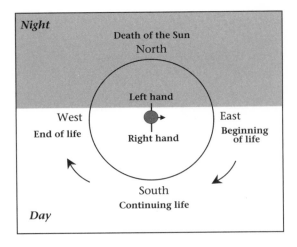

Figure 2.3 Schematic representation showing the relationships of right and left to the points of the compass and the movement of the sun relative to a person, shown in the centre, facing due east at sunrise. The diagram is only correct north of the Tropic of Cancer.

here? As so often with symbolism, as the anthropologist Lévi-Strauss emphasised, the pattern only really becomes clear when we look at the entire *set* of myths or symbols. Any one on its own can sometimes be explained away; the complete set cannot.[8]

To understand these burials one needs to think about the natural and cognitive world of early Europeans. The heavens were of vast importance to Neolithic peoples in a way that is almost inconceivable nowadays, few people today being aware of the phases of the moon or the position of the sun. For Indo-Europeans, living far north of the Tropic of Cancer, the sun, even in mid-summer, was always to the south. It would rise towards the east, swing round through the southern sky, and set in the west. Then, almost miraculously, it would rise again the next day, once more in the east. The sun's position literally provided a way of *orienting*, of finding east. If they looked toward the rising sun, and followed its movement throughout the day, it was always to the right – on their right-hand side. There is an intimate link between the dominant right hand and the movement of the sun, that giver of life and warmth (Figure 2.3). Night and day, and the four points of the compass, are part of a continuously turning circle – a circle that turns to the right. Attaching symbolic meanings is then straightforward: east denotes the birth of the sun and the beginning of life; south, warmth and the continuation of life; west, the setting of the sun and the end of life; and north, the death of the sun itself, to be reborn in the east the next morning.[9]

Such symbolic associations probably explain many Indo-European burial patterns. In each case, the body is placed facing either to the east or to the south, with the connotations of rebirth or the continuation of life. If the theory seems a little vague, with its use of either east or south as the direction in which to face, then it is worth knowing that in many Indo-European languages the words for right and south are partly inter-changeable, so that in Sanskrit, for instance, *dakshina* means 'right hand' and 'south', and *puras* means 'in front' and 'eastwards'. Likewise, in Old Irish, *deas* and *ders* mean 'on the right' and 'southward' and *jav* means 'behind' and 'west'; and so we could go on.[10]

The symbolisms we are talking about here are not unique to the early Indo-Europeans. The cultural systems in which we find ourselves immersed today reflect many similar ideas. The Christian tradition, for example, incorporated many earlier symbolisms that have now become an intrinsic part of its iconography. One enters a church through the west door, facing the altar, which looks towards the rising sun. Similarly, in Christian churchyards, bodies are traditionally buried facing to the east, 'in sure and certain hope' of resurrection, or rebirth.

The movements of the sun have not only become associated with right and left, but also with rotation. The sun rotates in the same direction as the hands on a clock, and in many situations it is regarded as polite, or good form, to rotate in that direction. In ancient Greece, it was particu-larly auspicious for omens to go 'to the right'. After dinner it is tradition-al to pass port clockwise (that is, to the left), and there is even a seventeenth-century expression 'Catharpin-fashion', which means, 'when People in Company Drink cross, and not round about from Right to the Left, or according to the sun's motion'. In Old English, the terms *deasil* and *widdershins* mean 'sun-wise' and 'counter-sun-wise' (that is, clockwise and anti-clockwise) and in many situations it is regarded as unlucky or inappropriate to move widdershins. As an example, when waltzing, couples mostly turn clockwise, 'reversing' when they change direction, the implication being that clockwise is the proper or natural direction.

Many medieval and early modern machines such as windmills, water-mills and grinding machines have rotating parts (and a millstone that rotates clockwise is called a 'right-handed mill'). Surprisingly in the majority of such machines the visible parts rotate in a clockwise direc-tion, even though the prime movers are as likely as not to rotate anti-clockwise. The existence of machines with additional idler wheels, which act only to reverse direction (and thereby reduce the mechanical efficien-cy), suggests that there is a symbolic rather than a mechanical reason for machines mostly turning clockwise. The concept of a proper direction of turning was nicely illustrated in the thoroughly modern and secular context of an interview with the sculptor Richard Serra, who, when asked about a giant sculpture, 'Double Torqued Ellipse II', through which the viewer can walk, replied, 'I feel that it is less destabilising to walk the passage clockwise. But that might just be my preference, although I think it's a natural impulse to walk to the right.' A very old tradition is being continued here, with little sense of why, the justification being reduced to an inchoate feeling of being 'natural'.[11]

In the Indo-European graves shown in Figure 2.2 it is clear that south and east are preferred, and that this relates to the side on which the body is laid; left or right. However, in three of the four styles of burial, men and women are treated differently. Differently, yes, but females are treated in a way that is obviously related to the way that men are buried, with some sort of symmetry emerging, so that in Kurgan III–IV and Beaker III the women are the mirror-image of the men around the north–south axis, and in Beaker I they are the mirror-image around the east–west axis. The symbolic system of left–right is also tied in with another symbolic system of male–female. Such linking of symbolic systems is far from unusual; indeed, as will be seen, it is almost the norm. It is found not only in Europe but in almost all cultures. Sometimes it is very explicit, as in the Gogo of central Tanzania, who describe the right hand as *muwoko wokulume*, 'the male hand', whereas the left hand is *muwoko wokucekulu*, 'the female hand'. The languages of southern Africa typically refer to the right hand in such terms. In one survey of thirty-seven Bantu languages, the right hand was known as the male hand in sixteen cases. Other Bantu languages mostly talked about the 'eating hand' and, occasionally, 'the throwing hand' or 'the great hand'. In one or two cases the meaning is obscure. In Swahili, for example, the expression *mkono wa kuvuli* might be derived from *uvuli*, the word for 'shade' – perhaps 'the hand that carries the umbrella'.[12]

The association of right and left with male–female differences, and with sexual behaviour and reproduction, seems to be worldwide – indeed Freud commented, 'that right and left should mean male and female seems quite obvious'. In many parts of the world, the right hand is used for eating and for activities above the waist, whereas the left hand is used for cleaning, particularly at toilet, and for handling the genitals. The Gogo extend this distinction to intercourse itself, so that for sexual fore-play a man lies on his right side and uses his left hand to stimulate the woman's genitalia. Among the BaSotho, it is believed that if a woman lies on her right side during coitus then she will produce male children. In China, doctors predicted the sex of an unborn child on the basis that a girl would lie to the right and a boy to the left. The Mohave of Arizona and California wipe their buttocks with the left hand, which is believed to be the maternal side of the body, and eat with the right hand, which is paternal, the father's sperm having symbolically fed the developing fetus. The Kaguru of Tanzania believe that a person in the womb is made of two separate halves, the right half deriving from the father, the left from the mother.[13] The same idea also appears in Shakespeare's *Troilus and Cressida*, where Hector explains why he will fight no more with his cousin Ajax:

> This hand is Grecian all,
> And this is Trojan; the sinews of this leg
> All Greek, and this all Troy; my mother's blood
> Runs on the dexter cheek, and this sinister
> Bounds in my father's.

Right–left differences as the cause of sex determination are found throughout Western scientific thought. The sex chromosomes were only discovered in 1902, and before it was realised that the sex of the child depended on whether it received an X or Y sperm from its father, a myriad of speculative theories had been proposed on the origins of maleness and femaleness. Not surprisingly, right and left became tied in to those theories. Anaxogoras in the fifth century BC suggested, in line with modern science, that a child's sex depended upon the father, but, unlike modern science, suggested male children came from the right testicle and female children from the left. This idea led fairly naturally, even in ancient times, to the suggestion that a ligature around the left testicle would ensure sperm came only from the right testicle and hence a male child would be conceived. The idea was espoused by the medieval physician, Giles of Rome, and is found as late as 1891 in *The Essentials of Conception* by a Mrs Ida Ellis, who claimed, 'It is the male who can progenate a male or a female child at will, by putting an elastic band round the testicle not required. The semen from the right testicle progenates male, while that from the left female children.' Not all Greek philosophers thought that the male was important in determining sex. Empedocles, for instance, suggested that sex was determined by the female, the womb being hotter on the right and therefore producing male children on that side.[14]

Right and left are not associated only with male and female. Indeed, at times it seems as if there is almost nothing that they have not been associated with. Consider the Purum, a small tribe who live on the Indo-Burman border. They have been studied both in anthropological field studies and at the theoretical level, particularly by Rodney Needham of the University of Oxford, who has probably done more than anyone else to revive interest among modern researchers in Hertz and his studies.[15]

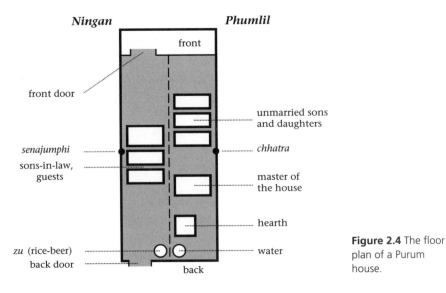

Figure 2.4 The floor plan of a Purum house.

The Purum social system is divided into two very separate groups or clans, which Needham terms *wife-givers* and *wife-takers*. The Purum also divide material objects, which are exchanged in a form of barter, into *masculine* and *feminine*: examples of masculine include pigs, buffaloes and rice beer, and of feminine include cloth, weaving looms and domestic articles. The feminine also includes one other article of exchange; women themselves. The rules of exchange are straightforward: feminine objects go from wife-givers to wife-takers, and masculine objects go from wife-takers to wife-givers.

A Purum house (see the plan in Figure 2.4) is divided lengthwise into two parts, a right-hand, *phumlil*, and a left-hand, *ningan*. It is also divided into a front and a back by two posts, *chhatra* on the right, which is built first, and *senajumphi* on the left. Viewed from the back of the house, looking towards the door, the right-hand side is the private part, belonging to the master of the house and his unmarried sons and daughters, and it is here that one finds the fire-place. In contrast, the left-hand side, which is also the side of lower status, is a public area, including the place for suitors and married daughters visiting their parents. The right-hand side is therefore the side of the wife-givers, the left-hand side of the wife-takers.

The Purum system is typical of what is known as *dual symbolic classification*. The world is divided into pairs of opposites which are then tied together, so that given one member, the others are also known. Table 2.1 allows one to see the system quite straightforwardly.

Right	Left	Right	Left
Male	Female	Kin	Affines
Masculine	Feminine	Private	Public
Moon	Sun	Superior	Inferior
Sky	Earth	Above	Below
East	West	Auspicious	Inauspicious
Life	Death	South	North
Good death	Bad death	Sacred	Profane
Odd	Even	Sexual abstinence	Sexual activity
Family	Strangers	Village	Forest
Wife-givers	Wife-takers	Prosperity	Famine
Gods, ancestral spirits	Mortals	Beneficent spirits	Evil spirits, ghosts
Back	Front		

Table 2.1: Purum dual symbolic system

A system like this is a sort of calculus for assessing symbolic values. As an example, when selecting a site for a new village, a cock is strangled. If the bird lies on the ground with its right leg over its left, the omens are taken to be good, whereas if the left leg is over the right the site is abandoned. A similar sacrifice is used in the child-naming ceremony to assess whether the future is auspicious. If the child is a boy, then a male bird is strangled by the priest, and if the right leg lies on top of the left that is a good omen. However, for a girl child, a female bird is used. Since male and female have been reversed, the interpretation of the legs is also reversed: for a girl it is auspicious if the left leg, the feminine leg, is on top. There is not, therefore, a simple equation of right = good and left = bad, but instead various combinations specify the outcome. In such a system, symbolic inversions can readily occur, right being auspicious in a male but inauspicious in a female.[16]

Symbolic inversions are commonplace in anthropology. For the Kaguru of East Africa, witches are the inverse of the normal world, even walking upside down on their hands, and one would not be surprised to find, as in medieval Europe, that one of the stigmata of a witch was being left-handed. The Ngagu of southern Borneo believe everything in the after-world is reversed, 'sweet' becoming 'bitter', 'straight' becoming 'crooked', and 'right' becoming 'left'. Likewise, the Toraja of Celebes (Sulawesi) believed the dead do everything backwards, even pronouncing words backwards, and because the dead therefore use the left hand all the time, the living should also use their left hand when they do something for them. Analogous inversions occur in Western culture, as in the Anglican

Church where processions move clockwise around the church, except during Lent when they move anti-clockwise.[17]

By comparing different societies with dual symbolic classifications, anthropologists have found remarkable similarities. For instance, the Gogo of Tanzania may be geographically far removed from the Purum in Burma, but their table of oppositions, shown below, has many conceptual overlaps with the Purum table shown earlier:

Table 2.2: Gogo dual symbolic system

Right	Left	Right	Left
Male	Female	East	West
Man	Woman	South	North
Clean hand	Dirty hand	Up	Down
Strength	Weakness	Ritual side of house	Side of house with midden
Superior	Inferior	Fertility, health	Death, sickness
Clever	Stupid	Cool	Hot
Side man lies on during intercourse	Side woman lies on during intercourse	Medicines	Poisons
Side on which men buried	Side on which women buried	Black	Red/White
Bow	Calabash, drum	Older people	Younger people
Bush-clearing	Seed-planting	First wife	Junior wives
Threshing	Winnowing, grinding	Father	Mother

One can see from this why archaeologists should feel happy interpreting the burial patterns of the Kurgans in terms of related systems, and there is also support given here for the Durkheimian notion that the distinctions of the natural world are closely related to social systems – so that the sun, moon and points of the compass are related to agriculture in the Gogo and to the exchange of pigs and women in the Purum.

From the theoretical perspective of this book, the most important thing is that locked into these oppositions is a *right–left* divide. There is no compelling reason why this should be. There is nothing in the nature of houses, or of hot and cold, or of the village and the forest, or of threshing or winnowing that makes one of them intrinsically right and the other intrinsically left. Right–left symbolism has no inherent basis in the natural world, and must therefore originate in our own minds.[18]

Before talking about symbolism in general, and about left–right symbolism in particular, we need to confirm how widespread is this way of using the terms right and left. In particular, we need to convince ourselves that it is not confined to so-called primitive or pre-literate societies. In fact symbolism is everywhere, even in modern technological societies, and often it is so familiar that we barely notice its existence.

We can start in the everyday details of modern life, with a multitude of actions which seem impossible to justify totally in a rational way (although people will try surprisingly hard to do so). Why, when we meet someone, do we shake their *right* hand? Why, when we swear an oath in court, do we place our *right* hand on the Bible or holy book (a 'left-handed oath' not intended to be binding)? Why, at the dinner table, do we place the knife on the right-hand side and the fork on the left (and in Europe we insist that the less dextrous left hand is used for picking up food with the fork)? In Britain and America, a wedding ring is worn on the third finger of the *left* hand; a fact that Sir Thomas Browne tried to explain by claiming (wrongly) that there was a blood vessel running directly from the heart to the left ring finger, though this could hardly account for why in Germany the wedding ring is typically worn on the *right* hand. Why also is a morganatic marriage 'a left-handed marriage'? For further associations of right–left with male–female we need look no further than any English church wedding, the bride's family sitting to the left and the groom's family to the right, just as the couple themselves stand at the altar.[19]

The Christian Church provides a mass of symbolic associations with right and left, and the Bible is full of expressions involving both. Perhaps the most famous is the image of the Last Judgement, when the peoples of the earth will be divided:

> And he shall set the sheep on his right hand, but the goats on the left. Then shall the King say unto them on his right hand, Come, ye blessed of my Father, inherit the kingdom prepared for you from the foundation of the world… Then shall he say also unto them on the left hand, Depart from me, ye cursed, into everlasting fire… (Matthew 25:33–34, 41; A.V.).

Above the altar in Italian Renaissance churches there is invariably a painting of the Crucifixion in which Christ turns to his right side, showing the left cheek. In paintings of the Annunciation, the Angel Gabriel almost always come from the left side and Mary faces the left, and in early Renaissance paintings of the Madonna and Child, the Madonna holds the Christ Child on her left, showing her right cheek and the child's left cheek. Left and right play a key role in the layout of all churches (see Figure 2.5), and there is a formal resemblance to the floor plan of the Purum house (Figure 2.4). The altar faces the east and the

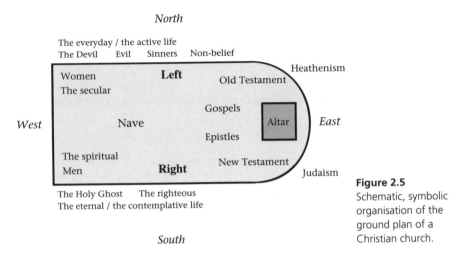

Figure 2.5
Schematic, symbolic organisation of the ground plan of a Christian church.

rising sun. On entering the church from the west, the northern, left-hand side has frescoes of the Old Testament, whereas the southern, right-hand side has frescoes from the New Testament. In many churches, the men would sit on the right (the south) and the women on the left (the north). The devil comes from the north in Milton's *Paradise Lost*, and in Dante's *Purgatory* the synagogue is to the north while the church is to the south. Other aspects of right and left in Christianity are more confused. Although it is conventional to talk of Eve as being formed from Adam's *left* rib, there is no biblical text to support this, and it perhaps represents a secondary symbolic association of male with right and female with left, or else comes from earlier Jewish traditions.[20]

Christianity is not the only religion for which right and left have important meanings. In the Jewish tradition there is also a clear division of right and left, perhaps originating in the fact that the early Israelites worshipped the sun, facing east and hence having the south to the right, a situation reflected in the Hebrew term for east. It was said of the Old Testament God of Judaism, 'The right hand of the Lord hath the pre-eminence: the right hand of the Lord bringeth mighty things to pass.' In the Talmud, the right is undoubtedly the position of honour. It was regarded as rude to walk on the right of one's teacher, and, should he be accompanied by two people, then the more important should walk on the right, with the teacher in the middle. The ceremony of *Halizah* should be performed on the right foot, with special rules for left-footers, footedness being determined according to the starting foot in walking. Jewish left–right symbolism is seen also in the mysticism of the Cabbala, where the Torah is the right hand and the oral law the left. At a more prosaic level, many people feel it lucky to put the right shoe on first, but a Jewish custom makes this far more sophisticated, firstly putting on the right shoe but not tying it, then putting on and tying the left shoe, and finally tying the right shoe, ensuring one begins and ends on the right side.[21]

As in Christianity and Judaism, so in Islam there is also a marked preference for the right. In the Qur'an, the elect are on the right of the Lord and the damned on his left, described in a contrast of great poetic beauty:

> *And those on the right hand; what of those on the right hand?*
>   Among thornless lote trees
>   And clustered plantains
>   And spreading shade
>   And water gushing
>   And fruit in plenty
>   Neither out of reach nor yet forbidden
>   And raised couches...
> *And those on the left hand; what of those on the left hand?*
>   In scorching wind and scalding water,
>   And shadow of black smoke
>   Neither cool nor refreshing...

The Black Stone in the Ka'ba, at Mecca, is 'the right hand of Allah upon the earth'. When God struck Adam's back and drew forth all his progeny, those destined for heaven came from the right side (as white grain), those for hell from the left side (as black grain). There are also strong injunctions about behaviour: 'One must neither eat nor drink with the left hand...for these are Satan's manners...[O]ne spits to the left and holds the genitals with the left hand.' On arrival at Mecca, one enters the Great Mosque with the right foot. Much of the meaning of right and left probably takes its meaning from the believer who addresses his prayers to the rising sun, thereby automatically situating right and left in relation to the points of the compass, with all their associated meanings. Within the Arabic language, left and right have similarly distinctive meanings to those in English, so that the right hand is used for taking an oath, *yamîne* meaning both 'right' and 'oath'. The left, *šimâl*, is the bringer of ill omens, and a left-handed person, *'a'sar*, is so called from the verb *'asara*, 'to render difficult, arduous, troublesome, irresolvable'. Although a strict interpretation of the Qur'anic texts might suggest that left-handers have problems, modern Islamic theology is far more liberal, emphasising that God knows who is left-handed, because it was he who created them that way, and that the rewards for the faithful come from those who have chosen to try to follow what is only a recommendation of the prophet: '[the] reward is not for the action itself but for the intention behind the action'.[22]

One intriguing reversal of conventional symbolism in Islam is that pilgrims at Mecca circle the Black Stone seven times *anti-clockwise*, the direction normally associated with funerals and sorcery. One explanation is that Mohammed chose to adopt the reverse of the pagan ritual previously present. An ingenious alternative hypothesis takes into account the fact that Mecca lies south of the Tropic of Cancer, at latitude 21° 25' N. For forty-six days a year, therefore, in June and the first fortnight of July, the sun goes round to the south rather than the north. If the first celebrations were in those months, it is possible that the direction became fixed as anti-clockwise. It is an attractive theory, except that rotation to the right is so very common, even in cultures that are well within the tropics or even within the southern hemisphere.[23]

In the great Eastern religions there is also right–left symbolism. In Buddhism, the path to Nirvana divides into two: 'The left-hand one is to be avoided, the one on the right is to be followed. This will lead successively to a thick forest, a marshy swamp, a steep precipice and, eventually, to a delightful stretch of level ground (Nirvana).' At Benares, the Hindu pilgrims circumambulate with their right hands towards the centre (that is, clockwise), as Krishna is alleged to have done at the sacred mountain. Likewise, a Buddhist *stupa* must be walked round clockwise, with one's right hand inwards to the *stupa*, and Tibetan prayer wheels must also be turned so that one's right side is towards the axle, turning in the opposite direction being to undo all that has previously been done. Pilgrims and prayer wheels both, therefore, turn in the direction of the sun, an exception being in the Hindu rites for the dead, where turning is in the opposite direction.

Historically, the founts of rational, philosophical, scientific thought in Classical Greece are everywhere touched with right–left symbolism. Pythagoras said one should enter a sacred place from the right, which is the origin of even numbers, and leave from the left, which is the origin of odd numbers. In his *Metaphysics*, Aristotle describes how the Pythagoreans identified ten first principles, which could be listed in two parallel columns:

Limited	Unlimited
Odd	Even
One	Plurality
Right	**Left**
Male	Female
At rest	In motion
Straight	Crooked
Light	Darkness

Good	Evil
Square	Oblong

It will hardly be a surprise to see 'right' associated with 'male', 'light' and 'good'. Several centuries earlier, the poet Hesiod, a contemporary of Homer in the eighth century BC, wrote the *Theogony*, in which he traces the origin of the world from initial chaos to the reign of Zeus, the king of the gods. In one of those complex mythologies involving much incest, plotting, monsters and revenge, there is a scene in which young Kronos, in one of the first ever Oedipal conflicts, plots with his mother, Earth, to kill his wicked, incestuous father, Heaven, who, hoping for a night of love-making, cuddles up to Earth at which point Kronos leaps out of hiding:

> ...the hidden boy
> Stretched forth his left hand; in his right he took
> The great long jagged sickle; eagerly
> He harvested his father's genitals, and threw them off behind.

And from these genitals arose, amongst others, the goddess of love, Aphrodite.[24]

Right and left in Classical Greece applied at many levels, even including food. James Davidson has elegantly described how the Greeks not only distinguished between food and drink, but divided food into two very separate categories, the staple foods, *sitos*, such as bread, and the relish or the flavour, *opson*, such as fish, meat or onion. *Sitos* was eaten with the left hand whereas *opson* was eaten with the right: 'The habitual differentiation at meal-times of left and right, bottom and top is easily translated into more ideological contrasts: substance and decoration, necessity and excess, truth and façade.' We can set this out in column form as follows:

Left	Right
Bottom	Top
Sitos	*Opson*
Substance	Decoration
Necessity	Excess
Truth	Façade

That centrepiece of Greek philosophical thought, the symposium, with its circle of couches around the room, also had its left–right symbolism, as Davidson again explains: 'Each couch could take two people, reclining on their left sides…Wine, song and conversation went around the room from "left to right", that is, probably, anti-clockwise.'[25]

Finally, in this brief account of symbolism in the modern and ancient world, there are also indications that symbolism plays an important role in the *unconscious* mind. Interpreting dreams has always been richly based in symbolism, and a recent *Dictionary for Dreamers* describes the left side as 'sinister, wrong, instinctive. Anything vicious and immoral: criminal tendencies, incest, perversion. The feminine passive principle; sometimes the faculty of intuition.' Much of the symbolism is similar to that of the symbols looked for in divination, as for instance when, in the *Odyssey*, Odysseus comments that 'the bird omens were favourable, wholly on his right, as he went away'.[26]

Movement from left to right is seen as bad, and movements in dreams have been similarly interpreted. An example was discussed by Freud, whose *Interpretation of Dreams*, first published in 1900, described dreams as 'the royal road to the unconscious'. The dream in question had actually occurred many years earlier, in the spring of 1863, being that of Bismarck, who was subsequently to become German Chancellor.

> I dreamed (as I related the first thing next morning to my wife and other witnesses) that I was riding on a narrow Alpine path, precipice on the right, rocks on the left. The path grew narrower, so that the horse refused to proceed, and it was impossible to turn round or dismount, owing to lack of space. Then, with my whip in my left hand, I struck the smooth rock and called on God. The whip grew to an endless length, the rocky wall dropped like a curtain and opened out a broader path, with a view over hills and forests, like a landscape in Bohemia…I woke up rejoiced and strengthened.

Freud interprets this dream in terms of sexual symbolism. The 'whip' that is held in the *left* hand, and which grows to an endless length, is a classic phallic symbol, representing the penis. The vista that is revealed is presumably orgasmic, and is followed by a sense of rejoicing. Freud here quotes from another psychoanalyst, Stekel, who claimed that 'left', in dreams, refers to those things that are forbidden or sinful, giving childhood masturbation as an example. That Bismarck seized the rod with his left hand indicated a forbidden and rebellious act. There is also meaning in the narrow path having the precipice on the right, since the horseman must be travelling anti-clockwise up the mountain – turning to the left. Again, Freud refers back to Stekel, who discussed the ethical component of right and left, the right being the path of righteousness, and perhaps of marriage, whereas the left is the path of crime and perversion, perhaps homosexuality or incest. Psychoanalysts have since developed such ideas, the contrast between right and left being two possible ways of resolving the Oedipal conflict: of identifying with the mother or with the father, or of resolving the castration anxiety through impotence or purity. The so-called left way eliminates the father, resulting in a matriarchal, property-less culture, with its implications of incest, homosexuality and impotence.[27]

The act of writing, the quintessential use of the right hand, has been of particular interest to psychoanalysts, Freud saying that, 'If writing – which consists in allowing a fluid to flow out from a tube upon a piece of paper – has acquired the symbolic meaning of coitus…then…writing… will be abstained from, because it is as though forbidden sexual behaviour were thereby indulged in.' Not surprisingly, the act of writing with the left hand then takes on a different meaning from writing with the right hand: 'the habit of writing with the left hand…[is] a symbolic gesture motivated, if not stimulated, by fantasies which are forgotten, repressed, and repudiated by the speech community'. One can begin to see why Freud might have had an aversion, albeit unconscious, to the suggestion of Wilhelm Fliess that he might be partially left-handed. Freud was honest enough, however, to admit that 'I [may] have been up to something that one can only do with the left hand. In that case the explanation will turn up some day, God knows when.'[28]

Wherever one looks, on any continent, in any historical period or in any culture, right and left have their symbolic associations and always it is right that is good and left that is bad. If we were not totally immersed within such a symbolic system we would yell out in incredulity and demand some explanation for this astonishing fact. It is time, therefore, to stop merely cataloguing and instead to ask why these right–left symbolic associations exist, how we explain them, and what they are for. It was Hertz who made the first serious attempt to make sense of these phenomena, and it is to him we now return.[29]

Hertz begins his essay as a biologist rather than a sociologist. He was in no doubt that the right hand is more skilful and that its preference has something to do with the brain. He had no time for the naive view that right-handedness solely represents only cultural pressures and social influence – a conspiracy of dextrals, as it were. Indeed, he neatly scuppered such a view by pointing out that since some left-handers retain an instinctive preference for the left hand despite a lifetime of the pressures of living in a right-handed world, so there is no reason why right-handers should not also have an instinctive preference for their right hand.

Having accepted that, though, Hertz was aware of a major problem. The right hand may be more skilful but, even in the strongest of right-handers, the left hand is capable of a wide range of activities and, as Hertz pointed out, can be trained to a high level, as in musicians or surgeons. This raised the question of why training is principally focused on the hand already endowed with a natural superiority, the right hand. Even if there is a difference between the hands, it is not absolute, but only relative; 'a vague disposition to right-handedness', as Hertz put it. Modern research, in which each hand moves a set of pegs from one row of holes to another, finds the right hand to be only about ten per cent faster than the left. It is a small difference, particularly for explaining why a typical right-hander will use their right hand for ninety per cent or more of tasks. So how did a 'vague disposition' become converted into 'an absolute preponderance'? The only way could be by something *outside* the individual organism, something which does not consist merely of instinctive preferences, and fairly weak preferences at that. To a sociologist such as Hertz, that something could only be society. The difference in the way we treat our two hands is not therefore merely a reflection of natural differences but is also an ideology, a set of ideas about how we *should* live rather than about how we are actually made. The left hand is not paralysed by cerebral inaction but by sanctions and prohibitions imposed upon it by society. The reasons for preferring the right hand, according to Hertz, are 'half aesthetic, half moral'.[30]

A key theoretical concept for Hertz was dualism, a division of the world into opposing groups of good and bad, rich and poor, tall and short, blonde and brunette, introvert and extravert, and so on. Even though we all tend to classify the world in this way, there are rarely actually two such distinctive and separable groups, but instead a continuum representing all possible intermediate values. Most people are neither strongly introvert nor extravert – they are somewhere in between, just as most people are not tall or short but of intermediate height, but that doesn't stop us thinking and talking in such terms. Psychologists who specialise in personality measurement have an old joke: 'People are divided into two types: those who divide people into two types and those who don't.' This dualistic tendency probably helps us to manage complexity. A full description of every object would be too rich for our cognitive systems to cope with. 'What did he look like?' 'Hair about 27 centimetres long, eyes 85 millimetres apart, between 190 and 195 centimetres tall, weight about 65 kilograms.' 'Ah, you mean tall, thin, long-haired with wide-set eyes.' Less can be more. Extra information impedes a fast, effective analysis.[31]

Hertz found dualism everywhere in primitive thought, ultimately linking it to those two great Durkheimian categories, the sacred and the profane: the things of this world and those not of this world; the domain of the gods and that of mortals, of the natural and the supernatural, of life and death, strength and weakness, good and bad. Symbolically tied into the sacred and the profane are light and dark, high and low, sky and death, south and north, male and female, and so on. As Hertz asked, how could man's body escape this polarity that applies to everything else? Humans will use anything to symbolise the sacred and the profane, and the right and left hands were crying out to fulfil such a role.[32]

Even if it was inevitable that the right and left hands would somehow come to be connected with the sacred and the profane, it is still of course not immediately obvious why the *right* should be associated with the sacred and the *left* with the profane. Here, Hertz returned to the anatomical and functional differences between the hands: the right is sacred *because* it is the stronger and more skilful. He was quick, though, to point out that this does not mean the symbolism is merely biological. The distinction between the hands would, on its own, be too slight to have any effect. It is only the social system of which we are a part, with its dualistic views on almost everything, that provides the power and energy to transform right to sacred and left to profane. When the tiny biological difference is coupled with this giant social force, it becomes transformed into a major effect. It is a bit like a police officer directing the traffic. The motive power of a 40-ton lorry could never be overcome by a policeman alone, who is far too weak, but because policeman and lorry driver are part of a social system, a tiny movement of the officer's hand changes the direction of the lorry. When carefully directed, a small force can produce large changes.[33]

Hertz's essay makes clear the relationship between the right hand as an anatomical piece of flesh and blood, with its fairly limited advantages of skill and strength over the anatomical left hand, and the symbolic right hand, which is almost infinitely powerful compared with the weak, inferior, symbolic left hand. Throughout this book, when looking at right and left we must remember to ask whether we are talking about substantive physical or biological differences, or about symbolic differences imposed upon the physical, biological and social world by the dualistic mode of thought that permeates everything we do. Symbolism can creep in everywhere.

Although symbolism is at the heart both of the social sciences and the humanities, it is difficult to find any universally accepted theory; indeed the semiotician and novelist Umberto Eco has despaired that 'a symbol can be everything and nothing', with as many definitions as there are authors. His own view is catholic, embracing 'the whole gamut of indirect and even of direct meanings: connotations, presuppositions, implications, implicatures, figures of speech, intended meaning, and so on'.[34]

A good starting point for thinking about symbols is Sir Edmund Leach's short, elegant book *Culture and Communication*. The central idea is that symbols are arbitrary and have meaning *only in relation to other symbols*. The 1's and 0's inside my computer, which effectively have produced the manuscript of this book, are arbitrary, make sense only as a set, and have meaning only in a specific word-processing program. They would be meaningless garbage in a graphics package or a spreadsheet program. Symbols have validity only within rules and bounds. However, people readily slide between different rules, resulting in metaphor. 'The lion is a beast' is valid in the natural world of animals, and 'the king is the most powerful person in the state' is a valid description of society. However, 'the lion is king of the beasts' is metaphor, merging two incompatible sets of rules. The lion may be the most powerful animal in the jungle, but it is only metaphorically the *king* of the jungle, jungles not being human societies. That the phrase is metaphorical can be seen by contrasting similar phrases such as 'the lawyer of the jungle', or 'the bus driver of the jungle' or 'the statistician of the jungle'; are such images meaningful or not? The only possible answer is that it depends on the context. Nevertheless, metaphors do make sense, and people have little trouble understanding them. Although, in principle, anything can be a symbol, Leach saw the structure of the human body as being particularly good for generating symbols, and in particular for creating opposites. Taking the navel as the centre of the body, there is a sense in which the arms 'match' the legs the genitals 'match' the head, the back 'matches' the abdomen, and the right side 'matches' the left, but they are not the same. They are, therefore, ideally suited for representing symbolic contrasts between paired ideas, such as good and bad.[35]

Although Leach carefully documents the logic of how symbols can validly be used, and also tells us something about where they come from, he does not really explain *why* humans use symbols so prolifically. The anthropologist Dan Sperber suggests that symbolism is an inevitable part of the functioning of the human mind. Despite continually struggling to make conceptual sense of the world, the mind often finds that no logical sense can be made of it. However, the mind doesn't stop working as a result. Instead, the symbolic system takes over, collecting, storing, putting information together, looking for patterns, in the hope that one day, somehow, meaning will emerge. Inevitably, each person approaches such tasks differently. Some symbols are nigh on universal, and everyone uses them in the same way – perhaps, light and dark, or male and female. Others are specific to particular cultures or belief systems, as with the symbolism of Christmas in a Western culture. Finally, some symbols are completely individual, so idiosyncratic that they can barely be explained to another. They are the stuff of dreams, paintings or poetry.[36]

Symbols share a feature of other types of human thought which are rule-bound in that we are often unable to specify the rules that govern them, yet can instinctively feel whether they are correct or otherwise. A classic example is that native speakers of a language often know whether sentences are grammatical, but cannot say why they are. Sperber describes examples of left–right symbolisms that work in the same way. Observing the Dorze in the Sudan, where he was doing field work, he noticed a ceremony in which a group of dignitaries circled the market-place in an anti-clockwise direction. He was told that the tour could not be done in the other direction. Why? The only answer he received was, 'It is the custom.' Explanation breaks down, although the system is clear enough to all who are using it. Even outsiders can learn such rules without knowing how, Sperber describing how he correctly used the Dorze division of objects into cold and hot, senior and junior, 'following principles I must have internalised intuitively since – I repeatedly tested this – I apply them as they do, without being able yet to make them explicit'. We in the modern Western world are no different. Sperber, a Frenchman, mentions a lateral symbolism in the *code de politesse*, the use of a knife and fork, which has been 'inflicted on each of us from infancy'. There are half-hearted explanations and justifications, often very superficial, of the type, 'It is polite to hold one's knife in the right hand,' without explaining why it is polite and what politeness means. Although rules underlie these behaviours that are acquired through learning, there is rarely explicit teaching and often no explanation is possible. For instance, notes Sperber, why is it that, 'when finished eating one puts the knife and fork together parallel towards the right rather than towards the left'? Think for a moment about one's surprise, and perhaps even the distaste or opprobrium that would follow, if the cutlery at a formal dinner table were laid the wrong way round or at random. Something deep is being stirred here but it is far from clear what it is.[37]

To complete outsiders, our cutlery symbolisms would be as pointless and as incomprehensible as the ones described at length by Lévi-Strauss, in which the Osage Indians tell us that 'the rising sun emits thirteen rays, which are divided into a group of six and a group of seven corresponding respectively to right and left, land and sky, summer and winter'. Why six, why seven, why is the group of six on the right, why does the right correspond to land or summer? Sperber's argument is that we must not seriously expect proper causal, functional explanations for such phenomena. This does not mean, though, that they are unimportant. They are there because our minds cannot accept otherwise, but they are likely to be as opaque to insight and reflection as are, say, the biochemical workings of our own internal organs.[38]

In this chapter, I hope it has become clear that left–right symbolisms are universal in human cultures, and that they are driven both by physiological differences between the two sides of the body and by social pressures. They gain much of their distinctive character from the tendency of the human mind to process the otherwise incomprehensible in symbolic terms. Since the first chapter of this book was about seemingly 'hard' biological and physical problems, to do with the layout of organs in the body, the way the brain is organised and the chemicals of which bodies are composed, this subsequent chapter's concern with the apparently 'soft' topic of symbolism might have come as a surprise. That, though, is to ignore the fact that even the hardest scientists (and my usage of 'hard' and 'soft' is itself deeply symbolic) need to give names to the phenomena they identify and these names inevitably acquire symbolic overtones that colour the way they are thought about. A thousand popular-science articles telling us that 'the universe is left-handed' or, even more revealingly, that 'God is a weak-left-hander' (the phrase of Wolfgang Pauli), show that symbolism will be ever present in a science carried out by humans.[39]

Throughout this chapter, we have taken for granted that the terms 'left' and 'right' have a clear and indisputable meaning. That assumption, though, is not quite so straightforward as it may seem, and is the subject of the next chapter.

☞   ☜

3

ON THE LEFT BANK

In 1869, Thomas Henry Huxley (Figure 3.1), nicknamed 'Darwin's Bulldog' for his tenacious defence of the theory of evolution at the British Association meeting in Oxford, where he had verbally savaged the Bishop of Oxford, had yet again taken on rather more commitments than even his talents could properly handle. Huxley was not only a savage controversialist and a gifted, meticulous scholar, but also, like contemporaries such as Ruskin, Tyndall and William Morris, he believed strongly in the benefits of education, especially for the working classes of Britain, and in the pressing need for the popularisation of science – in his own words, 'to bring science down from the skies'.[1]

When asked to give a series of twelve lectures at the London Institution in 1869, Huxley readily agreed. Although it took nine years, his shorthand notes were eventually published as a book, *Physiography*, which was an instant success, selling 4000 copies in six weeks, reprinting for Christmas, and soon going into a third edition. The scope of the book was broad, from rain, snow, ice and sea through to glaciers, earthquakes, volcanoes, coral reefs, the movement of the Earth and the composition of the sun. 'It took children from their parish to the outer reaches of the solar system,' it has been said. Not only children – in the libraries of the Mechanics' Institutes in the north of England it was the most borrowed book.[2]

Huxley's lectures were given in London and his opening lines start at the heart of the city which was then the capital of the largest empire the world had seen, and one of the intellectual centres of the world.

No spot in the world is better known than London, and no spot in London is better known than London Bridge. Let the reader suppose that he is standing upon this bridge, and, needless of the passing stream of traffic, looks down upon the river as it runs below. It matters little on

Figure 3.1 T. H. Huxley in 1893 at the age of sixty-eight. The young Julian Huxley, who later did research on the embryology of the side of the heart, sits on the knee of his 'grandpater'.

which side of the bridge he may chance to stand; whether he look up the river or down the river, above bridge or below bridge. In either case he will find himself in the presence of a noble stream measuring, when broadest, nearly a sixth of a mile from bank to bank. The quantity of water under London Bridge varies considerably, however, at different seasons, and even at different hours on the same day.

From this simple, almost parochial start, Huxley builds his entire edifice. The book, which reads as well today as when it was written a hundred and thirty years ago, is still a *tour de force* of popular science writing and, indeed, can be seen as one of the founders of the genre.

Our interest in this book is for a different reason. After only half a dozen or so paragraphs, Huxley considered the seemingly uninspiring question of how one might describe the geography of a river. Despite the banality of the subject matter, his words have an almost Miltonic ring about them.

It is obviously convenient to have some ready means of distinguishing the two banks of a river. For this purpose, geographers have agreed to call that bank which lies upon your right side as you go down towards the sea the *right* bank, and to call the opposite side the *left* bank. All that you have to do then, in order to distinguish the two sides, is to stand so that your face is in the direction of the mouth of the river, and your back consequently towards its source, when the right bank will be upon your right hand and the left bank upon your left hand. At Gravesend, for example, the right bank is that which forms the Kentish shore, while the

left bank is on the Essex side. With reference therefore to the rivers tribu-
tary to the Thames, it is said that the Churn, the Colne, the Leach, the
Windrush, the Evenlode, the Cherwell, the Thame, the Coln, the Brent,
and the Lea empty themselves into the Thames on the *left* bank; and the
Rey, the Cole, the Ock, the Kennet, the Loddon, the Wey, the Mole, and
the Darent, open into the river on its *right* bank.

This method for describing rivers seems to be used everywhere. Hence, in
Paris there is the left bank of the Seine, and so on. Although the method
is straightforward enough, Huxley points out that it has the problem of
assuming the reader is already familiar with the Thames, 'but to a perfect
stranger, one who had never seen the river and knew nothing of London
Bridge, such a method of description would be unintelligible'.[3]

Huxley contrasts this description with the advantages of flying over the
Thames in a balloon, when the river below would be laid out like a map.
With the map in the conventional direction, north pointing upwards,
and since 'the east lies on the right hand of the person who looks at the
map, and the west lies on his left hand', it is then possible to work out
that the Thames flows from its source near Cirencester in the west of
England to its eventual destination on the east coast of England. Huxley
stresses the utility of using north, south, east and west, since they

are terms which have a meaning quite independent of local circumstances, and indicate definite directions which can be determined in any
part of the world and at all times. When, in the early part of this chapter,
we used the local expression 'up the river' and 'down the river', 'above
bridge' and 'below bridge', it was assumed that the reader was familiar
with the Thames...By employing, however, the terms north and south,
east and west, we are using expressions that are familiar to all educated
people, since they refer to standards of direction universally recognised.

Huxley is right that it would be highly desirable to find terms 'which are
quite independent of local circumstances', but then, quite uncharacteris-
tically, he misleads himself, which is surprising to say the least. The key
to the problem is that phrase *universally recognised*. 'Universally' is a
strong word when used in the context of geography and astronomy, as
Huxley was doing here. The nature of the error, however, is subtle, and
not immediately apparent. Ultimately it will lead to some deep questions
of biology and physics, both of which Huxley would have found
fascinating.[4]

How did Huxley propose to find north and south? His first method was
the sort that a Boy Scout might use. On a sunny day, place a stick

vertically in the ground, measure the length of the shadow, measure it again a little later, and keep doing so until the shadow is at its shortest, which it will be at true noon. The shadow is then pointing due north.[5]

What is wrong with this method? Principally, that it is so very obviously far from *universal*; indeed, it would not work across a substantial proportion of the Earth's surface. Admittedly, he was mainly looking for a technique that would work for his audience in London and that would identify the terms north and south in a way that would be 'familiar to all educated people'. However, any reader in the southern hemisphere will instantly recognise the flaw in the method. In the same way that people in Australia who want sunny gardens prefer houses with north-facing gardens rather than those facing south as preferred in Britain, so the stick in the ground in Australia would tell the direction of *south*, not north. As long as the Australian reader was south of Alice Springs the stick would at least give a consistent answer the whole year round. However, for a reader between the Tropics of Cancer and Capricorn the method would give different answers at different times of the year.

Can the method be salvaged? Huxley, himself, recognises the practical problem that the method only works if the sun is shining, and he describes two other obvious methods: the use of the Pole Star or of a compass. In fact, both of these methods also fail in a general sense. The compass fails because although magnetic 'north' is currently in the direction that we call north it has not always been so. Geologists have found that the magnetic poles of the earth have 'flipped' on very many occasions in the past. The method therefore will not work in perpetuity, and that, from a scientific point of view, is a problem.[6]

The stars look as if they are a much better bet and, short of some dreadful accident to Earth such as colliding with a huge asteroid, the method of identifying north from the Pole Star (or in the southern hemisphere, south from the Southern Cross) should probably go on working into the far foreseeable future. The actual pattern of the stars as seen from Earth will change slowly (and we know that the night sky already looks somewhat different than it did even a few thousand years ago, at the time the Ancient Egyptians were building the pyramids). But there should always be some star or pattern of stars that will consistently identify the direction of the Earth's north pole.[7]

The deeper problem lies in generalising our concept of north to some other planet in some other part of the universe, for if the word north has any universal meaning, then any other planet should also have a north and a south pole. But which is which? Clearly the Pole Star is going to be of no use now, for all the stars will look totally different. Likewise, we can't make the assumptions that this other planet will also have some

vast internal magnet oriented in the same direction as Earth. To com-
pound our difficulties, we do not know how the planet will orbit around
its sun. There will, however, still be two poles on the planet around
which the rotation seems to take place. So which of them will be the
north pole?

Astronomers have a simple rule for this situation, which they call the
'right-hand screw rule'. Occasionally, people other than astronomers
need to solve this problem. One such might conceivably be Father
Christmas, at least according to a Christmas issue of the *New Scientist*. In
an article asking whether the north pole of some other planet or moon in
our solar system might provide a more congenial abode for Santa Claus
than the Earth's, Justin Mullins succinctly described the rule:

> Make your right hand into a 'thumbs up' shape. If the planet direction of
> rotation matches the way your fingers curl, your thumb points towards
> the north pole. Try it with the way the Earth rotates (the Earth's rotation
> is from West to East, which is why the sun appears to move from East to
> West).

This means, for instance, that relative to Earth, the north pole of Venus is
'underneath' the planet, since Venus, alone among the planets of our
solar system, rotates in the opposite direction.[8]

The surprising thing about the right-hand screw rule is those words
'right hand'. It seems that unless we know our right hand from our left
hand we will never be able to meet someone on the correct hemisphere
of an unknown planet. We might be able to navigate all the way to 'a
small planet somewhere in the vicinity of Betelgeuse', as Ford Prefect
describes home in *The Hitch Hiker's Guide to the Galaxy*, but the final stage
of meeting at some specific location in the northern hemisphere of that
small planet would be doomed to failure unless we were certain that the
concept of right on that planet was the same as ours.

This problem of defining left and right underlies numerous areas of
science and everyday life, and ultimately all have the same solution,
lying, as it were, in our own hands. Even such seemingly non-handed
navigational terms as 'port' and 'starboard' find their origins in the right
hand. Before the invention of the ship's wheel, ships were steered by a
paddle or rudder which was held in one hand. Since it typically had to be
controlled by a right-hander, it was on the right-hand side, this 'bord' or
side consequently becoming known as the 'steorbord', the steering side.
The opposite side was the 'ladebord', or lading side, perhaps because the
steersman found it easier to navigate this side against the quay. Either
way, it later became known as 'port', since 'starboard' and 'larboard' were

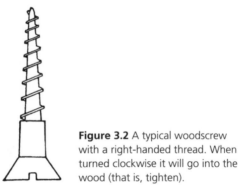

Figure 3.2 A typical woodscrew with a right-handed thread. When turned clockwise it will go into the wood (that is, tighten).

too easily confused. Why, though, the starboard light is green and the port light red is another matter.[9]

To discover additional complications in defining right and left, we have only to consider the humble woodscrew, a surprisingly late invention. Like so many everyday objects, there is a wealth of engineering, both formal and informal, behind this deceptively simple piece of metal without which so many useful things would simply fall apart. Look carefully at the screw in Figure 3.2 and it is obvious that it is highly asymmetric. It is, in fact, what we call a 'right-hand screw' with a 'right-hand thread'.

Most screws in everyday use have a right-hand thread. Occasionally there are exceptions, an example being the thread holding the left-hand pedal on a bicycle, which is left-handed so that it tightens as one cycles, rather than slowly coming undone. Left-hand screw threads were also used for the electric light bulbs in railway trains so that people could not steal them to use at home. Coffin screws are also traditionally left-handed.[10]

A right-hand screw is one in which the screw is tightened clockwise, turning a screw clockwise being much easier for right-handers since it uses the powerful forearm muscles of supination rather than the much weaker ones of pronation. It is worth noting here how another description of asymmetry, the direction a clock's hands turn, has also been related to the right hand. Since spirals (or helices) occur throughout the biological world, principally because they are the most efficient form of packing long thin molecules into a small space, it is important that there is a consistency in the way that they are named. Many of the large molecules that make up cells, such as DNA and proteins, are right-handed spirals in precisely the same sense that an ordinary screw is right-handed.[11]

Although a right-hand screw happens to be that which is most

Figure 3.3 A simple necklace found beside the body of Ötzi, 'the man in the ice'. The simple twist in the cord is a left-hand spiral, and has probably been made by one end being held fixed by the left hand, while the right hand twists in a clockwise direction.

Figure 3.4 Two Gaulish, Iron-age torques from the British Museum, perhaps made about 500 BC. Each has a clear left-hand spiral.

conveniently screwed in by a right-hander, it is a very confusing way of defining spirals in general. Look, for instance, at the twisted necklace in Figure 3.3, or the torque in Figure 3.4. The direction of twist in both spirals is quite clearly opposite to that of the right-handed screw in Figure 3.2. However, these are typical of the objects *produced* by right-handers, who fix one end with their left hand and then twist the free end clockwise with their right hand. If you don't believe it, try it. As is so often the case, we are surprisingly ill-observant about right and left. Potentially, however, there is a serious problem here, since some scientists – botanists, in particular – have chosen to call a 'right-hand spiral' one that would be produced by a right-hander making a twist – that is, like the necklace and the torque above, which most people would like to describe as left-handed in the sense that it is the opposite of the screw shown in Figure 3.2. Of course, none of this matters so long as everyone knows what they are talking about, but that isn't always the case. The risk of confusion is great. Remember *Misalliance*, that charming song by Michael Flanders and Donald Swann in which the 'right-handed Honeysuckle' and the 'left-handed Bindweed' – the Romeo and Juliet of the plant world – fall in love, despite the opposition of their families ('We twine to the right, and they twine to the left')? Eventually tragedy results ('Together they found them the very next day. They had pulled up their roots and just shrivelled away'). In fact, no-one seemed to know which way they were twining. At the beginning 'The fragrant Honeysuckle spirals clockwise to the sun' (that is, a left-hand screw thread, which is the correct description), but later we read 'Said the *right-hand thread Honeysuckle* to the left-hand thread Bindweed' (emphasis added). If one family had been advised by a botanist and the other by an engineer, they would have thought the couple ideally suited.[12]

The situation with shells is both simpler and more complex. Conchologists, who study shells, imagine the path of a small insect crawling in at the bottom of a shell and climbing to the top, and ask whether it is turning to its right or left side; if to the right, they call the shell *dexiotropic* (right-turning) and if to the left, *leiotropic* (left-turning) (see Figure 3.5). The complication in the story is that a right-turning (dexiotropic) shell has a left-hand spiral, and a left-turning (leiotropic) shell has a right-hand spiral. That can be seen by imagining an ant crawling up the thread of the screw in Figure 3.2 – it will continually be turning to *its* left. The same situation arises in spiral staircases, a left-hand spiral meaning we are climbing to our right and, as a result, have our right hand on the inner bannister (see Figure 3.6). The only blessing in all this is that conchologists, at least, make use of technical terms, dexiotropic and leiotropic, so there is less danger of confusion.[13]

Figure 3.5 The top figures show the common or leiotropic form of *Voluta vespertilio*, which has a right-hand spiral, whereas the lower figures show the rare, dexiotropic form of *Voluta vespertilio*, which has a left-handed spiral.

Figure 3.6 An oak staircase in St Wolfgang's, Rothenburg, in the form of a left-handed spiral, which is ascended with one's right hand on the inner bannister.

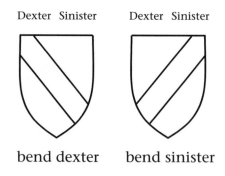

Figure 3.7 The *bend dexter* and *bend sinister* in heraldry.

Lurking behind these terminological difficulties are several deep problems. One is straightforward, but produces much confusion. Do we look
at something with respect to *our* right hand or the *object's* right hand. If
you and I stand opposite one another, then your left hand is on my right-
hand side, and so on. Does one use a *viewer-centred* or an *object-centred*
view? Different groups of people have chosen different answers to that
problem. When, many years ago, I studied whether painted portraits
faced in one direction or the other, I described them as showing their
right cheek or their left cheek – an unambiguously object-centred
description. I was following fairly standard medical nomenclature where
'the left leg' means the patient's left leg and not the leg on the surgeon's
left side as the surgeon looks at the patient. Nevertheless, one can see the
potential for occasionally disastrous confusion. Even doctors get themselves in a mess. Chest X-rays invariably are looked at as if the patient
was lying in a bed in front of the doctor, so the left-sided heart is on the
doctor's right-hand side. Fine. The only trouble is that when cross-sec-
tional brain scans were invented, doctors chose instead to imagine they
were hovering above the patient's head, looking down on them. The
result is that the right side of the brain is on the doctor's right side.
Again, great potential for confusion, particularly for radiologists who are
looking at both chest X-rays and brain scans from the same patient.
Fortunately, there is a straightforward ergonomic solution – a large letter
R on one side and **L** on the other.[14]

Doctors are not the only ones with problems. Actors have it the whole
time. The producer sits in the theatre auditorium looking at the actors,
who are looking back. 'Come in from the right-hand side' is then totally
ambiguous. Whose right? Theatrical usage therefore says 'Enter stage
right' which is understood as the right-hand side of the actor on the
stage, looking out at the audience. Art historians use the terms 'proper
right' and 'proper left' in a similar way, to avoid confusion in interpreting

manuscripts and paintings. Heraldry has similar difficulties – how does one refer to the two sides of a shield? The answer is with respect to the person carrying the shield, so that a *bend dexter* runs from the knight's top right to bottom left, and a *bend sinister* from the knight's top left to bottom right, as in Figure 3.7. The *bend sinister* is conventionally the sign of a bastard son, whereas the *bend dexter* is for a legitimate son.[15]

Such problems, though, are merely from convention – do we mean the same thing when we talk about right and left? Lurking rather deeper are some bigger questions. In the 1830s, a young student realised the potential theological problems of not being able to distinguish left and right:

> [D]efinition is lacking…who shall define, who shall determine, which is the right side and which the left?…Oh, vain is all our striving, our yearning is folly, until we have determined which is right and left, for he will place the goats on the left hand, and the sheep on the right. If he turns round, if he faces in another direction, because in the night he had a dream, then according to our pitiful ideas the goats will be standing on the right and the pious on the left…I am dizzy – if a Mephistopheles appeared I should be Faust, for clearly and every one of us is a Faust, as we do not know which is the right side and which the left.

The student, rather surprisingly, was the young Karl Marx, and it is more than possible that he realised the problem after reading a paper by Immanuel Kant, the eighteenth-century philosopher who first properly considered the nature of right and left. Kant, one of the greatest philosophers, spent almost his entire life in the East Prussian city of Königsberg, reputedly never travelling more than forty miles from it. His was not a gripping life; 'wholly uneventful' was Bertrand Russell's description. Woken at five, he worked until twelve, ate his single meal of the day in a restaurant at lunchtime, walked in the afternoon – when one may fancifully imagine him contemplating the impossibility of the famous Königsberg bridge problem – and then read until ten when he went to bed.[16]

If Kant's life was outwardly unmomentous, inwardly he was grappling with some of the deepest and most difficult questions that science and philosophy have tried to attack, including the question of whether space is absolute in some sense or merely relative. The problem is an old philosophical chestnut. Sir Isaac Newton added a new twist to it by not only making it a question of metaphysics but also relating it to empirical science. Since then, as Lawrence Sklar puts it, 'the roster of names of those who have attended to this problem seems like a roll call of scientific genius in the Western world – Newton, Leibniz, Huyghens, Berkeley, Mach, Einstein, and Reichenbach are just a few'. The problem can be put fairly simply: is position in space defined in some absolute sense, or can we merely know the position of any particular object relative to others? Thus, at present I can say that I am about one metre away from the wall directly in front of me, one metre from the wall to my right and four metres from the wall to my left, and so on, defining those walls in relation to other objects until eventually the position of every object in the universe is specified. Those specifications, however, are only relative. If somehow the entire universe were to be moved one metre to the left, would it be possible to tell? Would anything be different?

Modern interest in the argument was started by Sir Isaac Newton, who found himself at odds with Gottfried Wilhelm Leibniz, with whom he also had a vicious and long-lasting dispute about who should be given priority for having invented the calculus. Whereas Newton argued strongly for an absolute version of space, Leibniz argued strongly for the opposite: 'I hold space to be something merely relative, as time is; that is, I hold it to be an order of co-existences, as time is an order of successions.' A flavour of the nature of the entire argument can be gained from a thought-experiment of Newton's. Imagine that the universe contains nothing but two weights, joined together by a piece of string. If the weights were not moving, then the string would be slack as the two bodies hovered in space. However, if the weights were rotating around a point at the middle of the string, then the string would tighten as what is loosely called centrifugal force pulls them apart. The question is, how, in this rather limited universe, would one know these objects were rotating? There would be nothing against which to measure the rotation. Nevertheless, one would still be obliged to argue that space was absolute and not merely defined by the relative position of one weight in relation to the other, which would be unchanging. This is no place to dabble any further in the arguments, which have already filled many books. Suffice it to say that the problem very much concerned Kant, for whom the right and left hands provided a possible answer to the problem.[17]

Kant's 1768 essay, with the unedifying title *Concerning the Ultimate Ground of the Differentiation of Directions in Space*, was brief, particularly in comparison with the vast, dense tomes he was later to produce, including the *Critique of Pure Reason* and the *Critique of Practical Reason*. It comprises only seven and a half pages of the 3000 that make up his complete works. And yet it claims to solve a major philosophical problem; indeed, to provide 'clear proof' to a question for which 'Everybody knows how unsuccessful the philosophers have been in their efforts to place this point once and for all beyond dispute.' That said, much of the essay is not directly relevant to the question of absolute space, interesting though it is. Kant does, however, mention the problem with which we began this chapter: of how to tell north from south without first distinguishing right from left. His passing comments on the right and left hands are of particular interest. Even in 1768, he was struck by the universal right-handedness of humans ('It is everywhere the right hand which is used for writing'), although he acknowledges occasional cases of left-handedness: 'If...we set aside individual exceptions which...cannot disturb the generality of the rule according to the natural order, all the peoples on earth are right-handed...Everywhere men write with the right hand.' Although not the first time that the universality of right-handedness has been noted, it seems to be the first time in the modern era by a major philosopher.[18]

The main theme of Kant's essay is the nature of the difference between right and left. Kant considered other objects which are asymmetric, such as right- and left-hand screws, but then concluded, 'the most common and clearest example is furnished by the limbs of the human body, which are symmetrically arranged relative to the vertical plane of the body'. The right and left hands are similar in so many ways and yet are fundamentally different in one crucial respect. As Kant put it, 'the glove of one hand cannot be used for the other'. There is the crunch. What is the nature of the difference between our two hands? To use technical jargon, our two hands are *incongruent counterparts*. What does that mean?[19]

Going back to the time of Euclid in the third century BC, mathematicians have studied the congruency of geometrical objects. At school we are taught some simple rule such as, 'If two triangles have the same angles and the same length of sides, then they are congruent.' So, in Figure 3.8, triangle A is congruent with triangle B, as can be shown by sliding triangle B across the page until it exactly overlies triangle A, as in Figure 3.9. What, though, about triangles C and D in Figure 3.10? The angles and the sides are the same, but one is the mirror-image of the other. So are they congruent? Can triangle D be slid so that it overlies triangle C? No; however much one tries, it cannot be done. For that reason, C and D have to be regarded as different, and are therefore known as 'incongruent counterparts', in contrast to A and B which are 'exactly congruent'.

Although triangle D cannot be slid across to overlie triangle C, there is nevertheless a way of making C and D exactly congruent. All one needs to do is to pick up triangle D, turn it over in mid-air, and then put it down on top of triangle C, as in Figure 3.11. The problem is solved by an important trick. The paper on which the triangles are printed is two-dimensional, and the triangles are also two-dimensional. Picking up the

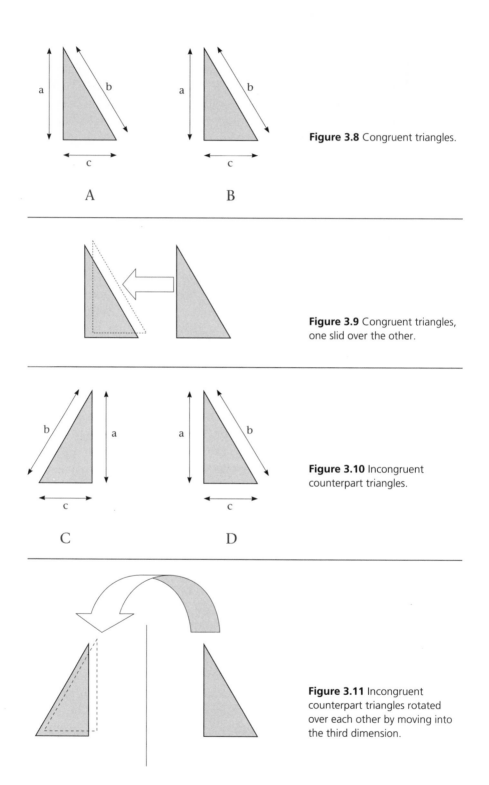

Figure 3.8 Congruent triangles.

Figure 3.9 Congruent triangles, one slid over the other.

Figure 3.10 Incongruent counterpart triangles.

Figure 3.11 Incongruent counterpart triangles rotated over each other by moving into the third dimension.

Figure 3.12 The train at the left, on a single track, cannot be turned around so that it is in the position shown on the right.

Figure 3.13 A reversing loop can be used to turn the train around, in effect by moving into another dimension.

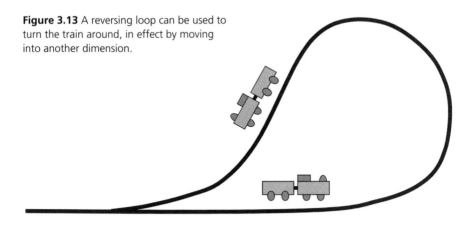

triangle rotates it through the *third* dimension, the one that hovers above the page. Incongruent counterparts can always be made exactly congruent by moving them into a higher dimension. We can see this in a simpler one-dimensional situation.

The twentieth-century philosopher Wittgenstein, in his only comment on right and left, pointed out that Kant's argument is true even in a one-dimensional space. Think about a very simple model railway, which has a train on a single straight track. Geometrically, the system is one-dimensional, being a line, and since the train can only roll from one end to the other, its position can be specified entirely by a single number describing its distance from the beginning of the track. Is it possible to turn around the train at the left-hand of Figure 3.12 so that it faces the other way down the track, like the one on the right? Anyone who has had a model railway knows that the answer is no. Or, rather, not if, like a real train, one has to keep the train on the rails. However, if we can lift the train off the track, into a higher dimensional space that is, then it can be turned in mid-air and put down the other way. Those with more experience of model railways will also suggest two other ways of turning the train around. One is to use a turntable. More subtle is to use the track layout shown in Figure 3.13. After running the train round the loop it returns facing the other way. These methods work because the railway system is now two-dimensional rather than one-dimensional – a single number no longer describes the position of the train, two numbers being needed, such as the distance north and east from a reference point.[20]

If this trick, of rotating into a higher dimension, works with one and two dimensions, would it also work with our two hands? Could the right hand be made exactly congruent with the left hand by rotating it through a higher dimension? Undoubtedly, yes. *If* one could pick up a right hand, lift it into the fourth dimension, rotate it, and put it back down again, then it would be a left hand. There is perhaps a sense in which a mirror does that.[21]

What does this matter to Kant and the argument about absolute space? The most important point is not that the difference between our right and left hands incontrovertibly demonstrates that space must be absolute, but that it provides very serious problems for those who, along with Leibniz, suggest that space is relational – in other words, space can only be described in terms of the interrelations between objects themselves. However, if the relationist position fails, then, in the absence of an obvious alternative, the absolute position seems more likely to be correct.

If space could be described adequately in terms solely of the relationships between objects, as Leibniz and the relationists argued, then objects that are different could be distinguished by different interrelations of their components. My right hand is different from the right hand of a child, say, because the tip of the index finger is further from the knuckle, and so on. That, though, is not the case with my own right and left hand. *All* of the angles and lengths are the same in my two hands, yet still the hands are indisputably different. I cannot put my right glove on my left hand or my right shoe on my left foot – they simply don't enjoy a fit 'as loose and easy and yet as close as a good glove's with your hand', as Ford Madox Ford put it in *The Good Soldier*. For Kant, the conclusion was inescapable: there must be something else against which the right and left hands can be compared – and that could only be space itself: 'Our considerations…make it clear that differences, and true differences at that, can be found in the constitution of bodies; these differences relate exclusively to *absolute* and *original space*.' Even empty space must have some absolute structure against which it can be said that our right hand is not the same as our left hand.[22]

An interesting question arising from Kant's argument is the possibility of taking my three-dimensional left hand, turning it round in a fourth dimension, and putting it back into our three-dimensional world as a right hand. If that could be done, then Kant's argument for an absolute space would have potential problems. But it is only a *theoretical* possibility. The spatial world we live in is *three*-dimensional, and there is no empirical evidence of a fourth spatial dimension. Furthermore, we can be extremely confident that no humans have ever been turned around in a fourth dimension. If they had been, they would be instantly recognisable. Not only would they have right and left hands reversed, and their heart on the right-hand side, but all of their amino acids would be dextro-rotatory and their sugars laevo-rotatory – which would swiftly become very obvious, being incompatible with life on this earth.[23]

Of course there is no argument so simple that philosophers will not dissect every one of its assumptions, corollaries and implications; and rightly so, that is their job. Philosophy is about seeing the complexities behind an apparently straightforward world. That doesn't necessarily, however, help the ordinary thinker – 'Like stirring mud', was Samuel Butler's description of philosophy. Kant's argument about left and right has been the subject of dozens of scholarly articles and at least one book, and the topic remains as alive and confusing today as in 1768. Kant, himself, did not help matters. A few years later, he rejected his own argument concerning absolute space, also rejected all arguments for a relational space, and instead substituted a 'third way', a transcendental view seeing space primarily as constructed by the human mind rather than passively observed. Space could, therefore, take any form in principle, but our minds have created it with three dimensions. This is not the place to think further about the huge and technical philosophical literature on Kant's idea. What we need to do is consider the implications for our own question of how the right- and left-hand sides of the body can differ so dramatically.[24]

Although it may seem like another detour, some of the problems relating to the right and left hands can be understood by thinking about the arcane, impractical problem of communicating with extra-terrestrials. If E.T. were to call, how would we explain the distinction between left and right? Hardly surprisingly, Kant did not consider the problem directly, but his 1768 paper suggests a serious difficulty for anyone trying to talk to E.T. about hands and gloves. Martin Gardner, who wrote the 'Mathematical Games' column in *Scientific American* for many years, called it *The Ozma Problem*, and it involves communicating with a remote, distant intelligence on a faraway planet by radio:

> Is there any way to communicate the meaning of 'left' by a language transmitted in the form of pulsating signals? By the terms of the problem we may say anything we please to our listeners, ask them to perform any experiment whatever, with one proviso: *There is to be no asymmetric object or structure that we and they can observe in common* (emphasis in original).

The problem can be made more explicit by imagining a Martian in a spaceship, with whom we only have radio contact. The Martian has made a symmetrical pair of gloves by following our verbal description. We now ask the Martian to pick up the right-hand glove. Can it be done? No, is the generally agreed answer (or at least it was before 1957 – and we will return to that in chapter 6). Why it is impossible goes back to Kant, who said that the difference between right and left hands cannot be stated 'in terms intelligible to the mind through verbal description'. Words, be they in a radio message, a book, Morse code, or binary digits are essentially one-dimensional messages and so cannot be used to describe the difference between the two gloves. The problem can, though, be solved by some form of comparison. On Earth, talking to another human by telephone, we might say, 'Lay the gloves in front of you with their palms down and fingers pointing away. The glove with the thumb pointing towards the side of your heart is the right-hand glove.' But this breaks the Ozma rule because it refers to the heart, an asymmetric structure common to both the transmitter and the receiver. Since we don't know whether Martians have a heart on the left side, or indeed have a heart at all, the method fails completely for the Ozma Problem. So do all the other ingenious methods that have been suggested, described by Gardner in detail: cunning methods involving the north and south poles of magnets, the induction of electrical current in wires, the optical rotation of crystals, D- and L-amino acids, the rotation of planets, or whatever. Each, at some stage, requires one to know which is left and which is right before it can be used. Kant himself said it clearly, 'the difference between similar and equal things which are not congruent (for instance, two symmetric helices) cannot be made intelligible by any concept'. Philosopher Jonathan Bennet made very clear what could and could not be done, when he said 'that one could explain the meanings of [right and left] only by a kind of showing – one could not do it by telling'. And telling is all that is available in the Ozma Problem. Showing is simply not possible.[25]

Now this is all very nice and interesting, and provides amusement for those of a philosophical bent, but how can it possibly help to solve Dr Thomas Watson's problem of why the heart is on the left side of the body? The answer is that, in terms of information, a developing fetus is just like a Martian in a spaceship. The messages it can receive are very limited in form, but it will also, at some point, need to know which side is right and which side is left. Most people have their heart on the left side, and there are good reasons to believe that having the heart on the left is, in some sense, under genetic control. If so, then the genes have to tell the developing body where to put the heart – in other words, to put it on the left-hand side and not the right-hand side. Inside spaceship 'Fetus', however, spinning along inside its mother's womb, there is only

very limited communication with the outside world. Most of the messages on how to specify the layout of the body are stored inside the
genetic material, the DNA, but the messages in DNA are just like Morse
code, or language, or any other sequence of pieces of information – they
are one-dimensional. There is, therefore, simply no way that the information stored in DNA can on its own tell the developing fetus where to put
the heart. This is not an empirical question, but a logical necessity, and it
is Kant's 1768 paper that makes the point.

Logical necessity and impossibility are fine things, but of course we *do*
mostly have our heart on the left side and somehow the heart must have
been told to be there. So how? Not by DNA alone, is the answer (and that
is why I was so careful to use the phrase 'in some sense, under genetic
control'). 'In some sense' means that, even though DNA is part of the
story, there must also be some other mechanism. The nature of that
mechanism ultimately has to be 'by a kind of showing...not...by telling'.

If there are people who believe that everything of importance about an
organism is encoded in the genes and the genes alone, then they will
have problems with what I have just said. Genes alone cannot make
one's heart be on the left side. Some extra-genetic information, what in a
broad sense we can call the environment, has to be involved in the story.
The problem can be seen in the attempts currently being made to bring
back to life the mammoth and other extinct organisms using preserved
DNA, or to clone endangered and near-extinct species. If attempted
purely from DNA sequences, perhaps in a virtual computer reconstruction of the actions of all genes, there would be a 50:50 chance that the
mammoth would have its heart on the left or the right side. Of course,
zoologists do not recreate animals only from DNA *in vitro*; that is, in a
test-tube of simple fluids. DNA can only develop into an organism in the
proper environment, which is the fertilised egg of the mother, inside her
womb; and that is a very complex environment. Those trying to recreate
the mammoth will therefore place their carefully obtained DNA inside
the egg of a not-too-distant relative, such as an elephant, hoping that
somewhere in that complex mixture of biochemical substances and sub-
cellular organelles there has to be the information that ensures that
mammoths have their heart on the left.[26]

How the body manages to put the heart on the left is a difficult
problem, to be returned to in chapter 5. Before that, however, we must
think about the words 'right' and 'left', and why it is that so many people
have difficulties with them, making them some of the most confused and
confusing words in everyday use.

4
KLEIZ, DREPT, LUFT, ZESO, LIJEVI, PRAWY

The Russian Imperial Army had such difficulty with its ill-educated, rural recruits not knowing left from right that to get them marching in step their drill instructors would tie a bundle of straw to the right ankle, a bundle of hay to the left, and shout, 'Straw, hay, straw, hay, straw, hay.' The well-trained Roman armies may have had similar problems, since 'spear' meant right and 'shield' meant left.[1]

Problems of left and right are not restricted to the illiterate. The German poet Schiller wrote to a friend, 'during the whole of the evening I could not make out which was right and which was left'. Even the best-educated sometimes resort to devices to help them remember. The physicist Richard Feynman used a mole on the back of his left hand to tell him about left and right, and Sigmund Freud used a method many will find familiar:

> I do not know whether it is always obvious to other people which is their own right and left and where right and left are in others. In my case (in earlier years) it was rather a matter of having to think which was my right; no organic feeling told me. I used to test this by quickly making a few writing movements with my right hand.

This trick, almost a manual pun, works particularly well in English since 'right' and 'write' are homophones – words that are spelt differently but pronounced the same.[2]

Distinguishing right from left is clearly essential in a technological society, for how else would we 'place the right-hand end of the ball wrangler inside the left-most of the glitch wurzlers'? Equally, simple directions have always needed to distinguish right and left ('Don't go into the left-hand cave where the sabre-toothed tiger lives'). Nevertheless, right and left are often confused and, according to Jean Aitchison, 'Muddling up *left* and *right* is possibly the commonest semantic tongue-slip of all, closely followed perhaps by the confusion of *yesterday, today* and *tomorrow.'*[3]

To speak of right and left means, firstly, that language must have words for them. A more subtle requirement is that our brains must be asymmetric. Even then, we still have to learn to use right and left reliably and appropriately. Let us begin by considering the two words in more detail. Where do they come from? English is part of the Indo-European family of languages, in which there are many other words for right and left, as Table 4.1 shows.

Table 4.1: Indo-European words for 'right' and 'left'		
	Right	**Left**
Ancient Greek	δεξιός	ἀριστερός, εὐώνυμος, σκαιός, λαιός
Modern Greek	δεξιός	ἀριστερός, ζερβός
Mycenean	de-ki-si-wo	[not available]
Latin	dexter	sinister, laevus, scaevus
Italian	destro	sinistro
French	droit	gauche
Spanish	diestro, derecho	izquierdo, siniestro
Portuguese	direito	canhoto
Rumanian	drept	stîng
Old Irish	dess	clē, tūath
Modern Irish	deas	clē, (tūath)
Welsh	de, deheu	aswy, chwith
Breton (modern)	dehou	kleiz
Gothic	taihswa	hleiduma
Old Norse (Old Icelandic)	hœgri	vinstri
Danish	højre	venstre
Swedish	högre	vänster
Norwegian	høgre	venstre
Old English	swīþra	winestra
Middle English	riht, swither	lift, luft
Dutch	recht	linker
Old High German	zeso	winistar, slinc
Middle High German	zese, reht	winster, linc
Modern German	recht-	link-
Lithuanian	dešinas	kairias
Lettic	labs	kreiss
Old Church Slavic	desnŭ	šujī, lěvŭ

	Right	Left
Serbo-Croatian	desni	lijevi
Czech	pravý	levý
Polish	prawy	lewy
Russian	pravyj (desnoj)	levyj
Sanskrit	dakṣiṇa-	savya-, vāma-
Avestan	dašina-	haoya-, vairyastāra-
Tocharian A	pāci	śālyās
Tocharian B (Kuchean)	śwālyai	saiwai
Luwian	išarwili-	ipala-
Hittite	kunna-	GÙB-la-
Akkadian	imnu, imittu	šumēlu
Ugaritic	ymn	(u)sm'al
Hebrew	yamin	semo'l
Arabic	yamîne	šimâl
Albanian	djathtë	majtë
Armenian	aǰ	jax

Some of these words are very similar – say, *pravý* in Czech and *prawy* in Polish, or *sinistro* in Italian and *sinister* in Latin – but then the regions where Czech and Polish are spoken are geographically close, and Italian is a modern descendant of Latin. That is why we talk of a language family, since, like blood relatives, the words tend to look similar. Having said that, some of the words are pretty different. Can they really be related?[4]

One of the jewels in the intellectual crown of linguistics has been working out how languages are related, so that it has been possible to reconstruct proto-Indo-European; the linguistic great-granddaddy of most of the languages in the table. That languages might be related was first suggested at the end of the eighteenth century by Sir William Jones, the Chief Magistrate of Calcutta, which was then capital of British India. Jones learned Sanskrit, the ancient language of India in which the Vedas had been written in about 1000 BC, and he was struck by its 'wonderful structure'. Anyone forced to battle with the conjugations and declensions of Latin, and the more complicated ones of Greek, might feel their heart sink with Jones' description of Sanskrit as 'more perfect than the Greek, more copious than the Latin'. What struck him, however, was more than the aesthetics of the language. In 1786, he commented on the strong affinity of Sanskrit to Latin and Greek, 'both in the roots of verbs and in the forms of grammar, [more] than could possibly have been produced by accident; so strong indeed that no philologer could examine them all three, without believing them to have sprung from some common source which, perhaps, no longer exists'.[5]

Jones was prescient, and his hunch of a common source has been more than vindicated, even if the sheer variety of modern languages might at first seem to be against it. An early insight was provided in 1822 when Jakob Grimm, who with his brother Wilhelm collected the famous Fairy Tales, realised that sounds in one language were systematically modified in other languages. Grimm's law described how words that in Sanskrit and Latin begin with the sound /p/ tend to begin with the sound /f/ in Germanic languages (so that *pater* in Latin becomes *father* in English, and *piscis* in Latin becomes *fish* in English). There are many other such laws. The implications were profound, and showed how Indo-European languages were related.[6]

Of course one does not only consider the sounds of words, but also grammar, morphology, and a host of other details. Similarly, we need to look not only at modern languages but at their predecessors; English coming from Middle English, which is related to Anglo-Frisian, which is related to Gothic, and so to East Germanic, and thence to Germanic and proto-Indo-European. Archaeologists have also rediscovered languages such as Hittite, from cuneiform tablets excavated at Boğazköy in Turkey, and Tocharian A and B, written on manuscripts found in the deserts of Sinkiang, in China, on the old Silk Road. Putting it all together is, as the linguist Robert Beekes has said, 'one enormous floor-puzzle' – and a puzzle that unfortunately has no picture on the lid of the box.[7]

The end result is a reconstruction of a language spoken in Europe before about 3000 BC, the linguistic offspring of which would eventually be spoken by half the people on Earth. Where the proto-Indo-Europeans lived is still controversial, though some subtle hints are given in the vocabulary itself. The word **mori*, for a sea or large inland lake, and **neh₂us*, for boat, suggests they travelled by water, and a word for snow, **snoig^{wh}os*, suggests the climate was fairly inclement (the asterisk in front of a word means that it is a reconstructed, hypothesised word, and the symbols such as h_2, and g^{wh} indicate ancient sounds reconstructed by linguists). Candidates for the ancestral home-range include the Southern Caucasus, the north of Mesopotamia, Anatolia, and southern Russia.[8]

Given the richness of proto-Indo-European, it would seem inevitable that there were words for right and left. One can hardly imagine any language which does *not* have those words. How would one give directions for hunting or agriculture, instruct in the use of a tool, or discuss tactics? A glance at the table shows that the various words for *right* and *left* exhibit sufficient similarities to have had common ancestors. The evidence, however, is mixed. Proto-Indo-European undoubtedly had a word for *right*. As in all technical areas, experts disagree as to its precise details, and even on the best way to write it down, but it is something like *dek̂s(i)-*, *t'ek^h-s-*, or *deksinos* / *deksiwos* / *deksiteros*. The commonality behind those reconstructions is clear enough. The astonishing thing, though, is the near total lack of a word for *left* in proto-Indo-European. In a recent encyclopaedic reconstruction, Gamkrelidze and Ivanov are quite dogmatic: 'the unanimous agreement [regarding] *t'ek^h-s-* "right" is in striking contrast with the impossibility of reconstructing a protoform for "left"'. Impossibility is a strong word, but coming from those who have spent their lives trying to reconstruct proto-Indo-European, it must be taken seriously.[9]

How could the proto-Indo-Europeans have a word for right but not for left? Adjectives and adverbs usually come in pairs, which are used as opposites; if some objects are hot then others must be not-hot, or cold; and so on. If the proto-Indo-Europeans had a word for right then they had to have the concept of right and left, and just as surely must have had a word for left. So what happened to it? Gamkrelidze and Ivanov are very clear: 'the reason…must lie in the symbolic meaning of "left" in Indo-European and its tabooing and partial replacement in individual dialects'. This is very much what Hertz had said in *The Pre-eminence of the Right Hand*: 'While there is a single term for "right" [in Indo-European languages] which extends over a very wide area and shows great stability, the idea of "left" is expressed by a number of distinct terms.' There is a hint of this fact in the table, more languages having multiple terms for left than for right – for instance, Latin, which has *sinister* and the older term *laevus*. But the real proof is in the inability to reconstruct a common origin for the words for left – they have mutated so much, with new words being created or coming from outside, that any commonality has long disappeared. It is a vindication of Hertz's intuitions about the symbolic pre-eminence of the right hand, the rampant replacement of left indicating stigmatisation. We can even see the process in English, there being no modern descendant of the Anglo-Saxon *winstre*.[10]

The different linguistic status of *right* and *left* is also seen in the phenomenon that linguists describe as being 'marked'. An unmarked form is a word such as *happy*, whereas a marked form is *unhappy*, the *un-* indicating a deviation from the normal state. Unmarked words are typically more frequent, are linguistically older, and are more neutral (so that the answer to 'Is he happy?' could be either yes or no, whereas 'Is he unhappy?' implies the person is indeed unhappy). Words for *left* are typically marked linguistically, *right* being regarded as the norm, and *left* being indicated as 'other than right'. Right is normal; left is abnormal and is stigmatised.[11]

Before leaving the origins of words for right and left, it is worth asking why, in English, *right* means both 'the opposite of left' and 'the opposite of wrong', as well as having meanings to do with duties, expectations and obligations. Is there any proper connection between these different meanings? The answer seems to be no. Certainly some have argued for a link, but the general feeling seems to be that it is no more than coincidence. There is a word in proto-Indo-European, *$h_3$regtos*, which means just, proper or correct, but it clearly is not related to *deks(i)-*. From *$h_3$regtos* has come a series of words, most notably through the Latin verb *regere*, 'to direct, guide or rule', and thence a host of words, including many Indo-European words for king or ruler. The coincidence of the two meanings of right occurs only in Germanic languages, and is what is known as a 'reverse semantic shift'. If the several meanings of *right* are not equivalent, it hardly needs saying that the fact that in English people mostly *write* with their *right* hands is inconsequential, the pun occurring in no other European language (although that doesn't stop it being useful for helping children to remember which hand is which).[12]

Given that proto-Indo-European can be reconstructed so effectively, some readers may be wondering whether even earlier proto-languages can also be reconstructed, of which proto-Indo-European is but one descendant. Such a thesis is controversial. In 1903, a Danish linguist, Holger Pedersen, pointed out similarities between the Indo-European, Semitic, Uralic, Altaic and Eskimo-Aleut families of languages, and suggested they were descended from a common ancestor that he called Nostratic. Since then, the idea has gone in and out of fashion, and is currently back in fashion, with attempts being made at a reconstruction. As yet, no Nostratic words have been identified for right or left.[13]

Since left and right seem such important concepts, and since the words go back so far in our linguistic history, it is surprising how bad so many of us are at distinguishing right and left. Before discussing that, though, let us try a little memory experiment. Look at Figure 4.1, depicting a British postage stamp. Which of the two images shows the correct orientation of the Queen's head? Similarly, in Figure 4.2, which is the version of George Washington found on an American dollar bill? Finally, Figure 4.3 shows one of the most dramatic comets of recent times, Comet Hale-Bopp, which appeared in 1997 and which many people still recall. Can you remember which way round it appeared? The answers are in the notes at the end of the book.[14]

If you had difficulty with these, don't worry; you are in good company. In a study of Oxford University students, carried out six months after the comet Hale-Bopp had disappeared, only about sixty per cent of those who had seen the comet could remember whether it was pointing left or

Figure 4.1 Which of the British stamps shows the Queen's head as it is actually portrayed?

Figure 4.2 Which of the pictures of George Washington is the way that it actually appears on a US one-dollar bill?

Figure 4.3 Comet Hale-Bopp seen from Chandler's Ford, Hampshire, England on 28 March 1997. One of the photographs shows the comet as it actually was, and the other is left–right reversed. Which picture shows the comet in its correct orientation? The answers can be found in the notes.

right. That was despite the majority having seen the comet on between two and nine occasions, with many seeing it more often than that.[15]

How do we learn to use the words 'right' and 'left', and why do some people seem to have such difficulty with them? As is so often the case, one can understand how adults correctly do something by seeing children do the same task wrongly but then eventually develop the skill. Left and right are not understood until quite late in childhood. The doomed

Figure 4.4 A test of right–left awareness in children and adolescents by Sonja Ofte and Kenneth Hugdal. Which 'hand' is the right hand in each of these figures? Because some of the figures are facing and some have their back to the viewer, and some have their arms crossed, the question can only be answered correctly if one fully understands the nature of right and left.

seven-year-old André in Sebastian Foulkes' novel *Charlotte Gray* demonstrates the problems young children have with left and right: 'When he arrived at the road, panting, he hesitated for a moment. Left, right…He was still too young to know the difference, but he knew the school was that way, up the hill.' Winnie-the-Pooh, that bear of little brain, could also never get it right: 'Pooh looked at his two paws. He knew one of them was the right, and he knew that when you had decided which one of them was the right, then the other one was the left. But he never could remember how to begin.' Like others with problems sorting out left and right, Pooh also had troubles with reading: 'Winnie-the-Pooh read the two notices very carefully, first from left to right, and afterwards, in case he had missed some of it, from right to left.' Children's left–right knowledge can be tested more systematically using tests such as the one shown in Figure 4.4.[16]

Much of modern child psychology was developed by the Swiss psychologist Jean Piaget, who in 1928 described his studies of how children developed the idea of right and left. He started with some very basic questions: 'Show me your right hand. Your left. Show me your right leg. Now your left.' By five or six, most children answered correctly. That, though, doesn't mean they understood the difference between right and left. Child psychologists are a seasoned, sceptical lot, like psychologists in general, who often wonder if what looks like a high-level skill is actually something less sophisticated. Piaget, therefore, asked another set of questions which was very like the first. Sitting opposite the child he said: 'Show me my right hand. Now my left. Show me my right leg; now my left.' Whereas most five-year-olds got the first set of questions right, hardly any answered the second set correctly. Five-year-olds, therefore, do *not* fully understand what right and left mean. In fact, they only answered these second questions correctly when they were about seven years old.[17]

Piaget's next test was somewhat more complicated. The researcher sat opposite the child across a table on which were a pencil and a coin; the coin to the child's left and the pencil to the right. He then asked: 'Is the pencil to the right or the left of the penny? And is the penny to the right or left of the pencil?' Although this set of questions seems approximately the same difficulty level as the previous set, children did not get them right until they were about six months older, at the age of seven and a half.

The next question was the really crucial one, although even Piaget himself didn't realise this when he originally designed the experiment, commenting after an earlier pilot study, 'One ought to have asked the child – this did not occur to us until after the experiment had been completed – to go to the other side of the table after having said that the coin was to the left of the pencil and to have added, "Now is the penny to the left or to the right of the pencil?"' That question did indeed make the task much harder, children not getting it right until they were about nine years old. One little boy, Pi, who was exactly seven and a half, showed what goes wrong:

Is the penny to the right or to the left of the pencil?
The left.
And the pencil?
The right.
[Pi then walked around the table and sat beside the experimenter. The penny and the pencil were left untouched.]
And now is the penny on the right or on the left of the pencil?
The left.
Really?
Yes.
And the pencil?
The right.
How did you do it?
Easy, I remembered how they were before.

Pi's confidence is striking, although both the error and the confidence are typical of children of this age. His error was to think that right and left are fixed properties of objects rather than relationships that depend on where they are viewed from.[18]

Even by age nine, the child has still not fully mastered the nature of right and left, as yet another experimental variation shows. This time the researcher puts *three* objects on the table – a pencil on the left, a key in the middle and a coin on the right of the child. Again, questions are put, firstly while the child sits opposite the researcher: 'Is the pencil to the left or the right of the key? And of the penny? Is the key to the left or to the right of the penny? And of the pencil? Is the penny to the left or to the right of the pencil? And of the key?' It sounds confusing even to adults, and we have the added difficulty of imagining the scene, whereas the child has everything in front of them. Still, though, they only get it right at about nine and a half to ten years old. Now, though, they finally have a proper understanding of right and left, which they demonstrate convincingly in the final test, when they are able to come to the other side of the table and still get the answers correct.

Why do children take so long to learn about right and left, and why do they have such problems with the concepts? A complete understanding involves the coordination of three separate skills – understanding right and left, carrying out a mental rotation, and seeing the world from a different perspective. Piaget recognised three stages in children's understanding of left and right. In the first stage, children see left and right only *in relation to themselves*. This is what Piaget called 'egocentrism': seeing the world from the child's perspective rather than that of other people. The difficulty is not a moral or selfish one, but a cognitive problem – an inability to turn the problem around. That means the terms left and right are seen as absolute, applied by the child to their own left and right, and then projected on to the rest of the world. This process can be seen in a subtle rewording. When asked if an object was '*to* the left' a child often replied that it was '*on* the left'; the vocabulary of relative position replacing the vocabulary of absolute position.[19]

In the second stage, the child is more socialised, understanding that left and right, as they see them, are not necessarily the same as how other people see them. They can now distinguish *their* left foot from *my* left foot. However, it is only in the third stage that right and left are seen as properties of relationships *between* objects, and not of objects themselves. An object such as the key can both be to the left of the coin and the right of the pencil from the child's point of view, and the other way around for someone else. Only in this final stage, which Piaget calls 'complete objectivity', has a mature understanding been achieved.

Even though the vast majority of children successfully solve left–right problems by the age of twelve or so, left and right can continue to provide problems for many people throughout their life. Before looking at the data, answer this simple question by choosing one of the five answers:

As an adult, I have noted difficulty when I quickly have to identify right versus left:

- All the time
- Frequently
- Occasionally
- Rarely
- Never

Lauren Harris asked this question of 364 professors at Michigan State University. Two per cent said 'All the time', six per cent 'Frequently', eleven per cent 'Occasionally', thirty-six per cent 'Rarely' and forty-five per cent 'Never'. It is striking that one in five of these university academics had troubles 'Occasionally', 'Frequently' or 'All the time'. The study is not unique; a similar result was found in doctors, and another study also found the same in university graduates and members of high-IQ societies such as Mensa and Intertel. Some people are particularly likely to be confused about right and left, and in all the studies women and left-handers report more confusion. The following typical case history is reported by a left-handed doctor, practising as a neurologist:

> I was first aware of right–left confusion in about the third grade when I had curious difficulty following simple instructions such as to do something on the right or the left side of the page. In order to orient myself I had to first begin writing to determine which was my left hand and develop a point of reference…I still [as a thirty-three-year-old] need to consciously make the right–left decision, mentally check my decision and will make an error if preoccupied, tired, or have had a drink. I have no difficulty with direction otherwise.

Like this neurologist, Shakespeare was aware that alcohol made right–left confusion worse, Cassio in *Othello* protesting his sobriety by saying, 'Do not think, gentlemen, I am drunk…this is my right hand; and this is my left hand.'[20]

Even people who say they are not confused about left and right show problems when they are tested in the laboratory. To give you an idea of the tests, look at Figure 4.5, and quickly say out loud whether each hand is pointing *up* or *down*. Now look at Figure 4.6 and say out loud whether each hand is pointing: *to the right* or *to the left*. Most people take half as long again to do the second task. Now comes the most difficult test of all. Look again at the hands in Figure 4.6 and say out loud whether each one is a *right hand* or a *left hand*. This is much more difficult and usually takes

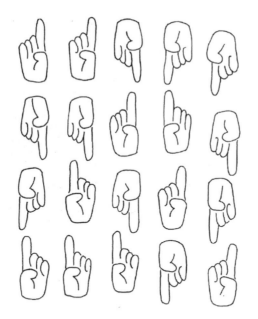

Figure 4.5 The Hands Test of Left–Right Confusion. This is the control condition, in which one has to say whether each hand is pointing 'up' or 'down'.

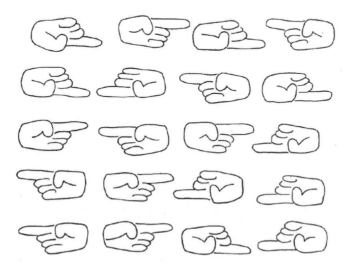

Figure 4.6 The Hands Test of Left–Right Confusion. This is the main condition, in which one has to say whether each hand is pointing 'to the right' or 'to the left'.

about two and a half times as long as saying that the hand points to the right or to the left.[21]

Even for adults, then, there is something more difficult about identifying right or left rather than up or down. To understand why that should be, psychologists separate two different sorts of explanation, called *perceptual encoding* and *verbal labelling*. When we distinguish two objects, we first have to *see* a difference (the perceptual stage), and then have to *describe* that difference (the labelling stage). In terms of right and left these processes can be distinguished by careful experiments. The typical task for the subject is to distinguish symbols such as ∧, ∨, > and <, arrow heads pointing up, down, right and left. The subject sees one of them flashed up for about a tenth of a second. The subject responds by saying the word 'right', 'left', 'up' or 'down'. Typically, the response for 'right' and 'left' is slower than for 'up' and 'down', as in the previous experiment. In the current experiment, however, the subject has both had to see the difference and describe it. However, in a slight experimental variation, known as a 'Go/No Go' design, the subject just says the word 'Go' when a particular type of arrow appears (∨ or >), and says nothing at all if the other type appears (∧ or <). In other words the subject distinguishes the arrow heads but does not label them. In a carefully designed experiment, left and right arrows can be compared with up and down arrows, and a significant difference appears from the previous result. Subjects are now just as quick and accurate with > and < as they are with ∧ and ∨; right and left are no more difficult than up and down. Since in the second experiment the stimuli have been distinguished *perceptually encoded* but have not been *verbally labelled*, the problem with left and right must result from verbal labelling.[22]

Something about the *words* 'right' and 'left' makes them more difficult to use than, say, 'up' and 'down', or 'above' and 'below', or 'front' and 'back'. Think firstly about 'above' and 'below'. To decide if one object is above another, we need to know what is 'up' and what is 'down'. That is easy – hold something in the air and let go. The way it falls is 'down' because gravity means things go down and not up. Likewise, 'near' and 'far' or 'front' and 'back' are easy – one can touch the front of an object which is nearer but can't touch the back, which is further away. But what about left and right? We are back to Kant's problem – there is nothing obvious in space, such as gravity or the length of our arms, which tells us which way is right and which left. Right and left will, therefore, be harder to use.

Intriguingly, even paired terms such as 'above' and 'below' are not entirely equivalent to one another in their usage. Subjects are quicker to say an object is above another object than below it. 'Above' is the unmarked form of the description, whereas 'below' is marked – in effect it means 'not above' – and takes extra time for the brain to process. A straightforward experiment can be carried out to test which of 'right' or 'left' is marked. Once again the stimuli are flashed on a screen for a very short time. On each trial, the subject sees a single box with a word in it, and a spot alongside the box. The subject has to say if the relationship between the word and the spot is true or false, so that if the spot is actually where the word says it is, the correct reply is 'true'. The stimuli and the correct answers are shown below.

• Left	Right •	Left •	• Right
'True'	'True'	'False'	'False'

The response times for this test were about one tenth of a second faster when the word was 'right' than when it was 'left', suggesting that 'right' is unmarked and 'left' is marked. If you have tried the task above, you might have realised that there is another way of running the experiment, so that instead of left and right being defined from the subject's point of view (subject-centred) they are defined relative to the box (object-centred). Now the correct answers are different from before:

• Left	Right •	Left •	• Right
'False'	'False'	'True'	'True'

Overall, subjects are slower on this version of the experiment by about one sixth of a second. Describing 'right' and 'left' from the subject's point of view is therefore more natural, or better learned, than describing it from the object's point of view, and that fits with Piaget's finding that children assess right and left from their own viewpoint before they assess it from the object's point of view. Despite this difference the experiment still gives similar results to the previous one, 'right' being faster than 'left', confirming that right is unmarked and left is marked.[23]

If you are left-handed, you may be wondering whether 'right' is also unmarked for left-handers, or whether 'left' is now the unmarked term. Unfortunately, the experiments were only carried out on right-handers, and although the authors of the study said they were studying left-handers, they do not seem ever to have published the results. So we don't know, I'm afraid, interesting though it would be to find out.

There is one final variant that can also be carried out with this experiment. We have already seen that problems of right and left seem to depend on verbal labelling. We can check whether that is the case by replacing the words in the boxes with an arrow pointing to right or left, as below. Now there is no confusion between the arrows pointing right and arrows pointing left. Problems arise only with the *words* 'right' and 'left'.

    'True'   'True'   'False'   'False'

It is clear that left and right are difficult for people to distinguish, that children have many problems with them, and many adults have difficulties throughout their life. If humans have problems, what about animals? Can they tell left from right? Before answering that, we must be clear what it means to 'tell left from right'. If a cat scratches its left ear when a fly lands on it, that doesn't mean the cat can tell left from right, merely that it can respond on the same side as an irritant stimulus. Proper differentiation of right and left requires an arbitrary, non-symmetric response to mirror-symmetric stimuli, as in the earlier experiment, for instance, with the spoken response 'right' and 'left' to the arrows pointing right or left, > and <, or with mirror-symmetric responses to arbitrary non-symmetric stimuli, such as turning right or left on hearing the arbitrary sounds that comprise the spoken words 'right' and 'left' in English. Can animals do such things? The simple answer usually seems to be no; or, if they do show some ability, it is only with great difficulty. Experimenters have tested octopus, goldfish, pigeons, rats, rabbits, guinea pigs, cats, dogs, monkeys and chimpanzees, and most show no evidence of distinguishing right from left. Even though people are not always expert at such tasks, they are still much better than animals. The best explanation for this difference is that humans have brains that are asymmetric, whereas the brains of animals are symmetric, and a symmetric brain cannot distinguish asymmetric stimuli or make asymmetric responses – a theoretical insight first made by the Austrian physicist, Ernst Mach.[24]

Mach's ideas have been incorporated into psychology by Michael Corballis and Ivan Beale. Think firstly about a perfectly symmetric machine such as an aeroplane on automatic pilot, which responds differently to events happening on the right or the left – a crosswind, for example – by turning its rudder to right or left. Like a cat with a fly on its ear, such a machine cannot be said to know the difference between right and left, despite functioning very effectively. Imagine now a perfectly symmetric machine, an aeroplane on automatic pilot, or a perfectly symmetric brain, which can make an asymmetric response such as turning left when an asymmetric stimulus such as the letter 'p' appears in front of it. What happens if the mirror-image stimulus, the letter 'q', appears instead? The machine *has* to respond by making the mirror-image response of turning right. To see why, imagine both the letter 'p' and the machine being reflected in a mirror. We have already said the machine is perfectly symmetric and so it is completely unchanged on the other side of the looking-glass. The stimulus, however, is reversed, the 'p' becoming a 'q', and hence it has to be the case that the response also is reversed, from a left turn to a right turn. If that were not the case, the machine could not have been properly symmetric in the first place. The final twist, though, is to turn the whole theory inside out. If the machine *does* turn in different directions to stimuli which are not mirror-images of one another, say left to a '+' and right to a '*', then that has to be because the machine itself is asymmetric. Only an asymmetric machine or an asymmetric brain can distinguish left from right.[25]

Although Mach's ideas were only thought-experiments, they have proved of great worth in understanding right–left distinctions and, like Kant's ideas, have set strong limits on what symmetric systems can do. Since most animal brains are, to a large extent, symmetric, animals will inevitably have trouble distinguishing left from right. Occasionally, animals do learn to distinguish them and they do so by a clever trick. For instance, some pigeons can be trained to distinguish the mirror-symmetric stimuli ╱ and ╲. What they do is tilt their head 45 degrees to one side, and that converts ╱ and ╲ into | and —, which are no longer mirror-images, and so are distinguishable. In effect, the pigeon has made itself asymmetric by tilting its head. The experimenter putting a patch over one eye of the pigeon can also produce the same effect.

Mach's theory explains why some adults have trouble distinguishing right and left. People who have a more asymmetric brain should distinguish left and right better than those with a more symmetric brain. Left-handers have less lateralised brains than right-handers, and women are less lateralised than men, which explains some of the differences in right–left confusion. Adults or children who are strongly right- or left-handed are better at distinguishing right and left than those who are weakly right- or left-handed. People with large asymmetries between the sides of their body – for instance, due to hemiplegia or paralysis – are also better at distinguishing left and right. A simple practical solution, therefore, for anyone with left–right confusion is to make the body more asymmetric. A judge in the Family Court in New York described his solutions to a long-standing problem:

> I can't tell my right from my left without looking at my wedding ring or my wrist watch. I tell my wife that our marriage is forever because, if I ever take my wedding ring off, I won't be able to find my way home. When I drive my car I notch my wrist watch just a little bit tighter so if

somebody says turn left, I know which way to go. Or when I come out of
an airport gate, you know, where they have all those signs, baggage to
the left, taxis to the right, go forward, go up, go down – I wear my watch
a little tighter.[26]

The cases so far of right–left confusion have been relatively mild, confined to normal adults who, although they *can* do a task involving
right–left differentiation, find it more difficult than other tasks. Severe
right–left confusion can, though, occur in adults because of brain
damage, these patients having the controversial syndrome named after
the Viennese neurologist Josef Gerstmann, who first described it in the
1920s. Gerstmann's syndrome has four separate components: inability to
write (agraphia), inability to count or do simple arithmetic (acalculia),
inability to name fingers (finger agnosia), and confusion about left and
right (left–right disorientation). It is a bizarre collection of symptoms,
described by the British neurologist MacDonald Critchley as 'this colliga-
tion of unexpected and unlikely phenomena'; so odd that many are sus-
picious as to whether it even exists. Although some neurologists regard
this as merely a chance association of symptoms rather than a true syndrome, it has nevertheless stood the test of time. That mainly is because
Gerstmann's syndrome is usually associated with damage to a specific
part of the brain – the angular gyrus in the parietal lobe of the left hemisphere. The relationship between a specific ability such as knowing right
from left and a tiny area of the brain is disconcerting, although it would
not have surprised Oscar Wilde, who is said to have commented on those
who 'found a curious pleasure in tracing the thoughts and passions of
men to some pearly cell in the brain'.[27]

HP is a typical case of Gerstmann's syndrome. Well educated, with a
degree in literature, and working as an insurance agent in Geneva,
Switzerland, at the age of fifty-nine he went to a casualty department
complaining of suddenly being unable to write, calculate, or dial telephone numbers. A brain scan showed a tiny area of damage, due to lack
of oxygen, in the white matter underneath the angular gyrus. He had few
other symptoms, but detailed testing revealed the other components of
Gerstmann's syndrome. In particular, there was right–left disorientation,
which showed not only because he made more errors on tests identifying
points on his own and others' bodies but also because he was far slower
on the test, taking almost ten minutes to carry it out compared to the
usual forty seconds or thereabouts. His writing had also become bad – he
said that it looked like that of a child – and he mixed up the letters 'b'
and 'd', and 'p' and 'q'.[28]

The question most concerning neuroscientists is whether the strange

tetrad of symptoms in Gerstmann's syndrome is due to a single underlying deficit. One suggestion is that there is a problem in understanding
space, which affects how right and left are seen as spatially organised and
how they change as objects rotate and move. Such problems can be identified by careful testing of mental rotation – the ability to imagine what
an object might look like if rotated into a new position. Could such a
deficit in visualising left–right differences explain the other symptoms of
Gerstmann's syndrome? It may. Inability to write is perhaps most readily
explained, since writing so clearly depends on moving in one particular
direction (left to right for English). Confusing right and left would,
therefore, disrupt writing dramatically.

Problems with counting and arithmetic seem more difficult to explain
until one sees the close relationship between counting and distinguishing left and right. With ordinary Arabic numerals, the left-most digits
represent the larger numbers, the hundreds and thousands, whereas the
right-most digits represent the smaller numbers, the tens and units. A
confusion of left and right could well impair mental arithmetic. Left and
right are also involved in 'number forms', mental maps of the numbers
that were first described in the nineteenth century by Sir Francis Galton.
About one in seven people say they have number forms, which can be
quite complex, as in Figure 4.7, although it is often the case that the
smaller numbers are arranged in a straight line from left to right. The
number form in Figure 4.7 is especially interesting because this patient
had suffered damage to the angular gyrus in the left parietal lobe – the
same area implicated in Gerstmann's syndrome – and the number form
had then disappeared.

Another mathematical difficulty relates to the fact that the final part of
the syndrome is finger agnosia. Perhaps it is not coincidence that we first
learn to count on our fingers (and hence use decimal numbers, with their
ten digits). Our fingers are also named and identified by counting – first
finger, second finger, and so on – and any disruption of arithmetic will
impair this. It is also the case that the way we identify our fingers changes
with the position of the hands: palms down, my thumbs are in the
middle and I count fingers outwards from the centre, with my fourth
(little) finger at the outside; palms up, my thumbs are at the edges and I
count fingers inwards from the outside. The direction changes according
to the rotation; the very ability said to be impaired in Gerstmann's syndrome.[29]

In Gerstmann's syndrome there is a problem with *knowing* about left
and right. Some patients have almost the converse problem, being unable
to distinguish an object from its mirror-image. RJ is a sixty-one-year-old
man who has suffered strokes in both right and left parietal lobes. In one

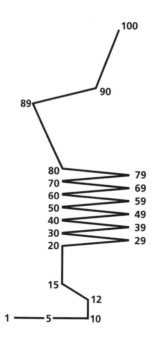

Figure 4.7 A number form.

experiment, he was shown a picture and its mirror-image, and the experimenter demonstrated the impossibility of superimposing one picture on
the other; in other words, demonstrated that they must be physically different in some way. Despite that, RJ said, 'I know from what you did that
they should be different...but when I look at one, then the other, they
look just the same to me.' RJ was even unable to choose the odd one out
in pictures like those in Figure 4.8 where two objects face one way and
the third faces the other. RJ's problem was not one of poor vision or
attention to detail. If the three bears all face the same way but one has a
tiny change to its head, as in Figure 4.9, then he notices that with no
difficulty.

The change to the bear's head in Figure 4.9 meant that for RJ it was
now a *different* bear rather than 'just the same'. He is partly correct here.
In a sense, the three bears in Figure 4.8 *are* the same. A bear viewed from
the left or from the right is still the same creature, and part of RJ's visual
system knows that. However, knowing that an object remains the same
object when seen from different views, and knowing that those views
nevertheless look different, are separate processes. RJ knows the former
but not the latter, and the parts of his brain that know about objects do
not know, or even need to know, about right and left.[30]

Confusion of left and right is a common process, and not even professional scientists or artists are immune. The anatomist and surgeon
Frederick Wood Jones found several examples of drawings with 'the most

Figure 4.8 Which of the three bears is the odd one out?

Figure 4.9 Which of the three bears is the odd one out?

curious anomalies' in which illustrators managed to mix up right and left (Figure 4.10). A famous example is Tischbein's portrait of Goethe, which in Frankfurt has the status of the Mona Lisa in Paris (Figure 4.11). Somewhere, though, the anatomy is wrong, the left leg being far too long, perhaps because the composition has been radically altered. Whatever the reason, there seems little doubt that the right leg has a left foot. Scientists are no better. For instance, the sub-atomic particles called neutrinos are all left-handed. Nevertheless a couple of years ago, *Scientific American* had to publish a correction after describing them as right-handed. The most consistent errors occur with that most famous of all molecules, DNA, the decoding of which won James Watson and Francis Crick the Nobel Prize in 1962. Its double helix is almost the definitive icon for science itself, being represented in thousands of places. DNA normally forms a right-handed spiral (although a rare left-handed variant can occur). In other words, it twists like a conventional screw. That, though, has not stopped it being reproduced wrongly in hundreds of places. Dr Tom Schneider has a website where he has collected hundreds of examples of incorrectly drawn 'left-handed DNA', most being found in scientific journals. Many are in advertisements, so we may perhaps charitably suggest the final copy was never seen by a scientist, but that doesn't quite explain them all. Certainly not the editorial comment in *Nature*, the place where DNA's structure was first described, which in 2000 mentioned the clues that 'led Watson and Crick to deduce the left-handed double helical structure of DNA'. Watson, in fact, has been particularly

a   c

b

Figure 4.10 a, A right foot is on the left leg. **b,** The right and left legs have left and right feet. **c,** A right hand is on the left arm.

Figure 4.11 Tischbein's portrait of Goethe in the Roman Campagna. The right leg has what seems to be a left foot.

badly served, his 1978 textbook, *Molecular Biology of the Gene*, having six different illustrations with left-handed DNA, and in 1990 the American journal *Science* printed reply cards for joining the American Association of Science quoting Watson as saying, 'I have to read *SCIENCE* every week', this being illustrated with left-handed DNA. Perhaps worst of all, a 1998 reprint of Watson's *The Double Helix* was illustrated on the front and back with left-handed DNA. Perhaps it is not a coincidence that Watson is left-handed.[31]

Despite left and right being very confusable, nature in general manages to avoid being confused, creating bodies in which the heart is consistently on the left side, with only occasional rare cases of the pattern being reversed. The biology underlying right- and left-handed bodies will be the subject of the next two chapters.

THE HEART OF THE DRAGON

In the second act of Wagner's *Siegfried*, the third of the four operas in the *Ring* cycle, the hero knows he has to kill the dragon, Fafner. Siegfried has already made his trusty sword, Nothung, and is talking to the blacksmith, Mime, about Fafner. Mime describes Fafner's lashing scaly tail, his vast jaws, and how,

> poisonous foam
> he will pour from his mouth;
> if you are splashed by one single drop,
> it shrivels your body and bones.

Siegfried plans his attack and asks Mime, 'Has the brute a heart?' 'A merciless, cruel heart,' replies Mime. And then Siegfried asks,

> And is that heart
> in the usual place,
> at the left of his breast?

to which Mime replies, simply,

> Of course; dragons
> have hearts just like men.

This tells Siegfried what he needs to know. Later in the fight, 'Fafner roars, draws his tail back quickly, and rears up the front part of his body to throw its full weight on Siegfried – thus exposing his breast. Siegfried quickly notes the place of the heart and plunges in his sword there to the hilt.' And so Fafner dies.[1]

The scene invites several thoughts, the main ones in terms of this chapter being whether a dragon would have its heart on the left side, and why indeed it is that the left of the breast is the 'usual place' for the heart in people and many animals. Questions about the asymmetry of the heart force the even more basic question of how so many organs in the body come to be *symmetric* – we automatically presume, for example, that the dragon will have similar, albeit mirror-imaged, right and left front legs, and right and left rear legs, as well as right and left eyes on each side of the head; for so dragons are usually portrayed. That word 'similar' begs the question of whether the two sides of the body are indeed exactly identical, or instead merely alike. If they do differ slightly, why should this be, and does it matter?[2]

The original libretto of *Siegfried*, which as usual Wagner wrote himself, doesn't mention the word 'left', saying only that the heart is in the same place as in man and other beasts. The English translator has tacitly assumed what most people would presume – that beasts, even mythical beasts, have their heart on the left. A similar assumption was made by the anonymous author of the Anglo-Saxon epic *Beowulf*, who presumed the dragon Grendel was right-handed. The location of the heart seems to have interested people since prehistoric times, at least if one believes an advertisement for the heart drug digitalis, put out in the 1920s by the pharmaceutical company Burroughs-Wellcome (Figure 5.1). It shows a cave-painting of an elephant or mammoth with the location of the heart clearly indicated (albeit a heart that looks uncomfortably like something from a St Valentine's Day card). One can imagine the intrepid hunters sitting talking by the fire, wondering, like Siegfried and Mime, precisely where the vital organ was located. Whether they knew the heart was on the left side is not clear from this picture, but my feeling is that primitive peoples who regularly butchered animals would have known where to find the heart.[3]

Many creatures familiar to us – mammals such as cats, dogs, sheep, cows and horses; birds; reptiles such as crocodiles and snakes; amphibians such as frogs and toads; and a myriad species of fish – all without exception have a body in which the heart is on the left side. Not only the heart is asymmetric, but so also is most of the associated plumbing; the delicate network of tubes that Sir Osbert Sitwell called 'that fragile, scarlet tree we carry within us' – the veins that bring blood to the heart and the arteries that return it to the body. The whole arterial system can be seen in the exceedingly beautiful medical image of Figure 5.2. Creating a body with a heart on one side has the same problems as building a car – once decided that it is a left-hand drive, then a mass of other things must follow, each part having its appropriate side, either from necessity or a simple lack of space elsewhere. Likewise, in the vertebrate body, not only

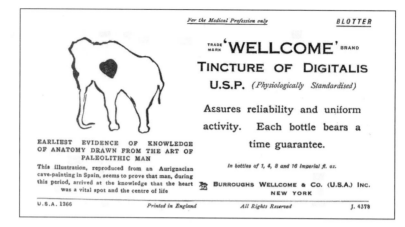

Figure 5.1 Advertisement by Burroughs-Wellcome in 1926 for tincture of digitalis, a drug which acts on the heart. The picture, presumably of an elephant or mammoth is captioned: 'Earliest evidence of knowledge of anatomy drawn from the art of palaeolithic man. This illustration by Henri Breuil, is of an Aurignacian cave-painting in Pindal, Spain, and seems to prove that man, during this period, arrived at the knowledge that the heart was a vital spot and the centre of life'.

is the heart on the left but there are also asymmetries of a host of other organs, most easily described in humans but found in most other vertebrates as well.[4]

Within the chest, the left lung, compressed by the heart, is smaller and has only two lobes whereas the right lung is larger and has three lobes. The aorta, the huge artery that takes the blood from the heart, curves upwards and over to the left, before passing downwards through the diaphragm and into the abdomen. In Figure 5.2 the most visible organs in the abdomen are the kidneys, the left as usual being slightly higher and larger than the right. The liver, the largest organ in the abdomen, is on the right-hand side, as is the gall bladder, and on the opposite side is the much smaller spleen, visible in the figure just above the left kidney. Not visible in the figure are the stomach, which is on the left-hand side, and the duodenum, which curls to the right, around the head of the pancreas, which is also on the right-hand side. The coils of the small intestine, although apparently confused and haphazard, have a clear, regular and very asymmetric structure. The small intestine ends low down on the right-hand side of the abdomen in the caecum, with its attached appendix. Then the colon ascends on the right-hand side, crosses over at the top of the abdomen, descends on the left to the sigmoid colon, and finally ends in the rectum and anus, the only parts of the gastrointestinal tract, apart from the mouth, throat and oesophagus, to be central and symmetric. In men, the two testicles are also asymmetric, the right-hand

Right   Left

Figure 5.2 The arteries in a healthy volunteer who has been scanned using magnetic resonance angiography. The arteries have been made visible by means of a contrast agent which appears white; the contrast has also started to collect in the bladder, which is visible in the centre of the picture.

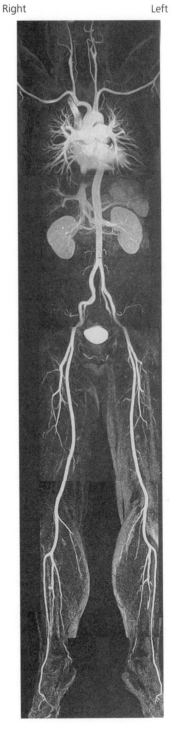

testicle usually being larger and higher, in part reflecting the need to pack these oval organs into the rather strange space left for them at the top of the thighs. In one of a number of lateral references, James Joyce, in *Ulysses*, alludes to the asymmetry of the scrotum in a comment about the side on which men 'dress':

> Zoe: How's the nuts?
> Bloom: Off side. Curiously they are on the right. Heavier I suppose. One in a million my tailor, Mesias, says.

The asymmetries inside the abdomen mean that surgeons are particularly interested in whether pain is on the left or right: if it is in the right iliac fossa, for example, then the diagnosis is likely to be appendicitis; if in the left iliac fossa, it is possibly diverticulosis of the sigmoid colon; if it is in the right hypochondrium, it is probably gall-bladder disease; and so on. Indeed, some diseases occur principally in only one of paired organs, the classic example being varicocoele; the vermicelli-like mass of enlarged veins around the testicle, which can cause infertility, almost always occurs on the left-hand side.[5]

The arrangement of organs I have described above seems so banally normal to anyone with a smattering of biology that we have to remind ourselves how atypical it is in living things in general. Gross asymmetry of this sort is mainly found in the vertebrates. Among other animals, the heart is not on the left and the cardiovascular system is not asymmetric. The earthworm, for instance, has a series of heart bodies in several segments of its body, exactly in the mid-line; and in other invertebrates, such as insects and crustacea, the heart is also symmetric and in the mid-line. Indeed, that workhorse of genetics, the fruitfly, *Drosophila*, is almost entirely symmetric throughout its body (with the curious exception that the penis rotates clockwise during development). Our very familiarity with those large organisms similar to ourselves, the vertebrates, makes it easy to forget the strangeness of having a heart on the left. How, why and when did it evolve?

Part of the 'why' is to do with the way fluids flow, the problems of growing big, and having an internal skeleton. Vertebrates differ from other animals in having a hard, internal skeleton. Tiny animals, particularly those living in water, do not need a skeleton at all – consider that blob of protoplasm, the amoeba. Once an organism lives on the land and gets above a certain size, support is essential to fight the ever present force of gravity. Simple animals such as earthworms can make do with a hydrostatic skeleton, a system of tubes into which water is pumped and which maintain their shape in the same way as 'bouncy castles' or balloons. For larger animals, a hard skeleton is essential, and insects, crustacea and the like have an external skeleton – a hard outside box made of the protein chitin, inside which the animal lives and to which muscles are attached. This arrangement works well up to a point. Growth, however, is tricky, with the need to moult at regular intervals.

More problematic, as J. B. S. Haldane pointed out in his famous essay of 1928 titled *On Being the Right Size*, is the increasing difficulty of getting oxygen to the various parts of the organism. Part of the solution is an internal skeleton, as found in all of the largest animals on the earth, from dinosaurs to whales to elephants. The most obvious component, which gives the vertebrates their name, is the backbone, composed of a series of vertebrae. The backbone forms alongside a more primitive form of skeleton, the notochord, a stiff rod that forms early in the development of vertebrate embryos. The notochord is found not only in the vertebrates but also in organisms such as sea-squirts, which are more primitive than vertebrates and which, together with the vertebrates, form the chordates. An internal skeleton allowed vertebrates to grow much larger bodies, first underwater as fish, and later, on land as amphibians, reptiles, birds and mammals. Attached to the skeleton are the many muscles that allow movement, and these all require blood, which is provided by an enlarged and more efficient heart.[6]

Although, in theory, it might seem possible to have an effective, symmetric, centrally placed heart, in practice becoming more efficient also means becoming asymmetric. The heart is a machine pumping liquid, and as fluid is pumped at a greater rate, so the constraints of fluid dynamics – the ways in which liquids flow – begin to apply. What must be avoided is the disordered, erratic flow known as turbulence, because it wastes energy and also traumatises fragile red blood cells which can cause blood to clot. As blood flows through the heart, it moves in a series of spirals that allow the streams of blood entering and leaving the heart to pass one another without colliding. Since spirals are chiral, being either right- or left-handed, hearts inevitably have to be asymmetric. Although that neatly explains the particular asymmetry of the heart, it does not explain why vertebrates have a heart to one side of the body, since either a left-handed or a right-handed heart would satisfy the needs of fluid dynamics. To understand the *left*-sided heart we must therefore look at the ancestors of the vertebrates, and find where body asymmetries originated.[7]

The fossil origins of the vertebrates have long been unclear. There are huge numbers of later fossils, from the Silurian onwards (about 430 million years ago), and there are a few specimens from the Ordovician (about 480 million years ago), but before that the fossil record is sparse.

Recently, though, fossil specimens have been uncovered in China which, at about 550 million years old, are the oldest known vertebrates. They show all the key identifying features of a vertebrate – a skull, gill slits, fin rays and a large heart – but they are also sufficiently primitive not to have jaws, as is also the case in a few living vertebrates. So where did these early vertebrates come from? That is a complicated and technical story, and here I am only able to give a flavour of it. It is also fair to say that the story is still being written, and some details will inevitably change over the next few years.[8]

Early animals probably started as single cells, and then became multicellular, as in the sponges. Next, they developed a mouth and a nervous system, forming organisms that sat glued by their bottom to a rock, sucking food from the water as it passed by, digesting it in the hollow body cavity, and spitting out the indigestible remains through the mouth. The next step is a crucial one in our story. At some point, an organism stopped sitting vertically motionless on a rock, passively waiting for food to come to it, and instead lay on its side and then crawled off across the bottom of the sea, actively searching for food. While sitting on the rock it had radial symmetry – that is, it was shaped a bit like a tin can or a bottle, the top and bottom being distinguishable but, because the cross-section is circular, there being no identifiable front, back, left or right. Once it lay down on the sea-bed, however, the old top and bottom became a new front and back, with the mouth at the front. Gravity also meant that the organism now had a new top and bottom (Figure 5.3). Once there is a top, a bottom, front and back, it is also possible for there to be a left and right, since left and right can only be defined once front–back and top–bottom have also been defined. At this stage, however, the left and the right sides are still identical, the organism exhibiting bilateral symmetry and being classified among the Bilateria.

The simplest descendants of the early Bilateria are flatworms, but other descendants include a variety of worms such as the common earthworm; arthropods, such as crabs, lobsters, insects, spiders and the extinct trilobites; molluscs, including oysters, slugs, snails, squid and octopus; and, of course, the vertebrates. However, and here is a twist in the literal sense, the bilateral symmetry that we see in the arms and legs of human beings and other vertebrates may not be the same symmetry as that present in early Bilateria and modern worms.[9]

At this point, it is worth noting some technical terms used in describing an animal. Think of a dog. Its nose is at the front, which we call *anterior*, and its tail at the back or *posterior*, these together defining the anterior–posterior axis. The legs and belly of the dog are on the underside, and are called *ventral*, in contrast to the dog's backbone, which is

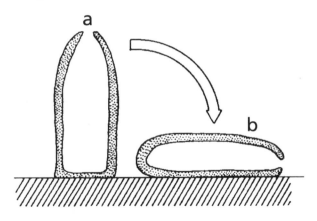

Figure 5.3 The process which Jefferies calls 'pleurothetism', in which a radially symmetric organism lies on its side, moves forward and becomes bilaterally symmetric.

uppermost or *dorsal*. These define the dorsal–ventral axis, which is perpendicular to the anterior–posterior axis. The third dimension, left and right, which only becomes apparent when anterior–posterior and dorsal–ventral are set up, is at right angles to both anterior–posterior and dorsal–ventral. Most animals, be they fish, frog, snake, bat or sloth, can be described in these terms. Although it might seem easier to use more everyday terms such as front, back, top and bottom, these present a problem when it comes to describing our own bodies. When human beings first stood on two legs, everything in the body turned through a right angle, with the exception of the head, which tipped forward on the neck. 'Forwards' is therefore anterior in a dog, but ventral in the human body, and 'upwards' is dorsal in a dog but anterior in the human chest and abdomen. Despite their somewhat unfriendly-looking appearance, I will therefore stick to the unambiguous terms 'anterior', 'posterior', 'dorsal' and 'ventral'.

The problems of palaeontology are immense. Specimens are few and far between; they are often crushed, distorted and broken by being buried for aeons, deep in the rocks; soft body tissues are very rarely properly preserved; the fossils found are probably only intermediate stages rather than the definitive common ancestor; and so on. Dick Jefferies of the Natural History Museum in London, who has spent his career as a palaeontologist trying to work out the origins of the vertebrates, describes his approach to such a mass of confusing evidence: '[It] is like solving a crossword puzzle. If sufficient numbers of things fit with these very complicated fossils, I hope I am getting it right. Any particular argument will only be presumptive, but they all fit together so closely that I hope the ultimate result is correct.' A crossword solver's mind, or perhaps

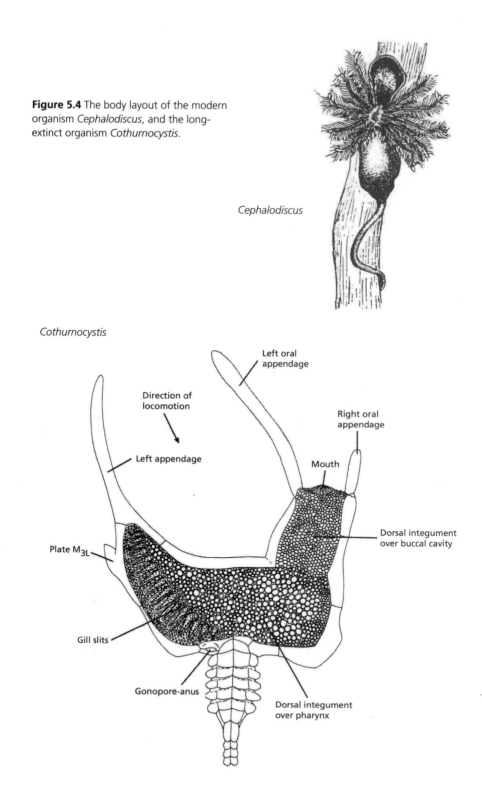

Figure 5.4 The body layout of the modern organism *Cephalodiscus*, and the long-extinct organism *Cothurnocystis*.

Cephalodiscus

Cothurnocystis

Left oral appendage

Direction of locomotion

Right oral appendage

Left appendage

Mouth

Plate M$_{3L}$

Dorsal integument over buccal cavity

Gill slits

Gonopore-anus

Dorsal integument over pharynx

that of an aficionado of the board game 'Go', is needed to keep track of the twists and turns in the story.[10]

Dick Jefferies has put together a remarkable and surprising theory, which it is fair to say is still controversial although it accounts for a wide range of otherwise extremely perplexing phenomena. Jefferies' key insight is that there exists a deep and ancient relationship between a living animal, the hemichordate *Cephalodiscus*, and a tiny fossil, *Cothurnocystis*, one of a group known as the cornutes (see Figure 5.4).

Whereas *Cephalodiscus* is symmetric, *Cothurnocystis* is highly asymmetric. The end of *Cothurnocystis* with the mouth and two spines is the head, although it seems that the animals actually moved backwards, the tail going first. Jefferies has long tried to reconstruct how *Cothurnocystis* moved, and had suggested that it was by stretching out the stiff but muscular long tail in a whipping motion so that it dug into the mud and anchored the animal, which could then pull itself along to the new position.

Although originally only an elegant conjecture, based on beautiful cardboard and rubber-band reconstructions of the tail assembly, Jefferies' idea received much support when, in 1995, Wouter Südkamp found a remarkable piece of slate in a quarry in Germany. About 390 million years old and a third of a metre square, it contains no fewer than four fossils of the mitrate *Rhenocystis*, a relative of *Cothurnocystis*. Two of the fossils have left behind them clear tracks that are visible in the slate, and these tracks support Jefferies' idea that movement was backward, as seen in Figure 5.5.[11]

The most striking thing about *Cothurnocystis* is its asymmetry. How on earth did it get like that if it evolved from the symmetric Bilateria? Jefferies' idea is inspired simplicity itself, but is also wonderfully outrageous. He calls it dexiothetism. The proposal is that an ancestor of *Cothurnocystis* simply rolled over on to its right side, and crawled off in a different way. In other words, what was originally left and right became dorsal and ventral, the old dorsal and old ventral becoming the new right and left. Figure 5.6 shows Jefferies' own diagram illustrating the process, and Figure 5.7 shows how right, left, dorsal and ventral have been rearranged.[12]

Lying on one's right side must have had its inconveniences, to say the least. The body openings on the right side would now be deep in the mud, and would better be closed up. Likewise, the tentacles on the right would simply be in the way, catching in the mud, and again would be better lost. Having asymmetrical limbs, arms or legs would also be a problem for any organism faced with the practical problems of crawling, walking, running, swimming or flying, because there would inevitably be

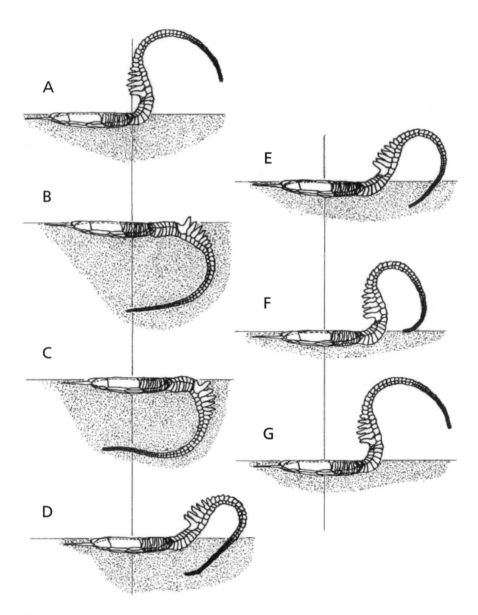

Figure 5.5 Reconstruction of the movements of the organism *Rhenocystis*, a relative of *Cothurnocystis*.

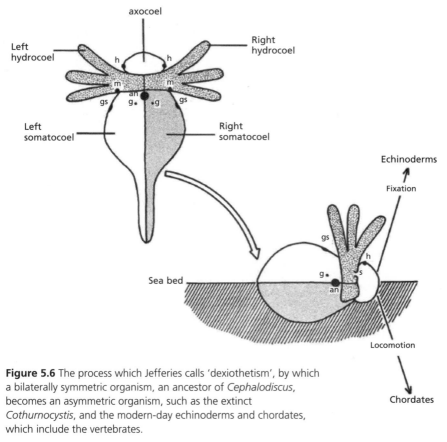

axocoel

Left hydrocoel

Right hydrocoel

Left somatocoel

Right somatocoel

Echinoderms

Fixation

Sea bed

Locomotion

Chordates

Figure 5.6 The process which Jefferies calls 'dexiothetism', by which a bilaterally symmetric organism, an ancestor of *Cephalodiscus*, becomes an asymmetric organism, such as the extinct *Cothurnocystis*, and the modern-day echinoderms and chordates, which include the vertebrates.

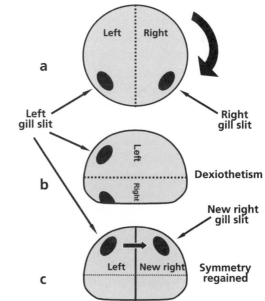

Left   Right

a

Left gill slit

Right gill slit

Left

Right

Dexiothetism

b

New right gill slit

Left   New right

Symmetry regained

c

Figure 5.7 Illustration of how, in Jefferies' theory, right and left are transformed into ventral and dorsal.

a tendency to go round in circles. Over millions of years, the descendants of *Cothurnocystis* therefore lost the characteristic 'boot' shape and instead became sleeker, more symmetric and more streamlined. Or, at least, they did so on the outside, because external symmetry mattered for survival. The asymmetries inside were another matter, being of little practical consequence to the organism, and so they continued in the descendants of *Cothurnocystis*. Bilateral external symmetry has therefore been superimposed upon an internal asymmetry. We are the descendants of an organism such as *Cothurnocystis*, and Jefferies' theory is that those innate asymmetries underlie our having a heart on the left, liver on the right, and so on.[13]

Jefferies' theory explains a confusing mass of fossil evidence, and shows how one body form can transform into another. What it does not do is provide a convincing explanation of why it should have been advantageous for *Cothurnocystis* to roll on to its right side, yet there must surely have been some advantage, for the immediate disadvantages seem immense. It seems fair to say that no-one, at present, has the faintest idea of a solution to the problem. Indeed, it might even seem dubious that organisms really can change their entire body layout so dramatically. Fortunately, we have a very good example of precisely such a change, involving the setting up of a new left and right, and it shows there is no reason in principle why Jefferies cannot be correct. Nor does it occur in a particularly obscure beast. Indeed, many people have had one in front of them on the dinner table – a flatfish, such as a sole or a plaice.

Organisms continually look for new ecological niches that have not previously been exploited and where the living is easier. Flatfish found that lying flat on the sea-bed, camouflaged to look like sand, had all sorts of benefits. However, making oneself entirely flat is not easy, particularly if one starts out, like most fishes, essentially tall and thin. It is possible slowly to modify the genes of a fish so that it compresses down and becomes very flat; rays and skate have managed exactly that. However, flatfish adopted a very different approach. What they did can whimsically be described as lateral thinking, for they turned themselves on their sides and lay flat on the bottom. They needed, though, to rearrange their eyes so that both looked upwards, for there is little use and there are many potential problems in having one eye deep in the mud of the sea-bed. The flatfish therefore moved one eye to the other side, so that both eyes were located on the same side of the body. More remarkably still, the flatfish did not do this at some remote time in the evolutionary past, but each and every flatfish continues to do it afresh during its early development.

Young newly hatched flatfish look indistinguishable from most other

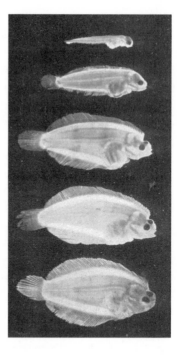

Figure 5.8 Five stages in the development of the plaice, a flatfish. In the first picture the two eyes are on opposite sides of the head, but as the fish grows one eye moves around the outside of the head until both eyes are on the same side of the body. Note also that the fish at the top swims in a normal upright position, whereas the one at the bottom lies flat on the bottom of the sea and swims in that position.

newly hatched fish (Figure 5.8). In particular, like other fish they have one eye on the right side of the head and the other on the left side. When the flatfish larva is about a centimetre long, a bar of cartilage above one eye dissolves, letting the eye on that side slide across the top of the head and on to the other side, the entire skull, twisting in the process. Other things change also; most noticeably the colouring on the two sides becomes completely different, so that the fish is fully camouflaged when viewed from above or below. Flatfish demonstrate that animals can undergo dramatic rearrangements in their body organisation, swapping left and right with dorsal and ventral. If that can take place over a few days within the development of an organism, it could certainly have happened during the evolution of organisms over millions of years.[14]

The body plans of animals have therefore evolved in three distinct phases. The first stage, the simplest, shows only radial symmetry. The second stage shows bilateral symmetry, the organs being arranged symmetrically to right and left of the mid-line. The third stage, characterised by the vertebrates, involves an external symmetry coupled with a gross internal asymmetry of the viscera. Perhaps this is the moment briefly to return to Siegfried's question concerning the position of a dragon's heart. Would this have been on the left side? It seems a fairly safe bet to say yes. Any animal as large as a dragon would need an internal skeleton and would almost certainly be a vertebrate, characterised by having a bony skeleton, a skull, four limbs, and so on. Further, since all other vertebrates are internally asymmetric so it would also be with the dragon Fafner, his heart being on the left side.

The grand evolutionary picture of radial symmetry followed by bilateral symmetry and then internal asymmetry, may be fairly clear, but it tells us very little about the mechanisms by which a developing organism manages to achieve bilateral *symmetry* (as with arms and legs) coupled with the *asymmetry* found in heart, lungs, liver and other internal organs. Something in embryos allows them to simultaneously develop arms that are the same and lungs that are different. The simpler problem for the embryo is producing bilateral symmetry; for example, developing right and left arms that are more or less mirror-images of one another.

In terms of developmental genetics it is easier to make two things that are mirror-images – say a right and left hand – than it is to make two things that are not mirror-images – say, a right and left arm each with a *right* hand on its end. The trick is in realising that the same set of instructions can produce both the right and left hands. Think about two people standing back to back at the old Royal Observatory at Greenwich on the meridian line, one facing east and the other west. It is midday, Greenwich Mean Time, and so the sun is due south. Each person holds a bag of flour that they pour on the ground as they walk, leaving a white trail behind them. Imagine shouting out a set of instructions:

> Walk two steps away from the meridian without dropping any flour;
> turn and walk two steps north leaving a trail of flour;
> turn and walk two steps away from the meridian leaving a trail of flour;
> turn and walk two steps north leaving a trail of flour;
> turn around and walk two steps south without dropping any flour;
> walk two more steps south leaving a trail of flour;
> turn and walk two steps towards the meridian leaving a trail of flour;
> walk two last steps towards the meridian without dropping any flour.

At the end, the two people should be together again, now face to face. The more easterly will have written a large letter 'd' on the ground, whereas the other will have written a large letter 'b'. One set of instructions has produced mirror-image results. Think now what the instructions would be like if each person were to write the letter 'b', one to the east of Greenwich and the other to the west. Either there would have to be different instructions for each person, or the instructions, instead of being relative to the Greenwich meridian ('walk towards the meridian') would have to be absolute ('walk in an easterly direction'). Either way, it is more complicated. Producing mirror-images, however, is easy, with neither person needing to know anything about left and right (or east and west, which are their equivalents); they only need to know the position of the sun (south), and the direction of the starting line, the meridian.

The evidence from embryology suggests that instructions for developing right and left hands work in precisely the way described above. It is more complicated in three dimensions but, as long as the embryo knows where to find anterior–posterior and dorsal–ventral (front–back and top–bottom) – equivalent to north and south in the Greenwich model – and also knows where to find its mid-line – the equivalent of the meridian – then producing mirror-image arms and legs is straightforward. The strongest evidence that the system works this way comes from experiments, usually on chick embryos, in which drugs and other manipulations produce all sorts of defects, including massively deformed hands, arms or fingers. Never, though, does one obtain an organism with two right arms, one on the right and one on the left. Similarly, although humans are occasionally born with varied and sometimes horrible abnormalities of their limbs, a left arm on each side is simply not one of them. 'They have two left hands' may be a good linguistic metaphor but it is not an embryological reality.[15]

The Greenwich model is a useful analogy for several aspects of the biology of left and right sides, because it has certain instructive weaknesses. The old Royal Observatory was chosen as the location because the meridian line is marked by a long straight piece of brass, where tourists are photographed with one foot in the eastern hemisphere and one in the western. Imagine carrying out our task a few hundred metres away in Greenwich Park. Now there is no actual line marked on the ground. Picture, also, the two people being asked to move not just two paces at a time but fifty. When they turn towards or away from the meridian, there is now nothing tangible to aim for, no line being visible, and the estimation of north and south is more approximate, relying on the sun. Furthermore, the ground is rougher, and error is consequently altogether more likely. There is even a risk that one or other person will accidentally end up on the wrong side of the meridian. The task worked well at the Observatory precisely because the line on the ground there was so clear, allowing each person to navigate accurately. The brass meridian acted as a mid-line, distinguishing one half of the world from the other. Embryos also need to know the exact location of the mid-line. It doesn't matter much if organs are developing far away from the mid-line, for nothing very serious can go wrong. But near the mid-line, where precisely co-ordinated events are happening, an exact knowledge of the border between right and left is vital for the embryo to assemble itself properly. It is a bit like the old Cold War border between East and West Germany – getting lost and being a few hundred metres out mattered little when in the middle of either country, but the same mistake at the border could be dangerous and perhaps even fatal.

Only in the past few years has control of the mid-line during development been studied at the molecular level. That something had to specify the midline had long been suspected because of occasional, rare and horrendous fetal malformations affecting mid-line organs such as the nose, or paired organs near the mid-line such as the eyes. An extreme defect, named after Homer's one-eyed giant in *The Odyssey*, is cyclops, in which there is a single malformed eye in the centre of the head, sometimes on a stalk or proboscis, and often associated with other life-threatening abnormalities. Cebocephaly is a less severe abnormality, in which the nose has only a single, small circular nostril, while in holoprosencephaly development of the face and the brain are so devastatingly disrupted that few normal features can be recognised. Understanding of these conditions advanced substantially in 1996, when it was found that a protein, the bizarrely named *Sonic hedgehog*, was crucial for defining the mid-line. Mice in which the gene for *Sonic hedgehog* had been 'knocked out' showed many problems, and the deformed fetuses did not survive pregnancy. The earliest abnormalities were at the front of the brain, where the two lobes were no longer separate but fused into a single organ, as were the eyes, which formed a single cyclopic eye rudiment in the centre. Although it is still early days in determining precisely how *Sonic hedgehog* and other substances interact, the key principle is becoming clear – left and right have to be defined with respect to something, even in paired organs, and that something is the mid-line, which must be clearly demarcated so cells know where they are, where they should not be, and what they should do. Not knowing where left stops and right starts is a developmental disaster.[16]

A well-defined mid-line is also needed for the development of that most asymmetric of organs, the heart. The heart is first visible as a single straight tube running down the middle of the embryo; a tube that then kinks to one side and produces the asymmetric heart. The cells forming that single mid-line tube originate to the right and left of the mid-line, and therefore have to find each other before fusing into one mid-line organ. Doing that depends on a protein called *miles apart*, which has been studied in the zebrafish, an organism beloved of developmental biologists because it is almost totally transparent, allowing development to be seen directly as it happens. If *miles apart* is not working, then, instead of a single mid-line heart tube, the fish develops two entirely separate hearts, a condition known as cardia bifida, which is inevitably fatal.[17]

Symmetry is good for things meant to be paired and symmetric, but disastrous for things meant to be asymmetric. Being too symmetric shows its problems in the human condition called 'isomerism defect'. The normal body pattern has the heart, spleen and stomach on the left, and the liver and gall bladder on the right. Occasionally, people such as John Reid and Susan Wright, whom Sir Thomas Watson described (see chapter 1), have everything completely reversed, with the heart, spleen and stomach on the right, and the liver and gall bladder on the left; the condition known as *situs inversus*. In most cases, these completely reversed individuals have no problems, despite their back-to-front body organisation. That is because *everything* is reversed in *situs inversus*, the body remaining asymmetric and hence able to function normally. There is, though, a much rarer group of individuals with isomerism defects (Figure 5.9). They are much more symmetric than normal, and can have serious problems with the heart and lungs. Instead of having distinct right and left sides to the body, people with isomerism have either two right sides or two left sides. In right isomerism, the right and left lungs both have three lobes, just as a normal right lung would have, the heart has two right atrial appendages, there is no spleen, which is normally a left-sided organ, the liver is on the mid-line, and, if the patients are male, the testes are at the same height, rather than the right being higher. Left isomerism is, essentially, the opposite of this condition, each lung having two lobes like a normal left lung, the atria of the heart having two left appendages, the liver being central, and, in contrast to right isomerism, the individual having multiple spleens. Those with either right or left isomerism can have serious problems with the plumbing of the heart, which functions wrongly. To give an analogy, there is little problem in making a motor car which has a left-hand drive or is completely flipped over to make a right-hand drive, but imagine a car made of two right halves of a left-hand drive car; there would be no steering wheel, pedals or controls. The converse, of a car made of two left halves of a left-hand drive car would be no better, with steering wheels, pedals and controls on both sides.[18]

Let us return to Greenwich Park, and think further about how writing the letters 'b' and 'd' might go wrong. Imagine photographing the giant letters on the ground from a helicopter, scanning the picture into a computer, cutting out the letter 'b', flipping it over left to right, and laying it on top of the letter 'd'. Does it fit? The answer will almost certainly be no, or, at least, not exactly. Although the instructions might imply that the 'b' and 'd' should be precise mirror-images (and they would be if programmed on a computer screen), the real world does not work as a formula or an algorithm. The people carrying the bags of flour are flesh and blood, their legs slightly different lengths, their strides different,

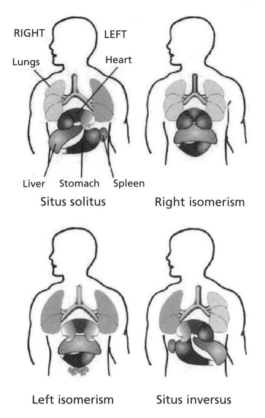

RIGHT   LEFT

Lungs   Heart

Liver   Stomach   Spleen

Situs solitus   Right isomerism

Left isomerism   Situs inversus

Figure 5.9 The normal layout of the heart, lungs and viscera (*situs solitus*), complete reversal (*situs inversus*), left isomerism and right isomerism.

their reaction times not quite the same, their opening the bags of flour unsynchronised and so on. That is why the 'b' and 'd' are only approximate mirror-images. There are methods which would make the 'b' and 'd' more similar, such as using identical twins, with identical height and stride, trained to open their bags of flour in a highly standardised way. At the other extreme, imagine the two people are very different in height, careless and in a hurry, with one wearing a personal stereo, and to make things worse the wind gusting so that the flour doesn't fall where it is meant to. The 'b' and the 'd' will now be much more different. The instructions remain the same, but, as Robert Burns said, 'The best laid schemes o' mice an men / Gang aft a-gley.'

Our teeth, eyes, arms and legs – indeed, virtually everything that is paired – show slight differences between the two sides, left and right not being precise mirror-images. Look in a mirror at your front two upper incisors, the big cutting teeth at the front of the mouth. Are they exactly the same width? They may look it at a glance, but with a pair of calipers, or better still a dental impression, one is found to be slightly larger than the other. For one particular person the right incisor may be one millimetre wider than the left; for another, it may be the other way round. Do that for several hundred people and plot a graph of the size differences, and you will find two things. First, there is the bell-shaped distribution that characterises so much in biology: many people have only a slight difference, fewer have a moderate difference, and very few have a large difference. Most importantly, in a large population almost exactly half will be a little larger on the right and half a little larger on the left. The right and left incisors have the same developmental plan for the two sides, the same DNA, so why are they not precise mirror-images of one another?[19]

The reason is the same as for the letters 'b' and 'd' in Greenwich Park not being precisely the same. Instructions for making an arm, a leg or a tooth are encoded in the genes, and the cells on each side of the body have identical copies of the instructions. But there is many a slip 'twixt cup and lip, and the same instructions do not result in precisely the same outcome. The cells forming the two sides are tossed around at the mercy of biological storms, outrageous fortune taking the form of heat, cold, viral infections, toxic chemicals, X-rays, background irradiation, the occasional cosmic ray, and so on; the thousand natural shocks that flesh is heir to. No developing organism is immune to them. Cell division, however reliable, is not infallible, and occasional failures mean the number of cells on the two sides of the body cannot be identical. Once small differences have started on the two sides, there is nothing the developing body can do to remove them, because one side of the body has no communication with the other. The question is not *whether* the two sides will be different but *how* different they will be. Such differences between right and left are called *fluctuating asymmetry*, the term coming from the fact that even if both parents have a larger left incisor, the child is as likely to have a larger right as a larger left incisor. The direction of the asymmetry fluctuates randomly from generation to generation.[20]

Fluctuating asymmetry results from variations in the environment of developing cells – what has been called 'biological noise'. Part of a cell's environment comes from the world outside, but it consists mainly of other cells, and within any particular cell the environment for a particular gene consists in turn of the mass of chemicals washing around it due to the effects of other genes. Genes produce a more consistent and reliable effect during development if they act within a predictable cellular environment, a situation known as developmental stability. If an organism maintains a well-buffered internal environment, then the same gene in different places – on different sides of the body, for instance – will result in the same outcome. In other words, the greater the developmental stability the more symmetric will be an organism's arms, legs, eyes, teeth, and so on. In yet another variant of the Greenwich Park model, think of the flour being dropped by two seven-stone weaklings or two fifteen-stone heavyweight boxers. If interlopers try to disrupt the task, then the seven-stone weaklings are far more likely to be knocked from their proper course than are the heavyweight boxers with their greater stability and resistance to external influences – the latter, in other words, are better buffered.[21]

Fluctuating asymmetry, therefore, provides an index of developmental stability. Individuals who have experienced fewer environmental insults, or who are better buffered against them, will be more symmetric. Low fluctuating asymmetry, particularly of limbs and face, ought therefore to be a measure of overall biological quality, and that indeed seems to be the case. People who are more symmetric have higher intelligence, better memory and are less impaired by drugs. Similarly, thoroughbred race-horses exhibiting the least fluctuating asymmetry demonstrate the best race performances. Highly symmetric individuals owe their symmetry to a low level of environmental stressors and hence have a high degree of developmental stability; a characteristic that their children are likely to inherit. They should, therefore, make attractive sexual partners, since they will have biologically fitter children. Since symmetry is so visible, then it could provide a basis for sexual attraction and sexual selection.[22]

Much of human sexual attraction depends on physical appearance and that elusive quality called beauty. Pinning it down has never been easy, but one theme keeps recurring. In *Left Hand, Right Hand!*, Osbert Sitwell tells of his grandmother, Louisa Hely-Hutchinson: 'As a girl she had been considered beautiful, and her features, in their aquiline mould, were *symmetrical* and distinguished' (emphasis added). Beautiful faces can be very variable in their overall characteristics, as Osbert recognises, mentioning, for example, the eagle-like nose seen in the Sitwells, which in the case of Edith conferred its own rather dramatic beauty. Proving that facial symmetry is a determinant of beauty is not easy, as one needs to take into account features such as nose shape, which are very variable and also affected by genes. This can be overcome, at least in part, by studying identical twins who, despite identical genes, are rarely *exactly* the same in appearance, not least owing to fluctuating asymmetry. Any preference for one twin over the other cannot then be due to differences in genes. In experiments asking which twin in a pair is the more attractive, there is a clear preference for the one who is physically more symmetric.

The importance of symmetry can also be demonstrated by computer manipulation of faces. Decide which of the two pictures in Figure 5.10 is

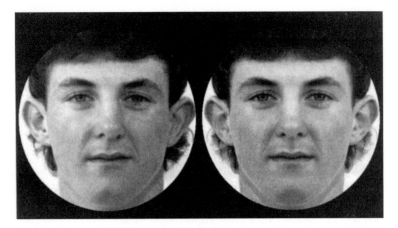

Figure 5.10 Normal photograph on the left and a version on the right that has been made more symmetric by computer manipulation.

more attractive. Although of the same person, the picture on the left retains the minor asymmetries often found in faces, whereas that on the right has been made exactly symmetric. Most people prefer the right-hand picture. So it does seem as if facial symmetry is related to facial beauty. But does symmetry contribute to increased sexual success? Certainly among animals with antlers or horns there seems to be an association between these being symmetric and greater breeding success. That, however, is only a correlation. Perhaps more convincing are the experimental studies of Anders Pape Moller on barn swallows. Using glue and a pair of scissors, he increased or decreased the length of the tail feathers to make them more symmetric or less symmetric. Swallows whose tail feathers had been made longer and more symmetric took less time to find a mate and had a greater breeding success. It does seem that symmetry is indeed a good index of developmental stability, and is related to sexual success.[23]

The Greenwich model for making a right and a left hand shows that producing mirror-symmetric organs is fairly straightforward, even if it is difficult to get the two sides to be precise mirror-images. What about the problem, though, with which this chapter started out: that of producing a heart that is entirely on one side and part of an asymmetric set of viscera? The only way that the Greenwich model can achieve such a result is if the instructions explicitly refer to left and right (or, equivalently, to east and west). One can imagine the instructions needed for the more easterly person to produce a letter 'b'. 'If you are east of Greenwich, walk two steps away from the meridian without dropping any flour; if you are east of Greenwich, turn and walk two steps north leaving a trail of flour,' and so on. That should do the trick nicely. Or rather, it will do so as long as each person knows whether they are east or west of Greenwich. How, though, would they know that? We are back to the difficulties we found in chapter 3, which ultimately ended up as the Ozma Problem. Unless there is some convention, some standard of agreement, some yardstick, some demonstration, one cannot know which side is left or right (or east or west). How, then, does a developing embryo put its heart on the left side? Somehow it must know about left and right. Surprisingly, during most of the twentieth century embryologists seem not to have been interested in this problem. Nevertheless, two of the defining experiments of classical experimental embryology, in the late nineteenth and early twentieth century, involved the heart and the sides of the body.

Life in an individual begins as a single cell, created when a sperm fertilises an egg to form the embryo. The cell divides over and over again to form the millions of cells making up the animal. In humans, the fertilised egg is tiny, a tenth of a millimetre in diameter, deep inside the mother's body, which makes studying it difficult. Things are easier in some other animals. After a hen's egg is laid, the developing chick can be studied easily. One of the first people to do so was Sir Thomas Browne in the seventeenth century, who removed the shell and observed 'how in the little *cicatricula* or little pale circle, formation first beginneth'. At the end of the nineteenth century, the most popular organisms for embryologists studying early development were amphibians such as frogs, toads and newts, which shed their fertilised eggs into water as the familiar spawn. The little black spot in the middle of each egg is the embryo, and the transparent jelly around it is the nutrient on which the embryo survives until large enough to wriggle free as a tadpole. With a microscope, it is easy to see this single cell dividing into two cells, then four, and then, by a third division at right angles to the first two, into eight, and so on (Figure 5.11). Within a day or so, the outlines of a recognisable tadpole can be seen, the beating heart is asymmetric, and the spiral of the gut is anticlockwise (Figure 5.12).[24]

In the late nineteenth century, little was known of how a single cell eventually forms the different organs of the adult body, or how some cells form the right-hand side and others the left. In 1885, the zoologist August Weismann contrasted two alternative explanations, both feasible at the time. The 'mosaic theory' said that at each cell division the genetic material also divided. At the two-cell stage, one cell may therefore have the genetic material for the left half of the organism, and the other the genetic material for the right half. At each division, the daughter cells have a more and more restricted set of genetic material, until eventually

Figure 5.11 The early stages of cell division in the embryo of *Xenopus*, the African claw-toed frog.

Right   Left

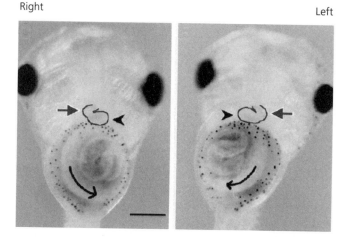

Figure 5.12 Normal tadpole (on left), looked at from below (ventrally), showing heart looping to the right and the gut coiling counter-clockwise. On the right is a tadpole with *situs inversus*, with the heart looping to the left and the gut coiling clockwise.

those making up the liver have just the genetic material for making liver cells. The opposing theory of 'regulation', said that at cell division each cell still had a copy of all the genetic material, and whether a cell formed heart or brain or liver was determined by how that cell interacted with its neighbours. In 1888, Wilhelm Roux, one of the first experimental embryologists, described an influential experiment that seemed to support the mosaic theory. Working with frog eggs, Roux took an embryo at the two-cell stage and then, with great difficulty, punctured one of the cells with a hot needle, killing it. How would such an embryo develop? According to the mosaic theory, the surviving cell had only half the genetic information and so only half the embryo would develop; in contrast, the regulation theory predicted a complete embryo. Roux's result seemed clear enough (see Figure 5.13). Only half of the embryo developed, exactly as the mosaic theory predicted.

Although Roux's experiment seemed strongly to support the mosaic theory, it was in fact flawed. Roux had made a critical error, for alongside

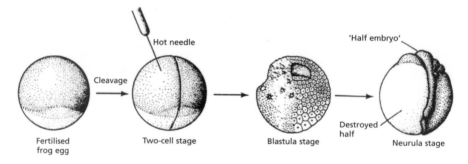

Figure 5.13 Roux's experiment in which one half of an embryo was killed by pricking it with a hot needle.

the living cells was 'the lifeless half…still attached as a decaying mass'. Three years later, in 1891, Hans Driesch carried out a better experiment, completely separating the cells at the two-cell stage to see how each would develop. Roux's results predicted one should form a right half and the other a left half, and Driesch himself was 'convinced that I should get the Roux effect in all its features'. He did not. 'Things turned out as they are bound to do and not as I had expected; there was a typically whole gastrula on my dish the next morning, differing only by its small size from a normal one.' That meant each cell had all the genetic information to make a complete organism, a fact amply confirmed a century later when a complete animal, Dolly the sheep, was cloned from a single body cell taken from an adult sheep, a cell that must have had all of the necessary genetic information.[25]

An outcome of such experiments was that they prompted people to investigate how to separate the cells of a developing embryo. Typical of the delicate manipulations required was a technique developed by Hans Spemann who, in 1935, was the first embryologist to win a Nobel Prize (Figure 5.14). In 1897, Spemann began a series of experiments on salamander embryos. He tied a loose knot in a fine hair taken from his own baby son, slipped the loop over the embryo and gently pulled it tight, either separating the cells completely or leaving them joined through a narrow isthmus. The result, in the latter case, was what are called conjoined or 'Siamese' twins, two organisms fused together at some part of their body. Of particular interest are ones such as those in Figure 5.15, fused at the tail but separate at the head, with each member having a separate heart.

One of Spemann's PhD students, Hermann Falkenberg, used the constriction method to study whether the heart of conjoined newt twins was on the right or the left side. He did not live to see his results published, since, like Robert Hertz, he died during the First World War, leading a

Figure 5.14 Hans
Spemann (1869–1941).

German night attack on 5 September 1916 at Belloy en Sauterre, in the battle of the Somme. In 1919, Spemann published the experiments posthumously. The results were striking and unexpected. Although both newts seemed to have a normal heart, and this was almost always on the normal, left-hand side in the left-hand newt, the right-hand newt had a heart which was mirror-imaged and on the wrong side (the right-hand side) in almost exactly half the cases. Whatever had caused this reversal, it was nothing to do with the mosaic theory, because if one carried on pulling the hair loop until the embryo was completely cut in half then *both* halves had a normal left-sided heart. Some interaction was occurring between the two partially separated embryos.[26]

The years after the First World War showed a massive interest in experimental embryology, with various researchers developing the techniques of Roux and Spemann. Take Lord Edward Tantamount, the second son in a noble line running back to the time of Henry VIII. Although younger sons were usually expected to enter the army, Tantamount's older brother was crippled and, in consequence, Edward was destined for a career in politics – a course in which, like much else in life, he had little interest. That all changed, however, on the afternoon of 18 April 1887 when, aged thirty and listlessly thumbing a quarterly magazine, he came across a quotation by Claude Bernard, the French physiologist: 'The living being does not form an exception to the great natural harmony which makes things adapt themselves to one another.' Suddenly, Edward knew what he wanted to do – to become a biologist. He started studying, and the next year he was working in Berlin under Du Bois-Reymond, the German physiologist of nerve and muscle, studying problems of assimilation and growth. Although a late starter, Tantamount rapidly became

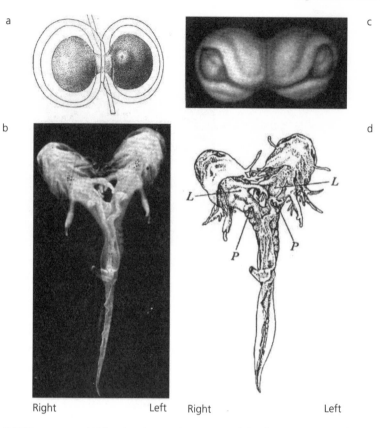

Right   Left   Right   Left

Figure 5.15 Spemann and Falkenberg's experiment on conjoined newt twins. **a,** The operation by means of a fine hair pulled tight to constrict but not divide the embryo. **b,** The conjoined twins at the neurula stage showing the neural folds which will form the brain and spinal cord. **c,** An original photograph of the conjoined twins by Spemann and Falkenberg in 1919. **d,** A line drawing of the same conjoined twins by Huxley and de Beer in 1934 which shows *situs inversus* in the right-hand member of the pair (which is on the left-hand side of the picture as the photograph is taken from the under surface, the ventral surface). L, liver, which normally is on the right; P, the centrally placed pancreas.

successful and in 1897, when the British Association met in Toronto, his paper on osmosis attracted much attention, not least from the much younger Canadian woman who discussed it with the forty-year-old Tantamount and, within a year, had become Lady Tantamount. They returned to London and, while his wife ran one of the great London salons, Edward continued to work in his private laboratory on the top floor of their Piccadilly mansion.

In 1922, Tantamount, with his assistant Illidge, had started work on a new and exciting area of experimental embryology, transplanting the developing tail-bud of a newt embryo into other parts of the body and seeing if and when it would develop as a tail or, perhaps instead, as a leg. Embryology was one of the great mysteries of biology. How could the single cell that comprised the fertilised egg keep growing and dividing endlessly, and yet not become that endlessly dividing cell, the cancer cell, forming instead the range of different body tissues and organs. As Tantamount put it, 'growing in a definite shape is very unlikely when you come to think of it'. Clearly something had to control and influence the process, not least making the two sides of the body end up being so different. One evening, although a party was in full swing at Tantamount House, the minds of Tantamount and Illidge were, as usual, elsewhere. ' "What about our tadpoles?" Tantamount asked Illidge. "The asymmetrical ones." They had a brood of tadpoles hatched from eggs that had been kept abnormally warm on one side and abnormally cold on the other.' It is unusual to have such a detailed insight into the very words spoken by working scientists, and it is time to come clean on Edward Tantamount. He is in fact a very rare bird indeed: a fictional scientist in a great novel, Aldous Huxley's *Point Counter Point*, a book described as the equivalent of Thackeray's *Vanity Fair* in its depiction of the richness of contemporary life.[27]

Huxley's novel, published in October 1928 and an immediate bestseller, is rare in its casual but informed awareness of ideas from science – there are passing references to Darwin, Pasteur, Mach, Kant, Haeckel, Newton, Mond and Faraday, as well as to Bernard and Du Bois-Reymond. Huxley was well placed to make such comments. His grandfather was T. H. Huxley, and his older brother was Julian Huxley, who in the 1920s was working as an experimental embryologist with newt embryos, and in 1934 had published, with Gavin de Beer, their massive *The Elements of Experimental Embryology*.[28]

Despite the efforts of embryologists such as Huxley and de Beer, little real advance was made in the next half a century or so, and few embryologists showed much interest in the problem of how the embryo distinguished left and right. When interest in the theoretical issues was revived, it came from a surprising quarter – psychology. During and after the Second World War, there had slowly been an increasing interest in how the brain differed on the two sides, and that interest accelerated during the 1960s. As a result, psychologists began asking about the biological underpinnings of the large psychological differences that were found in the two halves of the brain, and how they developed. What, though, was lacking was empirical findings that provided a handle on the biological processes by which left and right were defined in the organism. Then, in 1976, two research papers appeared that once more opened up the field, not least because the papers were mutually supportive, one about the laboratory mouse and the other about human patients with an unusual genetic syndrome.[29]

The Jackson Laboratory at Bar Harbor, in Maine, has specialised since 1929 in breeding mice, and particularly in breeding mice that have unusual genetic peculiarities. In 1959, two researchers described some mice that had their internal organs back to front. The mutation responsible was called *iv*, short for inverted viscera. In 1971, William Layton, an anatomist, obtained ten descendants of the first *iv* mice and started a breeding programme. By 1976 he could describe 441 descendants of the original ten mice, along with a further 507 mice that had been described in the earlier 1959 study. Fortunately, there is an easy trick to determine whether a tiny newborn mouse has its heart on its right or its left – wait twenty-four hours and then the tiny stomach, normally on the animal's right, will be full of milk and can be seen as a white blob through the pink, translucent tummy wall of the naked neonate. If the white blob is on the left, then the mouse is back to front and has *situs inversus*. Layton concentrated on those mice that were known from their pedigrees to have a double copy of the *iv* gene, the homozygotes, which had received a copy of the *iv* from each parent. Of the 948 mice, almost exactly half, 50.8 per cent to be precise, had their heart on the right side, and the rest had their heart on the left. Fifty per cent is what one would expect if chance alone determined whether the heart was placed to right or to left. And as Layton pointed out, it was also the same proportion of *situs inversus* that Spemann and others had found in the right-hand member of their conjoined frog, newt, salamander and trout twins.[30]

The second important paper of 1976 came from a Swedish medical researcher, Björn Afzelius, who was looking at patients with the condition called Kartagener's syndrome. Although usually detected in childhood, the condition can present as late as middle age, as in the case of a forty-eight-year-old London man who for four months had been coughing up large amounts of green sputum. Over the two days before being admitted to hospital he had suffered pains in the left side of his chest when he breathed in, due to acute pneumonia. For many years he had been coughing up at least a cupful of sputum daily, which was certainly not helped by his smoking twenty cigarettes a day. This wasn't, then, his first serious chest infection, and he also suffered from nasty bouts of sinusitis. When he was examined, and then X-rayed, it was clear not only that his heart was on the right side of his chest but that all of his body organs were reversed – a case of *situs inversus*. As well as administering a conventional battery of tests for patients with a chest infection, the doctors also asked the patient, in what must have seemed a very strange request, to provide a semen sample. It was, however, a shrewd diagnostic move, because under the microscope the sperm were seen not to be moving. This patient was a classic case of Kartagener's syndrome, having the unusual triad of problems of bronchiectasis (production of large amounts of infected sputum), sinusitis and *situs inversus*. Men with the syndrome are also infertile, although women are not.[31]

As with any syndrome in which there is a seemingly bizarre collection of symptoms, there must be some deep underlying feature that ties them together. Afzelius suspected that the bronchiectasis (the accumulation of infected sputum in the lungs) and the sinusitis were due to a problem with cilia. The bronchi in the lungs, as well as the sinuses of the nose, are lined with cilia – tiny little hairs that beat regularly and whose job is to sweep upwards and outwards all of the miscellaneous debris that would otherwise accumulate and cause infection. If the cilia are not working, then problems occur. Afzelius looked at the cilia of his Kartagener patients using an electron microscope. A normal cilium has a characteristic appearance, known as a '9+2' structure, with two tiny tubes, called microtubules, in a central sheath and nine microtubules surrounding it (see Figure 5.16). On one side of each of the nine microtubules are what are called dynein arms, dynein being a protein found in cells that generates movement. Afzelius found that the patients with Kartagener's syndrome had abnormal cilia that were missing the dynein arms. As the cilia could not move, the lungs and sinuses could not be cleared of any rubbish that had accumulated, and infection resulted. The male patients were infertile because dynein arms were also missing from the 9+2 structure found in the tail of their sperm, and, being immotile, the sperm were incapable of fertilising an egg.[32]

The abnormalities of the cilia elegantly explain most of the components of Kartagener's syndrome, but how does *situs inversus* fit into the picture? What can cilia have to do with the position of the heart? A clue is that among the brothers and sisters of people with full-blown Kartagener's syndrome there are others who also have bronchiectasis, sinusitis and, if male, infertility, but whose heart is on the normal left side, rather than the right. There seems to be one such patient without *situs inversus* for every one with the complete syndrome. In other words, the real syndrome seems to be bronchiectasis, sinusitis, infertility and *random* location of the heart, half of the patients having the heart on the right side and half on the left. It is that figure of fifty per cent once more. Maybe the normal role of the cilia is somehow to ensure the heart is on the left side, and that if the cilia cannot beat then the heart is as likely to be on the right as the left. That is what Afzelius suggested in his original paper. There were, however, problems with this theory. Not least, no-one had any idea of a mechanism by which cilia could determine the position

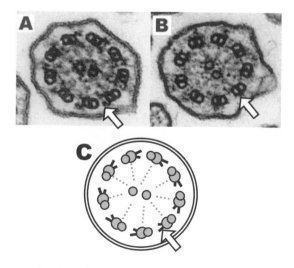

Figure 5.16 Cross-section through a normal human cilium (A), and a cilium from a patient with primary ciliary dyskinesia, a syndrome similar to Kartagener's syndrome (B). C shows a diagram of the 9+2 arrangement. The arrow in A and C points to one of the dynein arms, whereas in B the arrow points to where the dynein arm should be – the patient lacks them and so the cilium cannot beat. The cilium is viewed from below, looking up its length, and the dynein arms are arranged clockwise.

of the heart. More problematic was that patients with a related condition called Polynesian bronchiectasis had bronchiectasis, sinusitis and malfunctioning cilia but they did not have *situs inversus*. Furthermore, Layton's *iv* mice showed the opposite situation, since they did not have problems with their cilia; and likewise, there are human patients with *situs inversus* but no other features of Kartagener's syndrome who have normal cilia. By this stage, the cilia theory looked dead in the water, and was pretty well abandoned by everyone in the field, only being referred to occasionally as yet one more curiosity of lateralisation, of which there were already many.[33]

During the 1980s, the question of what causes the heart to be on the left was once more becoming of interest to biologists, although experimental analysis was hampered by the absence of a decent biological model. That changed at the end of the 1980s, with two joint studies by Nigel Brown and Lewis Wolpert. The first started by confirming an old but strange finding: that at high doses, drugs such as acetazolamide cause abnormalities in the legs of fetal mice, and that it was almost entirely the right side of the body which was affected. Intriguingly, however, as Brown and Wolpert went on to show, in *iv* mice with *situs inversus* the drug causes defects on the *left* side. Somehow, this simple chemical can distinguish on which side of the body the heart is located, a fact that suggests there is some fairly simple signpost or indicator. What could this be? In 1990, Brown and Wolpert proposed a theoretical model for how the developing embryo might tell its left from its right. At a philosophical level, the model proposed no more than what Kant had said two centuries earlier, that left and right can only be distinguished if there is some form of common signpost. However, Brown and Wolpert also translated that idea into a practical language that biologists would understand and want to explore further. Specifically, they proposed the existence of what they called the 'F-molecule', which could be oriented relative to the anterior–posterior and the dorsal–ventral axes of the organism so that the asymmetric molecule would then also indicate left and right. At the same time, Brown proposed a meeting in February 1991 at the Ciba Foundation in London of a small, carefully chosen group of scientists from a broad range of disciplines to thrash out the issues – a meeting that Wolpert was to chair and that lasted three days. Although called 'Biological asymmetry and handedness', the meeting might better have been called 'Twenty-nine scientists in search of the F-molecule'. Looking back, there is no doubt it was the moment at which left–right asymmetry became a reputable, tractable and interestingly deep problem that biology had to solve. Consequently, in the 1990s, after half a century of imperceptible progress, suddenly biologists were once more making headway, due to the advent of the powerful new tools provided by molecular biology.[34]

The field was really opened up by an experiment, published in 1995 by Mike Levin and his colleagues in Cliff Tabin's laboratory at Harvard University, which looked at the development of the heart in chick embryos, particularly in the period before the heart or any other organs are visible and when the embryo looks entirely symmetrical. At this stage, about sixteen hours after the egg is laid, equivalent to fifteen days old in a human embryo, the future chick consists of a slight ridge: the primitive streak. A structure called 'Hensen's node' develops at the front end of the streak, and then slowly passes backwards, leaving behind it the cells that will eventually form the head. The straight, mid-line tube that eventually forms the heart also develops in front of Hensen's node. Although it is initially symmetric, the first visible sign of the heart becoming asymmetric is a tiny movement to the left, called 'jogging', followed by the tube bulging out to the right (Figure 5.17). There then follows a long and complex sequence of other twists and turns before the final layout appears, in which the heart is principally in the left-hand half of the chest, and its beat can be felt to the left. Levin and his colleagues found that proteins such as *Sonic hedgehog* were present in very different amounts in the two sides of these early embryos. In Figure 5.18, at a stage

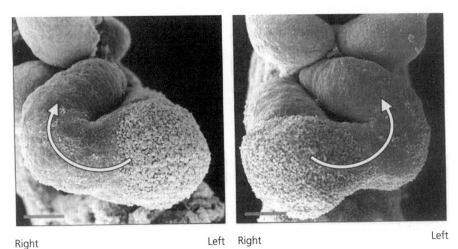

Right   Left   Right   Left

Figure 5.17 Nine-day-old mouse embryos seen looking towards what will eventually form the face. The two large blobs at the top are the head folds which will fuse together to form the head and the brain. Below these is the single tube of the heart. In the left-hand image the heart tube swings across to the right, and in this embryo the heart would eventually be on the normal side, the left. The right-hand image shows an embryo with *situs inversus*, the heart tube swinging across to the left side of the embryo.

long before the heart has developed, the dark blob indicating *Sonic hedgehog* is clearly visible on the left and almost entirely absent from the right.[35]

That the *Sonic hedgehog* protein was expressed more strongly on the left side was elegantly demonstrated, but it suffered from the recurrent problem in biology that correlation does not prove causation. The brilliance of the Levin study was that it answered such criticisms. The researchers put a tiny pellet of cells producing *Sonic hedgehog* next to Hensen's node on the *right-hand side*. They then waited until the next day to see what happened. The heart was developing normally, only not on the left-hand side, as was usual, but on the right. For the first time, a carefully placed chemical could determine whether an animal's heart was on the right or the left side. Put the pellet on the right and the heart is on the right; put the pellet on the left and the heart is on the left. Since then, a whole cascade of different signalling molecules have been found to be involved in the process, of which *Sonic hedgehog* is but one. Still, though, there was a key element missing from the story – what was it that distinguished the left-hand side of Hensen's node from the right-hand side?[36]

While Levin and others were looking at signalling molecules, Martina Brueckner and her colleagues, working in Arthur Horwich's laboratory at

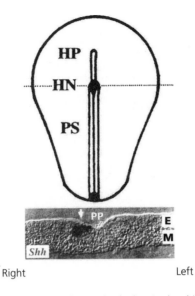

Right   Left

Figure 5.18 The asymmetric expression of *Sonic hedgehog* in the chick embryo. The top picture shows a schematic representation of the embryo: HP, head process; HN, Hensen's node; PS, primitive streak. The lower picture shows a cross-section through the embryo at the level of Hensen's node, indicated in the upper picture by a dashed line, with PP indicating the primitive pit in Hensen's node; E shows the outer ectodermal layer, and M the inner, mesodermal layer. The embryo has been stained with a probe for *Sonic hedgehog*, which shows as a large black blob, to the left of the mid-line on the upper surface of the embryo.

Yale University, were trying to work out what was wrong in the *iv* mouse by finding the faulty gene. It was a long, hard search laconically described by Horwich as 'a tedious job'. Finding the gene took a further six years, the sequence eventually being published in 1997. The *iv* gene was responsible for producing a protein, which Brueckner's team called 'left–right dynein'. Dyneins, the molecular motors of cells, had already been implicated in heart lateralisation because of their abnormality in Kartagener's syndrome. However, the only thing that was crystal clear about left–right dynein was that it had nothing to do with producing cilia in the lungs and sinuses of *iv* mice, for these were normal. So how was it involved in distinguishing right and left? Theories abounded, with some people suggesting that somehow the molecule acted as Brown and Wolpert's F-molecule, pumping chemicals from the left side of a cell to the right side, thereby telling the embryo which side was which. It seems, though, that the answer is probably nothing like that. In some ways it is far simpler, and much more similar to what Afzelius had suggested a quarter of a century earlier. The *iv* mouse *did* have abnormal cilia, but these were not where anyone had looked previously, nor were they of the same type as those which were abnormal in Kartagener's syndrome.[37]

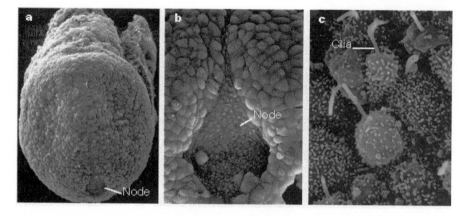

Figure 5.19 The node at three magnifications. **a,** The whole embryo, with the node at the bottom. **b,** An enlarged view of just the node. **c,** A further enlarged view of the monocilia in the base of the node.

The key discovery was made by researchers at the University of Tokyo, led by Nobutaka Hirokawa. They were looking at kinesins, which, like the dyneins, are molecular motors that transport molecules around cells by sliding along microtubules. One particular kinesin is made from smaller proteins, two of which are called KIF3A and KIF3B. Hirokawa and his colleagues studied the role of KIF3A and KIF3B in development by creating a mouse in which one or other of these proteins was knocked out. Both proteins were obviously very important, as the embryos were deformed in various ways and died early in pregnancy. Unexpectedly, about half of the embryos had their heart on the wrong side, the right, adding another gene abnormality to the list of those producing *situs inversus*. Knowing that the node was said to be crucial in determining left–right asymmetry, because kinesins were involved in building structures such as cilia, Hirokawa looked carefully at the nodal region under the microscope. The node can be seen in Figure 5.19, where it is a small pit, roughly triangular in shape, at the bottom of which are twenty or so cells each bearing what is called a 'monocilium'.[38]

Although distantly related to the ordinary cilia that are defective in Kartagener's syndrome, the monocilia are somewhat different. Instead of the usual 9+2 structure they instead show a 9+0 structure, lacking the central core of tubules. Most biologists also believed that monocilia did not function as cilia and were immotile. Hirokawa was surprised to find that the monocilia were entirely missing in the node of their knockout mice. The really big surprise, however, came on looking at the node of a normal, living mouse embryo under the microscope, when it became obvious that the monocilia were moving. Not only were they moving,

but instead of moving back and forth in the whip-like fashion characteristic of ordinary cilia, they were turning clockwise like propellers. That
immediately transformed all accounts of the origins of left–right asymmetry. The 'F-molecule' which Brown and Wolpert had been searching
for was not a molecule at all, but something far larger: a cellular
organelle, which could even be seen under the light microscope.
Furthermore, because it turned in only one direction, and was tethered at
the bottom end in the base of the node, which, in turn, was fixed relative
to the mid-line of the organism and its top–bottom (dorsal–ventral) axis,
then it could, in principle, provide a precise signpost telling the organism
which side was right and which was left. How?[39]

Although researchers can use a microscope to see the rotating monocilia, other cells in the developing embryo cannot, of course, see them in
this way. So how do the monocilia tell the rest of the embryo which is
right and which is left? The final surprise for Hirokawa's team came when
they dropped some tiny, fluorescent latex beads into the node and saw
them shoot at high speed from the right side of the node to the left – and
only from right to left. There was clearly a strong current flow, which
meant that any signalling molecule secreted by the node would be
pumped by the monocilia almost entirely to the left-hand side, where it
could trigger the cascades of signalling molecules such as *Sonic hedgehog*.
If that is happening in normal mice, then something must go wrong in
the knockout mice lacking KIF3A and KIF3B. Because those mice don't
have any cilia, the signalling molecule secreted by the node will appear
in about equal amounts on left and right. 'About equal' because here, as
everywhere, fluctuating asymmetry means that half the mice have a
slightly higher concentration on the left and hence will develop their
heart on the left, while half have a slightly higher concentration on the
right and so have *situs inversus*. Despite the *iv* mice also having a fifty per
cent incidence of *situs inversus*, at first it looked as if a similar explanation
couldn't work for them because, under the microscope, it was clear that
they did have cilia in the node. However, despite looking entirely
normal, those cilia didn't move – 'they were apparently frozen', as it was
described – and therefore the mechanism as with the mice lacking KIF3A
and KIF3B would mean half would have their heart on the right and half
on the left.[40]

The monocilia in the node seemed to explain most of what was going
on in one beautiful story. Beautiful except that, since 1992, there had
been a very large fly lurking in the ointment. A group of researchers at
Baylor College of Medicine in Houston had been inserting a new gene
into a mouse, a process that can be a very hit-and-miss affair. The gene
inserted by Paul Overbeek and his team must have landed right in the

middle of another important gene because the very abnormal mice died early in life. Most striking about these mice was that every single one had its heart on the right. Not just fifty per cent like the *iv* mice, but a full 100 per cent. The new mutation, somewhat confusingly, was called the *inv* mutation, short for 'inversion of embryonic turning'. What *inv* was doing was extremely mysterious, because the one thing researchers had accepted since the time of Spemann was that, however much one disrupted heart development – be it by mutations, trauma or whatever – the proportion of cases of *situs inversus* only ever went up to a maximum of fifty per cent. It seemed impossible to go beyond the chance barrier, but somehow the *inv* mutation had managed it.[41]

Once the rotating monocilia were discovered, there was an outside possibility that the *inv* mouse's monocilia were rotating the other way, though it seemed extremely unlikely. Indeed, when Hirokawa and his colleagues looked at the monocilia in the *inv* mouse, they saw that these were rotating at the same speed (about 600 times per minute) *and in the same direction* as in normal mice. However, although the *inv* monocilia looked normal in their movement, there was something wrong with the flow of fluid across the node. The latex beads, which had moved across the normal node so quickly, were moving at a fraction of the normal speed, and the flow looked turbulent. It seems that normal flow across the node depends not only on the rotation of the monocilia but also on the node having the proper triangular shape. However, the node in the *inv* mouse is smaller, longer and thinner than normal, and this disrupts the normal leftward flow. Nevertheless, even though the flow is slow it is still leftwards and not rightwards, so an explanation is needed of how the *inv* mouse always has its heart on the right. Hirokawa and his colleagues provided such an explanation with an elegant model.[42]

Most of Hirokawa's model is symmetric (see Figure 5.20). To both sides of the node are cells secreting a signalling molecule, and across the entire floor of the node are receptors which respond to that signal. The signalling molecule is not secreted in an active form but as an inactive precursor, which the nodal fluid activates and then, a few seconds later, deactivates and destroys. The molecule is therefore active only for a few seconds. During those critical few seconds, the nodal flow normally sweeps the inactive precursor away from the right-hand side, so that it becomes activated just as it approaches the left-hand side of the node. While it is active, the molecule is detected by the receptors on the left side of the node, triggering the cascade of signals that puts the heart on the left. In the *inv* mouse, the signalling molecule is released normally as the inactive precursor. The molecules enter the sluggish, turbulent waters of the node and move so slowly that they become active before reaching

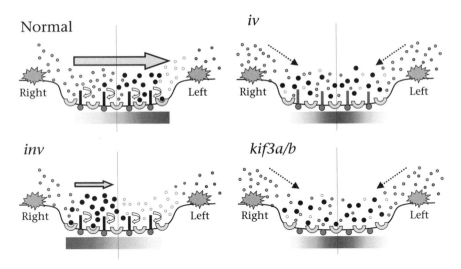

Figure 5.20 Schematic representation of the nodal flow in normal mice, and in those with the *iv*, *inv* and *kif3a/b* mutations. To left and right of the cleft the stars indicate cells secreting the signalling molecules, which are released as inactive molecules (small, grey blobs); they become activated (large, black blobs) and then become inactivated (small, white blobs). The monocilia shown in the normal and *inv* cases are rotating, whereas in the *iv* mouse they are frozen, and in the *kif3a/b* mice they are absent. The large leftwards arrow in the normal mouse shows the fast, linear flow of fluid to the left, whereas in the *inv* mouse the flow is slower and less strong. In the *iv* and *kif3a/b* mice the flow is by diffusion only, shown by the small, dashed arrows. The signalling molecules are detected by the receptors, the cups in the floor of the node, and the amount of signalling molecule detected is shown by the bars across the bottom, highest on the left in the normal mouse, highest on the right in the *inv* mouse, and highest in the centre in the *iv* and *kif3a/b* mice.

the mid-line – in other words, on the right-hand side of the node. They then trigger the cascade of signals, resulting in the heart of the *inv* mouse being on the right-hand side, the reverse of normal. Figure 5.20 shows a schematic diagram comparing what happens in normal mice, the mice lacking KIF3A/B, the *iv*, and the *inv* mice.[43]

It is an extremely elegant model, which explains how normal mice have their heart on the left side, and also shows how the process can go wrong in several very different ways. Another reason why I like it is that it not only provides plausible explanations of most of the major phenomena, but also feels like real biology, rather than just a list of DNA sequences. Hirokawa and his colleagues looked down the microscope and integrated what they saw happening there with the results of other sophisticated biochemical and genetic studies. At present, though, it is still a very new theory, and many biologists are asking a host of questions about its details. An important one concerns the seeming differences between frog, fish, mouse and man, and whether nodal flow is crucial in all of them. Whether it applies in some or all species, there are tricky evolutionary questions concerning how the node came to be organised in this subtle and sophisticated way. Such questions will be difficult to answer because fossil embryos are almost unknown, and there will certainly not be evidence concerning the rotation of their monocilia. Somewhere, somehow, however, the system must have evolved and for a purpose. The other major question is why the monocilia rotate clockwise, this being what makes the nodal flow go from right to left. That question actually has an answer, which, at one level, is rather straightforward. Pasteur would have understood it well, because amino acids – the building blocks from which bodies are constructed – are left-handed, and any motor built from asymmetric components will turn in one particular direction. That, however, merely prompts another, deeper question: why is it that almost all the proteins in our body are built from L-amino acids, the 'L' standing for laevo- or left-handed? The origins of that pervasive biological and physical asymmetry will occupy the next chapter.[44]

THE TOAD, UGLY AND VENOMOUS

It is one of the most famous scenes in all children's literature. Alice, in *Through the Looking-Glass*, holds her black kitten, Kitty, to the mirror, and wonders about Looking-glass House, which seems so much like the drawing room in her own house, except for some subtle changes – the books, for instance, 'something like our books, only the words go the wrong way'. When, eventually, she goes through the mirror she finds very different rules on the other side – the Red Queen keeps running faster in order to stand still, memory works forwards as well as backwards, and one needs to believe six impossible things before breakfast. However, nothing is as potentially strange as the thing Lewis Carroll merely hints at before Alice goes through the mirror. Alice asks Kitty if she'd like to live in Looking-glass House, and wonders if there would be milk for her there, suggesting that 'Perhaps Looking-glass milk isn't good to drink.' Undoubtedly, it wouldn't be – at least, not if drunk on this side of the mirror. To understand why, we must go back to the fundamental discovery of Pasteur, which Lewis Carroll might well have known about, that molecules can be right- or left-handed, although those produced by living things tend to be only one or the other. Molecules in living things would indeed be different in Looking-glass House.[1]

To be completely unpoetic, milk is a complex mixture of energy-rich fats, proteins such as casein, and carbohydrates such as lactose. It is easiest to think about the proteins, although the same considerations apply to lactose and many of the other molecules. Proteins are long chains of simple molecules called amino acids, at one end having an amine group (-NH$_2$) composed of a nitrogen atom and two hydrogen atoms, and at the other end having an acidic carboxyl group (-COOH) composed of a carbon atom, hydrogen atom and two oxygen atoms. The amine group of one amino acid can stick to the carboxyl group of another to make complicated molecules with several dozens, hundreds or even thousands of amino acids. Short molecules are called peptides, longer ones polypeptides and the longest proteins. The DNA which carries our genetic information acts almost entirely by specifying the sequence of amino acids in peptides and proteins. These then, for instance, form crucial parts of muscles, blood cells, teeth and the immune system; some proteins, such as the hormone insulin, act as messengers, and others form enzymes, complicated pieces of biochemical machinery that take one set of molecules, chop them up, rearrange them and convert them into others. Without amino acids, life on earth would hardly be possible.[2]

Amino acids, like most molecules in the body, depend heavily on the chemical properties of the element carbon. Atoms differ in their valency, the number of chemical bonds they can form with other atoms. Carbon has a valency of four, so that four other atoms can be attached to it. Other atoms have different valencies, that of hydrogen being one, oxygen two and nitrogen three. A carbon atom can therefore be attached to four hydrogen atoms and since the hydrogen atoms cannot then attach to anything else, the molecule is complete. It is written as CH_4, and is the gas methane, the substance comprising most of the natural gas used for heating and cooking. It can be represented by a simple diagram:

It is a convenient and traditional way of drawing molecules, but has a serious drawback. Molecules are not flat but three-dimensional, so methane's four hydrogen atoms are at the corners of a triangular pyramid, a tetrahedron. In 3-D, then, it looks rather more like in the diagram below, the two legs at the bottom right stretching forwards out of the page and backwards into the page respectively, like the legs of a tripod:

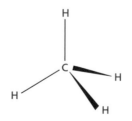

At the heart of every amino acid is a carbon atom. Attached to this are four *different* chemical groups, an amine group, a carboxyl group, a single hydrogen atom and 'something else' – the side chain, usually called R. It is R that makes the amino acids so different from one another – in alanine it is a methyl group, CH_3-; in methionine it is a complex chain of atoms, CH_3-S-CH_2-CH_2-, which includes sulphur (S); and in amino-acids such as phenylalanine there are complicated rings of carbon atoms. Although these details do not matter much for present purposes, the important thing is that if the four things attached to a carbon atom are all different, then there are two ways of arranging the molecule around the central carbon atom. We can see this best in three-dimensional drawings of the two different forms.

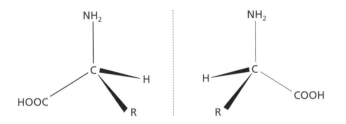

Even though such drawings are only a crude way of representing a three-dimensional molecule, it is nonetheless obvious that in some sense one is the *mirror-image* of the other. This is seen more clearly still using computer graphics to show the atoms in three dimensions as with the two types of valine in Figure 6.1, or even using the now wonderfully old-fashioned 'ball-and-stick' models of Figure 6.2. The key thing is that, however much one tries, the molecules of L-valine or L-alanine on the left cannot be rotated to look exactly the same as the molecules of D-valine or D-alanine on the right. Amino acids therefore come in two different forms, called L- (for laevo-, or left-handed) and D- (for dextro-, or right-handed), and they are left- and right-handed in precisely the same way as the two different forms of tartaric acid discovered by Pasteur (see chapter 1) – one rotates polarised light to the left and the other to the right. The two different types of amino acid are known as stereo-isomers, or enantiomers, and are said to be *chiral*.[3]

The term 'chirality' to describe objects that are mirror-images of each other was introduced by the physicist Sir William Thomson, subsequently Lord Kelvin, after whom degrees Kelvin (°K) on the absolute scale of temperature are named. On 16 May 1883 he lectured to the Oxford University Junior Scientific Club on crystals. Kelvin did not have a reputation for easy lectures and this was no exception. A few days earlier, at the Royal Institution, he had asked Lord Alverstone concerning another

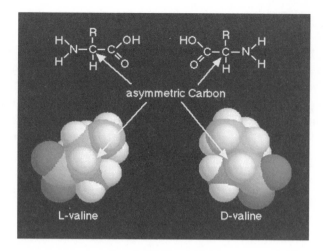

Figure 6.1 The three-dimensional arrangement of the atoms in the amino acids *d*-valine and *l*-valine.

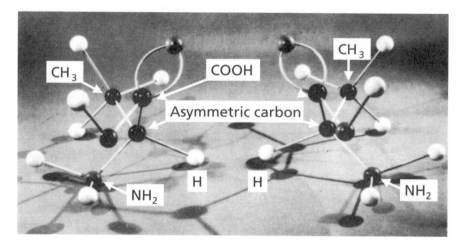

Figure 6.2 The ball-and-stick models of molecules used by Richard Feynman when he lectured on the difference between L-alanine (on the left) and D-alanine (on the right). Annotations are not in the original. Even though the molecular models are symmetric, the photograph itself is not symmetric because the illumination is asymmetric.

lecture, 'I hope you found my lecture interesting?' 'I am sure we should, my dear Lord Kelvin,' came Alverstone's reply, 'if we had understood it.' The Oxford lecture was forty-one pages long when printed. By page thirty-five Kelvin feared his audience was flagging and that he had sadly taxed their patience, even though he was barely halfway through what he had hoped to say. He battled on for a further five pages and then, despite his wish to present the poor students with 'a more thorough' account of the geometry of chirality, concluded 'but in pity I forbear'. Kelvin only defined chirality in a footnote to the printed lecture, and unless he spoke using footnotes, which even experienced academics find difficult, the students probably had little idea what he was talking about, this being the first time the word was used in English. 'I call any geometrical figure, or group of points, *chiral*, and say that it has chirality if its image in a plane mirror…cannot be brought to coincide with itself.' Our right and left hands are therefore chiral objects, since a right hand in a mirror cannot be brought into exact register with itself. The same is true of amino-acids, as can be seen in Figures 6.1 and 6.2.[4]

Although many carbon compounds in nature are chiral, not all are so, an example being alcohol, CH_3-CH_2-OH, in which one of the carbon atoms is attached to two hydrogen atoms and the other to three. That provides some answer to the question posed by W. H. Auden in an adult version of Alice's question in front of the Looking-glass. Quant, a right-handed widower who works in a shipping office near the Battery in New York, looks at his reflection in the mirror behind a bar and asks, 'What flavor / Has that liquor you lift with your left hand?' Without doubt, Looking-glass alcohol would get one just as drunk as the usual stuff, since the alcohol molecule is not chiral. The flavour of the liquor is another matter though. Many biological molecules are stereo-isomers, the L- and D-forms having very different properties. A well-known example is the molecule carvone, for which the D-stereo-isomer smells of spearmint, and the L-stereo-isomer smells of caraway. Likewise, one stereo-isomer of α-phellandrene smells of eucalyptus and the other of fennel. Many books also cite the molecule limonene, which is said in one chiral form to smell of oranges and in the other chiral form to smell of lemons. It is a lovely idea: 'Isomers and chirals, say the bells of St Clements.' Unfortunately it is wrong, a result of impurities introduced into the L- and D-forms while being synthesised. Oranges and lemons both actually produce D-limonene, which when pure has a sort of generically citrus smell. L-limonene is found in peppermint oil, and pure L-limonene smells of pine. Generally, D- and L-isomers have different odours because our sense of smell detects the three-dimensional shape of molecules, which engage like a key in a lock with receptors in the olfactory membrane in the nose. If two molecules have different three-dimensional shapes and engage different receptors, then they smell different. Since stereo-isomers are different three-dimensionally, they often smell different. Looking-glass liquor would therefore taste different.[5]

It is not only smell that works in this way. Hormones, such as thyroxine produced by the thyroid, are chemicals secreted into the bloodstream to modify far distant cells, which they do by binding to receptors on the surface. Like the receptors inside the nose, cellular receptors also look at the three-dimensional shape of molecules, and typically recognise only one of the chiral versions of a molecule. An analogy from everyday life is the Yale lock on a door, the key for which is usually chiral, meaning that the mirror-image could not unlock the door. For the hormone thyroxine, the natural version is the L- form and the D- form is inactive. A patient suffering from myxoedema, a shortage of thyroxine, has to be given L-thyroxine since D-thyroxine is not effective.[6]

Just as hormones and other natural chemicals are typically chiral, many drugs in common use are also chiral, one stereo-isomer being more effective than another. An example is salbutamol, used for treating acute asthmatic attacks. Although typically only one stereo-isomer of a chiral drug is effective, in the past most pills supplied by a pharmacist of this and other chiral drugs have been a racemic mixture of the D- and the L-form, the reason being that most drugs are synthesised chemically. It is the same situation that Pasteur found back in the 1840s, when he discovered that tartaric acid produced chemically was a fifty-fifty mixture of D- and L- forms, so-called racemic acid, whereas the tartaric acid produced by micro-organisms when wine goes off consists entirely of D-tartaric acid. Instead of being given a pure drug, therefore, patients have received two stereochemically different substances, principally because in the past it has been difficult to devise chemical techniques that produce one stereo-isomer of a drug but not the other.

Drug companies, in fact, are keen to produce drugs that are pure stereo-isomers, in part because drug regulatory authorities are very aware that one stereo-isomer can be therapeutic while the other causes unwanted side effects. For instance, bupivacaine, a longer-acting version of the lignocaine (lidocaine) used as a local anaesthetic in dentistry, is injected intravenously before surgery to produce a 'regional block' in an arm or a leg. Although it is generally safe, several cardiac arrests linked to bupivacaine were reported in 1979, and these were shown to be due to the otherwise inactive D- form. The active L- form did not have this side effect, and, as a result, L-bupivacaine, or levobupivacaine is now produced, to the obvious benefit of patients. A less glorious reason for drug companies wanting to carry out a 'chiral switch' – replacing a racemic mixture with a pure stereo-isomer – is purely commercial: if originally licensed as a racemic mixture, then introducing a pure stereo-isomer can extend the patent before generic competitors can produce the drug. Occasionally the strategy goes wrong, as the manufacturers of fluoxetine (Prozac) discovered. The company patented only the racemic mixture and not the enantiomers, which meant that eventually they had to come to a licensing agreement with a company producing the active stereo-isomer.[7]

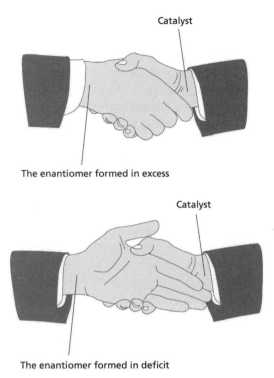

Figure 6.3 The principle of chiral catalysis. The catalyst, symbolised by the hand on the right, fits better and more naturally with a right hand (top) than with a left hand (bottom), and as a result has a lower energy, so that more of the right-handed product is produced than the left-handed product.

Recent years have seen a surge in the number of drugs marketed that are single stereo-isomers – fifty of the top one hundred by one estimate – due mainly to a quiet revolution in synthetic chemistry which now has a host of techniques for industrial synthesis of stereo-isomers. Many advances use techniques similar to those of biological enzymes, creating small clefts or holes in catalysts into which only the left- or right-handed molecule can fit, so that only one form then takes place in the chemical reaction (Figure 6.3). In 2001, William Knowles, Ryoji Noyori and Barry Sharpless were awarded the Nobel Prize in Chemistry for developing such methods.[8]

Stereo-selective drugs, although often useful, are not a universal panacea. A mythology has developed around one in particular, the notorious drug thalidomide, marketed originally to help pregnant women. A recurrent story is that D-thalidomide is a safe and effective hypnotic, whereas the appalling side effects in developing fetuses were entirely the result of L-thalidomide. The story has a fatal flaw in its logic. Although many chemicals do come in L- and D- forms, they are not always permanently fixed that way. Some are, but others, including thalidomide, have very limited stability, the process of racemisation converting them to the other stereo-isomer. A patient taking only D-thalidomide would find substantial amounts of L-thalidomide in their blood within six to twelve hours, because of racemisation. D-thalidomide would be as dangerous as the original version unless modified to prevent racemisation.[9]

Having wandered a little from the amino acids, it is time to return to them, and consider how they are put together into proteins. The amino acids in a protein are coded in DNA by triplets of the four bases, cytosine, guanine, adenine and thymine (C, G, A and T), any of the four occurring at each of the three positions to give a total of $4 \times 4 \times 4 = 64$ potentially different combinations, which together specify the set of twenty amino acids from which organisms are constructed. The first step in protein synthesis is to make an RNA copy of the DNA sequence. A ribosome then runs along the RNA, stopping at each successive triplet. A molecule of transfer-RNA brings the amino acid encoded by that triplet, and the amino acid is spliced on to the end of the growing chain of amino acids. The ribosome then moves to the next triplet, which in turn is read and its amino acid added; and so on until the complete protein has been built. Every time an amino acid is added to the growing protein chain it is an L-amino acid. The genetic code and the entire translation machinery are based only on L-amino acids. As far as we can tell, that is true for every organism on this planet. Such a complete dependence on L-amino acids raises many fundamental questions for biology, the answers to which stretch beyond biology itself, both outside our solar system and deep into sub-atomic physics.

Amino acids are not the only place where biochemical asymmetry can be found. The sugars that make up our bodies, such as glucose, are also found in just one form, this time the D- form. Again, it is a perplexing truth deeply in need of explanation.[10]

If our bodies are made entirely of L-amino acids and D-sugars, what would an organism be like that was instead constructed of D-amino acids and L-sugars? Would such a 'D-organism' work as well as the conventional 'L-organism'? Although we are unlikely to answer that question for a long while, it is now possible to synthesise peptides, polypeptides and even proteins entirely from D-amino acids rather than L-amino acids. If the protein is an enzyme, then that D-enzyme's three-dimensional structure is the mirror-image of the normal L-enzyme, and therefore it should act on D-amino acids rather than the usual L-amino acids, just as a mirror-image lock would open with a mirror-image key. That has been tested in one case, a protease enzyme from the human immunodeficiency virus (HIV), and it worked precisely as predicted. The enzyme was completely back to front, and the Looking-glass D-protease was as good at chopping up D-peptides into little pieces as was the ordinary L-protease at chopping up L-peptides. There seems every reason to believe the same would be true for an entire Looking-glass organism, as long as it lived in a Looking-glass world where it could drink Looking-glass milk and the like.[11]

Although our bodies are constructed almost entirely from L-amino acids, D-amino acids do occur in nature, and their presence often illuminates many biological processes. D-amino acids in fact occur ubiquitously because of spontaneous racemisation. When Pasteur heated D-tartaric acid crystals for six hours at 170°C he found they changed into racemic acid; that is, a fifty-fifty mixture of D- and L- tartaric acid. Natural L-amino acids also racemise, being converted to a fifty-fifty mixture of D- and L-amino acids, a process that occurs most easily in free amino acids but happens also to amino acids in proteins during cooking. Racemisation is potentially disastrous for proteins, because a D-amino acid has a different three-dimensional shape from an L-amino acid, altering the overall shape of the protein and preventing it from binding properly to other proteins. Cells usually avoid 'protein fatigue' – a phenomenon loosely similar to metal fatigue – by continually replacing old, decrepit proteins with fresh, new ones, synthesised by ribosomes and containing only pristine L-amino acids. Sometimes, though, continual replacement cannot happen. Some proteins in the body, such as dentin in the teeth, or crystallin in the lens of the eye, are extremely long-lasting because they form the physical structure of the organ, and with increasing age they inevitably accumulate D-amino acids.

A different problem occurs for the short-lived red blood cells. These cells, which carry oxygen in the blood, are extremely active cells that are unable to replace old proteins with new because they have none of the key equipment for making proteins, a nucleus or ribosomes. The proteins present when a red blood cell is made therefore comprise all those the cell will have for the rest of its natural life. Protein fatigue results in errors gradually accumulating, and there is no way of replacing the failing, geriatric proteins. The scenario is rather like an ageing space station, the crew having to make do with what they find on board since parts cannot be replaced, until eventually the ship catastrophically fails owing to an accumulation of errors. Although red blood cells may not survive long – about 120 days in humans – they make an ideal system for studying protein fatigue. Within forty days, about one per cent of a red cell's L-aspartic acid has converted to D-aspartic acid, a surprisingly rapid degradation. Fortunately, most amino acids do not racemise that quickly, particularly when forming a part of proteins. Most amino-acid racemisation is sufficiently slow to be used as a 'molecular clock' for estimating the age of biological specimens. Long-term protein racemisation can be seen in the man known as Ötzi, whose body was found in a glacier in the Tyrolean Alps in September 1991, and who was estimated, using radiocarbon dating, to have lived five thousand years ago, between 3350 BC and 3100 BC. In his hair, thirty-seven per cent of the amino acid hydroxyproline was in the D form, compared with thirty-one per cent in 3000-year-old hair, nineteen per cent in 1000-year-old hair, and only four per cent in recently collected hair samples.[12]

Proteins are powerful molecules and can be extremely dangerous if, as toxins or venoms, they are injected into the body. As a mild example, consider that slight soreness in the mouth after eating fresh pineapple, which is due to bromelain – a protein-digesting enzyme in the pineapple – digesting the surface of your lips and tongue; as you eat the pineapple, so it is eating you. Proteins may be dangerous, but they are also essential to the body, both as a source of energy, and as a source of the nine 'essential' amino acids, such as phenylalanine, which our body cannot make but must take in from outside. In other words, we must eat them. The potential risks of eating dangerous proteins are minimised by the intestines secreting large amounts of special enzymes, such as trypsin, pepsin and elastase, collectively called proteases, which chop proteins into their constituent amino acids. Before they are absorbed, the risky dietary proteins have been neutralised into safe, non-toxic, amino acids, which pass into the blood. Proteases, like all enzymes, work by looking for a particular three-dimensional structure which is specific for each enzyme. Trypsin, for instance, cuts proteins next to the amino acids lysine or arginine, and although the biochemistry textbooks rarely emphasise it, trypsin is looking for L-lysine and L-arginine. What happens then if trypsin encounters D-lysine or D-arginine? The answer is simple: nothing. It is like putting a left foot into a right shoe, or a mirror-image key into a lock – it does not fit. Proteins and peptides containing D-amino acids are not digested in the body, and when D-amino acids are given to a person they pass through unchanged into urine and faeces. This is the reason why Looking-glass milk would not be good to drink, for the D-casein and other D-proteins would not be digested but merely pass straight through the bowels, maybe causing diarrhoea as a side effect.[13]

The failure of proteases to break down proteins containing D-amino acids may sometimes underlie disease. One potentially important example is the protein called amyloid, which accumulates catastrophically in the brain of patients with Alzheimer's disease. Amyloid, somewhat surprisingly, contains a D-amino acid, D-serine. To find out its role, researchers created an eleven-amino-acid peptide from the middle of the amyloid protein, and included in it either the D-serine found in patients, or a more normal-looking L-serine. Both the D- and the L-serine versions were toxic to brain cells, but the brain's protease enzymes rapidly removed the L-serine version, whereas the D-serine version was not affected. It may be that the normal L-serine protein spontaneously racemises into the D-serine version, which cannot be removed, and thus slowly accumulates, damaging brain cells.[14]

Although proteins containing D-amino acids are not digested in the bowel, some D-amino acids are found in the blood, either from food (particularly cooked food) or as by-products from the bacteria living in the large bowel (it has been shown that rats raised in germ-free environments have lower levels of D-amino acids). Wherever they come from, D-amino acids have long been regarded as unwelcome phenomena, particularly since studies in the 1940s showed that D-serine and D-aspartate damaged the kidneys and slowed the growth of rats. The body, though, seems to have sorted out the problem, for in 1935 Hans Krebs, who later won the Nobel Prize for his work on the Krebs cycle, found an enzyme he called D-amino acid oxidase. Present in large amounts in the kidney, it neutralised the nasty D-amino acids so that they could be excreted in the urine. Until the 1990s, that would have been the end of the story for free D-amino acids. Then, in 1984, the Japanese researcher Atsushi Hashimoto and his colleagues at Tokai University School of Medicine discovered high levels of the amino acid D-serine in the brain of rats, and soon after D-aspartate was also found. Neither of these seemed to come from spontaneous racemisation or the rat's diet. Experiments led to the discovery of a new enzyme in the brain, serine racemase, which converted standard, boring L-serine into new, exciting D-serine. Further, D-amino acid oxidase, that old enzyme desperately looking for an interesting function, was found wherever D-serine and serine racemase were present in the brain, and seemed to be particularly effective at removing D-serine. If the brain has enzymes that make D-serine and other enzymes that break it, and if also there are large quantities of D-serine in the hippocampus and the cerebral cortex, then D-serine must have a pretty important job to do. Although not yet fully sorted out, D-serine seems to modulate one of the sexiest neurotransmitter systems discovered in recent years, NMDA, which is involved in memory and learning, epilepsy, various diseases and protecting the brain after stroke. Why, though, a D-amino acid should be such an important neurotransmitter is anyone's guess at present.[15]

The D-amino acids discussed so far, occurring as free D-amino acids in the brain or by mistake as a result of spontaneous racemisation, do not violate the traditional view that proteins in higher animals are made only of L-amino acids. The central dogma of molecular biology, which says that DNA produces RNA which produces proteins, has the corollary that since DNA only encodes L-amino acids and transfer-RNAs only carry L-amino acids, then proteins have to be made of L-amino acids. That is indeed true, but one needs to add the proviso that although proteins may come into the world made only of L-amino acids, they need not continue that way. If an organism finds it useful to replace an L-amino acid in a protein with a D-amino acid, then there is nothing in biology to stop it. In fact, that process is very common in bacteria and fungi, which have many proteins containing D-amino acids.[16]

The cell wall of bacteria acts as a physical protection and keeps the internal environment constant in hot or dry conditions. Bacterial walls often contain unusual amino acids, including exotic L-amino acids not included in the twenty encoded by the genetic code, and also D-amino acids such as D-alanine, D-aspartic acid, D-glutamic acid and D-phenylalanine. The advantage is probably that ordinary proteases, the protein-digesting enzymes of other organisms, cannot digest the unusual amino-acids in the cell-wall proteins, and so the bacteria are protected from attack. The trouble with such an efficient defence mechanism is that it inevitably starts an arms race. As bacteria with near-indigestible cell walls infect other micro-organisms, so those micro-organisms themselves have evolved a novel defence mechanism, producing chemicals that can inhibit the growth of bacteria. The most famous was discovered by Sir Alexander Fleming, in 1928, when he noticed there were no bacteria growing on a culture dish around a patch of mould. The mould produced penicillin, which had the unusual property of blocking the growth of the bacterial cell wall – in other words, it prevented the D-amino acids in the cell wall from being an effective defence.[17]

For simple organisms such as bacteria (the prokaryotes), it is normal to make proteins containing D-amino acids, usually for special defensive purposes, but what about multicellular organisms (the eukaryotes), which comprise plants, animals and, of course, us? Again, a clear dogma had evolved concerning this. In 1965, in his authoritative book on amino acids, Alton Meister wrote, 'At this time there is no conclusive evidence for the occurrence of D-amino acids in the proteins of plants and animals.' It might well have stayed that way but for the work of Vittorio Erspamer, who in 1962 discovered in the skin of a South American frog a peptide called physalaemin, which proved to be a potent stimulant of smooth muscle. Although pure serendipity inspired that first discovery, the next three decades witnessed an effort to screen amphibian skin from around the world for biologically active peptides. It was a labour of love

for Erspamer, but also a labour to cover the pain of loss. In a brief biographical detail, so rarely found in scientific papers, Erspamer describes
his life's work, 'which had given results surpassing by far our boldest
expectations. It was a long-lasting game, which made fleeting the months
and years, and was the only solace in the moment of agony when I tragi-
cally lost my eighteen-year-old daughter, Maria Luisa, heart of my
heart.'[18]

Erspamer's work stimulated many other researchers to study the
obscure, bizarre and fascinating peptides that he isolated. In the 1990s,
the work also prompted two massive review papers by the pharmacolo-
gist Lawrence Lazarus, both with Shakespearean titles, which extol the
frog and the toad, lauding their place in cultural history, folk medicine,
language and literature, and, of course, pharmacology. From one of these
papers comes the title of this chapter, taken from *As You Like It* (Act 2,
Scene 1).

> Sweet are the uses of adversity,
> Which like the toad, ugly and venomous,
> Wears yet a precious jewel in his head;
> And this our life, exempt from public haunt,
> Finds tongues in trees, books in the running brooks,
> Sermons in stones, and good in everything.

Slightly modified for Lazarus, the precious jewel is now the toad's skin.
And Vittorio Erspamer, in his adversity, had found tongues that speak
pharmacology in the amphibians that live in trees, and books of drugs in
the frogs and toads found in running brooks and under stones. Already,
new biologically active peptides from amphibian skin number into three
figures, with no end in sight. Only one family of compounds concerns us
here, the dermorphins and deltorphins, collectively known as the opioid
peptides.[19]

Meister's claim in 1965 that D-amino acids did not occur in the proteins of plants and animals was finally shown to be wrong in 1981, when
the first dermorphin was discovered, and since then ever more substances
containing D-amino acids have been discovered. The dermorphins and
deltorphins all contain seven amino acids, and in the second position all
have either D-alanine or D-methionine. The D-amino acid seems to be
crucial, since replacing it with the equivalent L-amino acid makes the
peptide completely inactive. The D-amino acid creates a sharp kink in the
peptide chain, which allows it to form the key that opens so many
biological locks.[20]

Dermorphins and deltorphins are opioid peptides because they act on the brain in the same way as natural opiates such as morphine and heroin. In fact, weight for weight, dermorphin is a thousand times more potent than morphine, and ten thousand times more potent than the proper neurotransmitter in the brain, enkephalin. From the dermorphins and deltorphins may well come morphine substitutes that are potent pain-killers but also non-addictive and without the side effects of sedation and gastro-intestinal stasis. Of course there is also the possibility for new designer drugs to feed abuse, the juice of the poppy being replaced with a simple peptide. Indeed, through luck or desperation, it is already known that the skin of toads contains powerful psychoactive substances, drug addicts in Australia reportedly licking the dried skin of the toad *Bufo marinus*, and those in America smoking the dried skin of *Bufo alvarius* (described by *Newsweek* as 'the granddaddy hallucinogen, so powerful it makes LSD look like a glass of milk'). It is hardly a new idea. During traditional shamanistic rituals, the Matses of the upper Amazon basin would induce hallucinations by putting dried skin secretions of the toad *Phyllomedusa bicolor* into skin wounds.[21]

So far, amphibian skin is the only place where proteins containing D-amino acids have been found in vertebrates, although no doubt there are others. Outside the vertebrates, there are many animals that use D-amino acids. For instance, the languid and beautiful marine snails of the genus *Conus* are sophisticated killing-machines that hunt fast-swimming fish by using their tasty-looking proboscis as a lure. Inside the proboscis is a disposable harpoon-like tooth, which injects a cocktail of toxins that cause immediate paralysis and convulsions, and allows the fish to be reeled in, engulfed and eaten. Part of the highly toxic venom is the nine-amino-acid peptide contryphan, which has a D-tryptophan in the middle of the sequence, a different amino acid and a different location from that found in the dermorphins and deltorphins, suggesting D-amino-acid proteins are more variable than previously thought. A final example of a D-amino-acid protein is found in the funnel-web spider, *Agelenopsis aperta*, which also paralyses its prey with a venom. At one end of this protein is a D-serine, which gives it the shape of a long, thin probe acting to block the brain's calcium channels and producing the devastating effect of stopping transmission between nerves. What is exciting about this venom is that, for the first time, we know where its D-amino acid comes from. An enzyme called peptide isomerase takes an ordinary L-serine in the protein sequence and converts it, on the spot, into D-serine. The enzyme is remarkably similar to the common protein-cutting enzyme, trypsin, showing how, as ever, evolution exploits an existing resource, rather than developing entirely new technology. It also suggests that many other animal proteins may contain D-amino acids.[22]

Although D-amino acids undoubtedly occur in a range of organisms, from bacteria through spiders and snails to frogs and toads, as well as in human brains, their scarcity re-emphasises the big question of why life on earth is composed so predominantly of L-amino acids and D-sugars. Two separate questions lurk here; one easy, the other difficult. The easier question is why amino acids and sugars should have the same chirality (that is, *all* the amino acids are L-amino acids and *all* the sugars are D-sugars). The reason is straightforward – it is easier to build objects using the same sorts of building brick, even asymmetric bricks, than to use a mixture of different types. RNA and DNA are a good example, being made of nucleotides containing a sugar (ribose for RNA and deoxyribose for DNA). Like all other sugars, they are D-ribose and D-deoxyribose. A mirror-image form of DNA could readily be built using L-deoxyribose, and it should function just as well. What is almost impossible is to produce a stable molecule that contains a *mixture* of D- and L-sugars. For DNA, stability is essential to its function, and so homochirality, having all the components of the same handedness, is essential. A similar argument applies to the amino acids that make up protein chains, these being much more stable if made entirely of one type of amino acid. That, though, is the easy part of the argument. The difficult bit is why life has D-sugars and L-amino acids rather than vice versa.[23]

One prevalent theory concerning the predominance of L-amino acids and D-sugars claims that there is nothing to be explained, the predominance resulting from chance and chance alone. The geneticist J. B. S. Haldane, the biochemist Leslie Orgel and the physicist Murray Gell-Mann have all suggested mere accident resulted in the first organism on earth using L-amino acids; an accident that could as easily have happened the other way around. However, once that accident had happened, then the use of L- rather than D-amino acids became a fixed, arbitrary but completely immutable characteristic of life – a frozen accident, carried on in perpetuity by all descendants of that earliest life form. Think of tossing a coin to decide which of two players should play first in a series of games. The two possible outcomes are equally likely in advance. However, a myriad of tiny factors mean that the coin must actually fall one particular way round, say as a head, and that outcome then determines the rest of the contest. Nothing special made the coin a head, and another coin on another day could as easily have shown a tail, but once that first decision is made, everything follows on. The difference from the game of life is that life on earth has been played once only, and is still being played. Although putting everything down to chance alone is an attractive and simple idea, there is a growing number of scientists who see it as insufficient for explaining the predominance of L-amino acids on earth, and that, as we shall see, is partly because of evidence originating many millions of miles away from planet Earth.[24]

The origins of life on Earth took place in very different conditions from those found now. It is tempting to imagine a benign environment where soft, vulnerable assemblages of nucleic acids, sugars, amino acids and membranes were able tentatively to explore how to become self-replicating – what Darwin described as a 'warm little pond with all sorts of ammonia and phosphoric, – light, heat, electricity, etc. present'. During the twentieth century, researchers visualised a more noxious world: J. B. S. Haldane envisaged a much hotter world, with drenching rain washing chemicals off the mountains into seas that 'reached the consistency of hot dilute soup', and in the 1950s the biochemist Stanley Miller simulated the early Earth with electric sparks passed through an atmosphere of methane, ammonia, hydrogen and water vapour. Even that is probably too salubrious, the chemist William Bonner describing how, 'during its first 700 million years or so the primeval Earth was subjected to intense bombardment by innumerable comets and asteroids as large [as] or larger than comet Halley, some of which impacted with sufficient energy to vaporise the oceans, form [a] rock-vapour atmosphere and sterilize Earth's surface to depths of tens of metres'. It hardly sounds the ideal location for some delicate chemistry and experimental biology, suggesting anything but a warm, wave-lapped beach. Nevertheless, complex organisms did somehow start to evolve in that world, and they used a genetic code based on DNA made of D-sugars and sophisticated enzymes constructed from L-amino acids.[25]

If chance alone is not the explanation for the supremacy of L-amino acids or, for that matter, of D-sugars, something must have given life based on L-amino acids an edge over life based on D-amino acids. Once again, there are two separate questions, one asking how L-amino acids are advantaged over D-amino acids, and the other how what was probably a small superiority became amplified to the extent that D-amino acids are almost unknown in living things. Most thinking has concerned the advantage of L- over D-amino acids, this accompanied by a vaguely stated hope, coupled with some arm-waving, that somehow everything else related to the theory will shake out. The devil, though, in all such theories is just how large an excess of L-amino acids might result. Too small (and it is usually exceedingly small), and surely it would have been swamped, literally washed out, in the chaos of the early Earth. Many theories have speculated about some asymmetry on Earth itself, perhaps in its magnetic field, or in quartz crystals, or in the gravitational fields, but all ultimately fail, either theoretically, empirically or both. Any feasible explanation, therefore, has to turn away from the parochial concerns of

Earth itself, and look inwards to the workings of the atom, or outwards to the depths of space.[26]

Physicists describe left–right symmetry as 'parity', and had always assumed that the universe was left–right symmetric – or, in their jargon, parity was conserved. 'Even parity' to physicists means things do not change when reflected in a mirror, whereas 'odd parity' means reversal in a mirror. Corkscrews have odd parity, changing from right-handed to left-handed in a mirror, whereas wine-glasses have even parity, being indistinguishable in a mirror or looked at directly. Conservation of parity meant, in the words of Lord Blackett, that 'Nature's hardware shop always stocked an equal number of right- and left-handed corkscrews.' Think of a game of billiards, either watched for real or reflected in a mirror. It would be impossible to tell which was the mirror version solely from the movements of the balls because the laws of physics dictating how one ball will move after striking another are the same in the mirror game and the real game. That though is not true for the artefacts of everyday human life. Imagine a right-handed friend opening a bottle of wine, pouring a glass of wine and sitting down to read *The Times*. A host of details betray the mirror version – the corkscrew would have a left-handed thread and would be turned anti-clockwise, the wine would be poured with the left hand, and the print on the newspaper would be back to front. Parity is not conserved in everyday life; human hardware shops do not stock equal numbers of right- and left-handed corkscrews, and humans have strong left–right asymmetries which they impose on everything they touch. Parity, however, did seem to apply to physics. As Chen Ning Yang put it in his Nobel Prize acceptance speech, 'the laws of physics [had] always shown complete symmetry between left and right'. In January 1957, however, there was great excitement about what Lord Blackett soon after described as 'one of the most curious and exciting events in the history of modern physics'. Chien-Shiung Wu carried out an experiment in Washington DC that took only forty-eight hours to perform and yet showed the opposite of what most physicists had expected; that parity was not conserved.[27]

The idea of parity conservation wasn't so stupid though. Physics recognises four very different types of forces in nature – electromagnetic, strong, weak and gravitational – and for three of them – electromagnetic, strong and gravitational – there were clear indications that parity was conserved. When, therefore, two young Chinese-American physicists, Tsung-Dao Lee and Chen Ning Yang ('Frank') suggested in a theoretical paper of October 1956 that conservation of parity might not apply to the weak force, and proposed an experiment to test it, there was much scepticism, despite the proposal being made precisely because Yang and Lee had found anomalies in previous experimental results. Persuading the US National Bureau of Standards in Washington DC to carry out the experiment was difficult because their theorists saw it as a waste of time doing an experiment that would inevitably confirm parity conservation. They weren't alone. In one of those classic statements of theoretical commitment that scientists come to regret afterwards, Wolfgang Pauli, the Nobel-Prize-winning theoretical physicist, wrote in a letter from Zurich, 'I do *not* believe that the Lord is a weak left-hander, and I am ready to bet a very high sum that the experiments will give symmetric results.' Whether he actually placed the bet is unknown, but if he did, he lost.[28]

The experiment itself involved the radioactive element Cobalt-60, which undergoes beta decay and emits an electron and an elusive subatomic particle called a neutrino. Cobalt atoms spin, much like the Earth spins on its axis, and so it is possible to talk of a 'north' and 'south' pole, although which is which is inevitably arbitrary. If cobalt atoms are cooled close to absolute zero in a powerful electromagnetic field, they line up so that all are spinning in the same direction (for instance, clockwise when looked at from above). They continue to emit electrons, and the key question is in which direction. Conservation of parity says the answer should be, 'equally, in both directions, north and south'. If that were not the case, then when looked at in a mirror the experiment would look different from the real version, just as the corkscrew looks different in a mirror, and the system would not have even parity. So it was to prove. More electrons came out from the 'south' than the 'north' end, meaning parity was not conserved. Electrons, it turned out, are left-handed. Pauli described the experiment as 'very dramatic', and news of the experiment flashed around the world of physics within days. The Nobel Prize committee agreed with Pauli and recognised the achievement by awarding the Nobel Prize for Physics to Yang and Lee that same year, 1957.[29]

The importance of Yang and Lee's discovery was well explained for non-scientists by Abdus Salam, the Pakistani physicist who himself won the Nobel Prize in 1979. He described to a classicist friend why the result was so exciting:

I asked him if any classical writer had ever considered giants with only the left eye. He confessed that one-eyed giants have been described, and he supplied me with a full list of them; but they always [like the Cyclops in Homer's *Odyssey*] sport their solitary eye in the middle of the forehead. In my view, what we have found is that space is a weak left-eyed giant.

The failure of conservation of parity has all sorts of ramifications, perhaps even explaining the excess of left-handed over right-handed galaxies. It also solves the Ozma Problem described in chapter 3, the challenge of which was to tell someone deep in outer space which of two gloves is the left-handed one. The solution is straightforward so long as one has the equipment to look at the rotation of a large sample of electrons, these predominantly being left-handed, or, even better, to look at some neutrinos, every single one of which is left-handed – as the physicist Otto Frisch put it, the flight of the neutrino is like a bullet spinning from a rifle with left-handed rifling. Once it is agreed that left refers to the direction of rotation of a neutrino, then there is no problem describing which glove is the left-handed glove. So, almost two hundred years after Kant's paper of 1768, the Ozma Problem was finally resolved in 1957.[30]

The real challenge for the asymmetry of the weak force is whether it might also explain the biological predominance of L-amino acids and D-sugars. Soon after the Wu experiment, J. B. S. Haldane pointed out that Pasteur had suggested the asymmetry of life might reflect asymmetries in the cosmos itself. Could that indeed be the case? Might there be a link between the handedness of sub-atomic particles and the handedness of sugars and amino acids? Certainly many theoretical physicists and chemists have suggested that, and many experiments have been carried out, although generally they have not been very successful, the effects either being too small to find, or not fully replicated by other experimenters. The problem is easily understood when one considers the numbers involved.[31]

Since 1957 theory has emphasised that although L- and D-amino acids look like mirror-images they are not *exact* mirror-images, since their atoms are not mirror-imaged; having, for instance, mostly left-handed electrons. D- and L-amino acids therefore have slightly different physical properties, with the D-amino acid being less stable and more likely to be broken down. This should mean eventually that an excess of the L-amino acid would appear. If this process could somehow be amplified, then living organisms might consist entirely of L-amino acids. It is an enticing theory and several theorists have put the case for it forcefully. Their calculations provide no doubt that there should be an excess of L-amino acids, but the figure is extremely small – about 1 part in 10^{17}. To realise how small this really is (1 part in 100,000,000,000,000,000), think about the circumference of the Earth, which is about 25,000 miles (40,000 kilometres). Imagine enlarging the Earth by one part in 10^{17}. How much longer would the circumference be? The answer is about a tenth of a millionth of an inch, half a millionth of a millimetre, or about four angstroms – the diameter of a couple of atoms. Extremely small, in other words. When put like that, it seems difficult to believe that such a small atomic excess could be of any practical biological consequence.[32]

To be important, such a tiny effect requires amplification. One possibility is what is called 'positive feedback', something everyone has experienced when an electric guitar or a microphone is put in front of a loudspeaker. The microphone detects a tiny sound, which is then amplified a little through the loudspeaker, so that the microphone picks up the slightly louder sound from the loudspeaker, which is again amplified through the loudspeaker, and so on, until very rapidly there is a deafening, whining sound at the maximum volume of the system. Something similar in chemical systems could amplify an initially very small difference. One possibility is that in some early chemical reaction the D form of the chemical inhibits the action of the L form and vice versa. Any tiny difference in concentration between the two then becomes amplified until one of the two completely dominates, the other having entirely disappeared. Should one of the two have a slightly higher concentration to start with, albeit as small as one part in 10^{17}, then eventually that is the one which will dominate. Such chemical systems have been demonstrated experimentally, showing that such a mechanism could indeed amplify the tiny differences arising from failure of conservation of parity.[33]

If the minuscule differences arising from the weak interaction are responsible for our L-amino acids, how on earth might one test or prove such a theory? Well there is a way, and, surprisingly, it does not occur 'on Earth'. Any theory based on the weak interaction makes a very clear prediction; namely, that the laws of physics apply throughout the universe, meaning that parity also fails to be conserved throughout the universe, so that wherever one finds amino acids the L form should be in excess. In other words, life, or at least amino acids, should be pretty much the same everywhere. This was the basis of the proposal in the 1960s for what was called the *Pasteur probe*, a space probe to search for signs of extra-terrestrial life by looking for optical activity.[34]

Although scientific theories have to make predictions, those predictions can seem rather empty if in practice they are not testable, so it might seem rather pointless speculating about amino acids from other worlds if there is no chance ever of seeing them. Well, no, actually. Material from deep space arrives regularly on Earth in the form of meteorites – meteors that have not entirely burned up in the atmosphere but crash into the ground. Most meteorites, being solid lumps of iron and minerals, are relatively uninteresting to biologists. There are exceptions, however, such as carbonaceous chondrites, one of which fell near Murchison, Victoria, in Australia at eleven o'clock in the morning of 28 September 1969. The meteorite broke up, and pieces were scattered over about five square miles. Its importance was recognised almost immediately, as the pieces of black rock consisted mainly of a 'sooty black substance' with a strong organic smell, something like methylated spirit. The bits were collected up and subjected to extensive research. Very soon it was found that the meteorite, about three per cent of which was carbon, was rich in amino acids. It was possible that these came from contamination when the meteorite hit the Earth, which is covered in amino acids, usually inside organisms. However, that possibility was soon excluded, both by looking inside the pieces of meteorite, and by finding amino acids such as methylalanine, sarcosine and methylnorvaline, which simply do not seem to occur biologically on earth. Initially, the amino acids seemed to be a fifty-fifty racemic mixture of L- and D-amino acids, but the first sample was small and the analytic techniques unreliable. Since then, though, studies have found a clear and definite excess of L-amino acids in the Murchison meteorite, up to ninety-seven per cent in the case of leucine. For the first time, there was solid evidence that amino acids from somewhere other than our planet are predominantly L-amino acids. Unfortunately, carbonaceous chondrites are rare, although the Murray meteorite that fell on Kentucky in 1950 also showed an excess of L-amino acids. More recently, on 18 January 2000, a large carbonaceous chondrite fell on Tagish Lake in British Columbia in northern Canada. Parts were quickly recovered from the frozen lake and immediately deep frozen, making them ideal for scientific study. Analyses of the amino acids may soon be available.[35]

Most theorising about the L-amino acid excess goes back to a principle put forward by Pierre Curie, who, with his wife Marie, won the Nobel Prize for discovering radioactivity. Curie said that 'asymmetry cannot arise from symmetry'. Since all the laws of physics except those involving the weak interaction show symmetry, the weak interaction seemed to be the only place worth looking. This, though, ignores the fact that the universe is an extremely large place. If in some local part of it there were an asymmetry, and in some other part the reverse asymmetry, then the laws of physics would still be symmetric, for the universe as a whole would be symmetric, even though the effects on amino acids in a local area could seem asymmetric. 'Local', in terms of the universe, could actually be vast. Such a principle forms the basis for the theory of William Bonner, a chemist at Stanford University, that circularly polarised light is the cause of L-amino acids on Earth. Whereas the waves of ordinary polarised light, the sort blocked by Polaroid sunglasses, are restricted to a particular direction such as up or down, the waves of circularly polarised light spiral either clockwise (to the right) or counter-clockwise (to the left). This might help explain the excess of L-amino acids, because circularly polarised light interacts differently with chiral molecules, with stereo-isomers in other words, making one predominate over the other. However, for this to be useful in explaining the excess of L-amino acids, one needs to find a source of large quantities of circularly polarised light. Bonner suggests that this is best found near a neutron star. These massively dense objects form after the explosion of a supernova, and are relatively frequent occurrences, three having occurred, for instance, in our galaxy in the past thousand years.[36]

Neutron stars are tiny, incredibly dense stars, having much the same mass as our sun but being only twenty to thirty kilometres in diameter, the size of a small city. They also have a vast magnetic field and their rapid rotation produces the characteristic radio signals of a pulsar, the rotating magnetic field producing a huge electrical field that makes the star's electrons move around the core in circular orbits and emit radiation. Part of that radiation is circularly polarised light. Above the star's equator, the light is polarised one way, say clockwise, while below the equator, it is polarised the other way, say counter-clockwise. Deep space is filled with clouds of molecules, frozen on to the surface of tiny interstellar grains of silicate dust. Those in the region of a neutron star are bombarded with circularly polarised light, and amino acids form from the water, ammonia, methane, carbon dioxide and other simple molecules in the ice. On one side of the neutron star, they will mostly be L-amino acids, and on the other side, mostly D-amino acids. The dust slowly agglomerates, forming large lumps of rock that move around the galaxy under the influence of gravity, eventually ending up in solar systems such as our own, to form comets or meteors. Should they come from one side of a neutron star, they will be laden with L-amino acids; should they come from the other side, they will be laden with D-amino acids. Here is an important departure from the theory that claims that L-amino acids occur because of failure of conservation of parity, according to which meteors *everywhere* in the universe should be laden with L-amino acids. The circularly polarised light theory says instead that only meteors on one side of the local neutron star will be awash with L-amino acids (the rest being rich in D-amino acids). It is an elegant distinction, but it will only be of importance when we can analyse comets or amino acids from the distant reaches of the universe.[37]

Meteors in our local patch of the universe may well be packed with L-amino acids, but that is a long way theoretically from our own bodies being full of L-amino acids. Certainly we are too far from any neutron star for circularly polarised light reaching the surface of the Earth to have produced the excess of L-amino acids. For Bonner, there is therefore only one explanation – the L-amino acids on earth must have come originally from meteors. Although this at first seems an outrageous explanation, there is actually much to be said for it. Firstly, it does not say that fully formed life came from meteors, merely bucketloads of L-amino acids (and if that seems unlikely, it is worth remembering that most of the gold and platinum found on Earth arrived in similar fashion). Secondly, it is far from clear that amino acids could have been created from simpler chemicals in the conditions found on the early Earth, so the proposal that they were brought in from outside has much in its favour. There seems little doubt that meteorites have crashed into Earth in large numbers since it was first formed, the most dramatic smashing into the Yucatán peninsular in Mexico at Chicxulub, sixty-five million years ago, and wiping out the dinosaurs. What would happen to the amino acids that these meteorites carried, once they reached Earth? If they merely dissolved in a vast hot ocean, they would be far too dilute to get life started, but how about if life did not start in the oceans but instead within the rocks of the Earth?[38]

A remarkable biological discovery in recent decades has been the bacteria known as 'extremophiles', lovers of what to us are extreme and intolerable conditions. Examples are hyperthermophiles, found inside hot springs at the temperature of boiling water; psychrophiles, found in the near-permanently frozen Antarctic sea-ice; alkaliphiles, found in the alkaline soda lakes of Egypt and Africa; acidophiles, found in sulphuric acid; and halophiles, found in salt lakes. Although it is easy to write these off as rare freaks of nature, Stephen Jay Gould has emphasised that the vast mass of such organisms, hidden deep in the rocks of the Earth, might together outweigh all other organisms on Earth. Molecular genetics has sequenced the genomes of these organisms, and compared them with other bacteria and with multicellular organisms, the eukaryotes. There is a mass of surprising findings. The extremophiles are distinct from traditional bacteria, being closer to the 'root' of the evolutionary tree than other organisms, meaning they are now classed as a separate domain of life, the 'archaea' – the ancient ones. The archaea may also be closer to multicellular organisms than ordinary bacteria, making them closer relatives to human beings than the everyday bacteria that cover our skin and fill our large intestine. However strange they may be, the extremophiles do, of course, share a crucial similarity to other life on Earth – they have the same genetic code and their proteins are assembled from L-amino acids. No other pattern is known on Earth. Intriguingly, though, and unlike conventional bacteria, the archaea have free D-amino acids inside their cells, in particular D-serine, the amino acid also present in large quantities in the brains of mammals such as rats and humans. It is another piece of evidence emphasising our shared ancestry with the very ancient archaea.[39]

If there is so much life in the rocks of the Earth, perhaps life started there rather than in the oceans. Although this may at first seem a strange idea, it grows more attractive the more one thinks about it. Warm, with plentiful supplies of energy coming from molecules such as hydrogen sulphide and nitrates; divided into tiny, interconnected compartments; undisturbed by oceanic or atmospheric storms; and affording a place where high concentrations of unusual chemicals could form and organisms could try out different solutions to the problems of metabolism and reproduction – it is an ideal testbed for novel life forms. Only when the complex biochemical systems were working well was it worth the organism's while to branch out into the really hostile environment of the Earth – the surface. The theory also provides a neat solution to the problem of the predominance of L-amino acids. When a meteor packed with L-amino acids crashed into the Earth, the impact would deposit large quantities of a rich mixture of carbon-based chemicals deep inside those warm rocks, and, since the precursors of life were rich in L-amino acids, it is not surprising that the living things that evolved from them were also based exclusively on L-amino acids.[40]

Where does all this leave us? One thing is clear. L-amino acids characterise life on Earth, and the occasional exceptions only re-emphasise the normally overwhelming dominance of the L-amino acids. A completely convincing theory of why L-amino acids predominate is still needed, but the idea that it is merely due to chance looks less and less likely, as also does any mechanism arising merely on our own planet. The evidence from meteorites suggests that some asymmetric process occurred over a much larger area of space, implying some handedness at the level of physics and chemistry. Circularly polarised light could produce large patches of space having amino acids of a particular handedness, and if Earth was in the catchment area of a region producing L-amino acids, then the meteorites bombarding Earth and the life on it would contain mainly L-amino acids. What, though, about the failure of conservation of parity and the asymmetry of weak interaction? Could that also account for the predominance of L-amino acids? There are many who would love to establish such a link between physics and biology, so enthusiasts for that theory are unlikely to give up quite yet. One possibility is that in the cold empty wastes of the inter-galactic molecular clouds, the failure of conservation of parity interacts with circularly polarised light and perhaps other esoteric phenomena such as the magnetochiral effect, each adding its two penn'orth to the overall effect.

If much of this seems idle speculation, then it is worth bearing in mind the hugeness of the biological problem to be solved – one that has been around since the time of Pasteur – and the fact that the different theories of the asymmetry of weak interaction and the effects of circularly polarised light make different predictions. Whereas the weak interaction theories predict that the *entire* universe will have an excess of L-amino acids, the circularly polarised light theory predicts that one half has an excess of L-amino acids and the other half an excess of D-amino acids. What is needed is to discover a way of finding the chirality of the amino acids elsewhere in the universe, not only within parts of our own solar system but in the remote molecular clouds distributed throughout the universe. If half are L and half D, then the weak interaction contributes nothing to the asymmetry on Earth, whereas if all are L, then the weak interaction is the explanation. If, say, three-quarters are L, then both theories are correct in part. At present, there is no way of knowing whether amino acids elsewhere in our solar system are L or D, although this may soon change. The SETH project, the Search for Extra-Terrestrial Homochirality, is producing small, versatile instruments that can assess the handedness of chemicals elsewhere in space, and it might well find a positive result long before SETI, the Search for Extra-Terrestrial Intelligence, does so. Advances in astronomy may also allow us to know the chirality of the amino acids in those giant molecular clouds, and then we will have an answer to one of the great questions concerning the origin of life.[41]

This chapter has looked at handedness at the smallest levels in which it is found in biology and physics. Now it is time to revert to the human level and ask about the asymmetry that is most obvious to us all; that which makes each of us literally either right- or left-handed.

☞   ☜

THE DEXTROUS AND THE GAUCHE

Charles Darwin was not yet famous in 1839. Three years earlier, in October 1836, he had returned from his epic five-year voyage around the world in HMS *Beagle*, where he conceived the ideas on natural selection and evolution that eventually formed the basis of *The Origin of Species*. Now he was settling down. In January he had married his cousin, Emma Wedgwood, and on 27 December 1839 their first child, William, was born (Figure 7.1). Darwin was clearly a besotted father, as is shown in a letter about William written in June 1840 (the spelling and punctuation in this and subsequent extracts is are in the original):

> – he is a prodigy of beauty & intellect. He is so charming, that I cannot pretend to any modesty. – I defy anyone to say anything, in its praise, of which we are not fully conscious. –
>
> He is a charming little fellow, & I had not the smallest concepcion there was so much in a five month baby: – You will perceive, by this, that I have a fine degree of paternal fervour.

There were ten children eventually, although one child, Mary, died in infancy, aged three weeks. Another, Charles, died aged eighteen months, and a third, Anne (Annie), the eldest daughter, died aged ten years of some form of gastro-enteritis. The death of Anne left what John Bowlby, the great psychoanalyst and a biographer of Darwin, described as a 'grief never to be wholly obliterated'. Darwin wrote thirteen years after her death, 'There is nothing in the world like the bitterness of such a loss.' As if to show the idiosyncrasies of fate and the capricious nature of infectious diseases in Victorian England, the seven remaining children prospered and lived to ripe old ages, dying at the ages of sixty-seven, seventy-five, seventy-seven, seventy-seven, seventy-nine, eighty-four and ninety-three.[1]

Figure 7.1 Charles Darwin and his son William ('Doddy'). From a daguerreotype, 23 August 1842, when William was aged two years eight months, and Charles was aged thirty-three.

Even before William's birth, Darwin was thinking of the relevance of child development to evolution, and in his scientific journal he had jotted down the heading, 'Natural history of babies', together with several questions that needed answering. When William was born, Darwin recorded his development in a notebook: 'During first week. yawned., streatched himself just like old person – chiefly upper extremities – hiccupped – sneezes sucked'. There are some lovely descriptions of how William (nicknamed 'Doddy' or 'Mr Hoddy Doddy') when he was just a year old invented words: for instance, *mum*, meaning food, soon generalised so that *shu mum* meant sugar, and *black-shu-mum* meant liquorice.[2]

Not only did Darwin observe; on 10 May, when William was four and a half months old, he also tried experimenting:

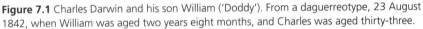

I made loud snoring noise, near his face, which made him look grave & afraid & then suddenly burst out crying. This is curious, considering the wondrous number of strange noises, & stranger grimaces I have made at him, & which he has always taken as good joke. I repeated the experiment.

'I repeated the experiment.' In that phrase is a marvellous insight into Darwin the father, naturalist and experimental scientist. Presumably, little Doddy obligingly burst into tears once again, to the consternation of nurse and mother.[3]

As with so many of the detailed notes and ideas that he was working on, Darwin did not publicise his thoughts until many years later. It was only in 1877, after reading an article by Hippolyte Adolphe Taine in the new philosophical and psychological journal *Mind*, describing language development in children, that he turned again to his notebooks, albeit now about an infant who was aged thirty-seven and a banker in Southampton, to write one of the most personal of his scientific works, entitled 'A biographical sketch of an infant'. The study of child development was still in its infancy, and Darwin's observations are some of the first systematic observations on a single child. It was yet another field he helped to create.[4]

When William was only eleven weeks old, Darwin was studying his handedness. The entry in the notebook reads:

Now eleven weeks old take hold of his sucking bottle with right hand, – when nursed either on right arm or left – He has no notion of clasping it with left hand, even when it is placed on body – this baby has had no sort of practice in using its arms. –

When one day under 12 weeks old took hold of Catherines finger with his right hand & drew it into his mouth. –

Now exactly 12 weeks & on following day old clasped his bottle with left hand just like he did formerly with right hand – Therefore his right hand is *at least* one week in advance of left. – I say at least for I am not quite sure, the first time he used his right hand, was observed.

These rough comments, thoughts and observations are smoothed out in the eventual scientific paper:

When 77 days old, he took the sucking bottle (with which he was partly fed) in his right hand, whether he was held on the right or left arm of his nurse, and he would not take it in his left hand until a week later although I tried to make him do so; so that his right hand was a week in advance of the left.

From such observations Darwin decided that his son would be right-handed. However, it was not to be, as he goes on to relate in the paper: 'Yet this infant afterwards proved to be left-handed, the tendency being no doubt inherited – his grandfather, mother, and a brother having been or being left-handed.' Elsewhere, Darwin comments in similar vein that left-handedness is 'well known to be inherited'.[5]

One can see why Darwin was interested in measuring handedness at such an early age. Although right-handed himself, his wife Emma was left-handed, as also was one of William's grandfathers. Although we do not know which, statistically it is more likely to have been Emma's father, Josiah Wedgwood II, rather than Robert Darwin. Either way, it was a strong family history of left-handedness, and thus it is no surprise that of the eight Darwin children who survived to an age where handedness could be determined, two were left-handed. Such a proportion is typical of the children of one right-handed and one left-handed parent.[6]

Darwin's interest in handedness and in particular the handedness of his children, would certainly not be seen as unusual nowadays. I have been researching into left- and right-handedness since the early 1970s and one of the most frequent questions parents ask is when they will know if their child is right- or left-handed. Perhaps, though, the most asked question is, 'How common is left-handedness?' Let's start with that, and begin by completing a brief questionnaire on handedness:

With which hand do you mainly carry out the following tasks? Ring *either* left or right, not both	
Write	Left / Right
Draw	Left / Right
Throw a ball	Left / Right
Brush your teeth	Left / Right
Hold scissors	Left / Right
Hold a knife (without a fork)	Left / Right
Hold a spoon	Left / Right
Hold a cup	Left / Right
Use a TV remote control	Left / Right
Open a can of fizzy drink (with a ring pull)	Left / Right

Although this is not the most complicated or sophisticated of handedness questionnaires, it is more than adequate for its task, and has recently been administered by Nigel Sadler of the Vestry House Museum to a large and representative group of nearly three thousand schoolchildren in Waltham Forest in North London.[7]

Scoring the questionnaire is straightforward. Count the number of times you have answered 'Left', which will be somewhere between nought (if you are a strong right-hander) and ten (if you are a strong left-hander). To find out how you compare with other people, look at Figure

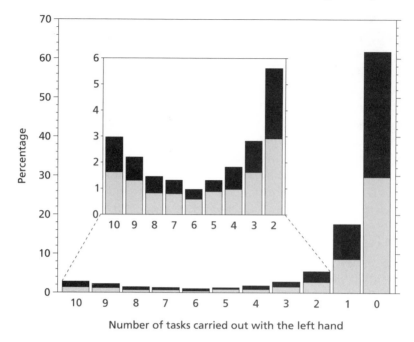

Figure 7.2 A study of handedness in schoolchildren in Waltham Forest, north London (based on unpublished data of Nigel Sadler). The proportional results of males to females are highlighted through shading: light grey for males and dark grey for females. The greater incidence of left-handedness in males can be seen most easily by comparing the '10' with the '0' column; the light grey is taller than the dark grey for the '10', whereas for the '0' column the dark grey is slightly taller than the light grey.

7.2. Begin by looking at the horizontal axis of the main graph, which goes from 10 to 0 and shows the number of left-hand responses. On the vertical axis is the percentage of people with each score. Several things are immediately apparent. Most people, two-thirds or so, don't carry out a single task with their left hand, and are consequently described as strong right-handers. Indeed, the strong right-handers so overwhelm the graph that it is difficult to see what else is going on. To make it clearer, the small, inner graph is a magnification of the left-hand and central area of the main graph.

The results, overall, form a 'J-shaped' curve – high at the right (strong right-handers), descending until it almost touches the horizontal axis, and then rising again at the left-hand edge (strong left-handers). Between the strong left- and right-handers are the people who are more intermediate, who are called weak left-handers if they use their left hand for six to nine of the tasks, or weak right-handers if they use their left hand for between one and five tasks. What about the very small group of people, about one in a hundred, who use their left hand for five of the tasks and

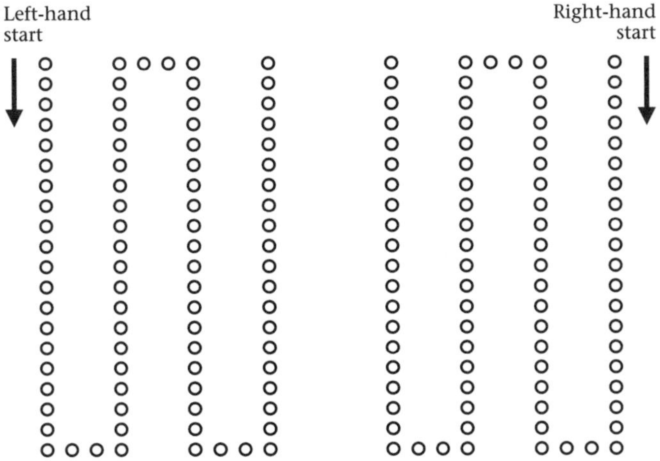

Figure 7.3 The Tapley and Bryden handedness measure. Make a dot in as many of the circles as possible in thirty seconds, using first one hand and then the other.

their right for the other five? Are they ambidextrous? Probably not. A longer, more detailed questionnaire reveals that most of these people still have a predisposition to one side; in other words they are actually weak right- or weak left-handers. Although there are people who claim to be truly ambidextrous, when they are rigorously tested in the laboratory each invariably favours one hand over the other. Unless people have practised very hard, perhaps as a party piece, then almost no-one can write equally well with both hands. If you think you might be ambidextrous, try the task shown in Figure 7.3. Set an alarm or timer to go off in thirty seconds and in that time use a felt-tip pen to place a dot in as many of the circles as possible. Then try it again with the other hand. Almost no-one does this task equally well with both hands.

How many people, then, are left-handed? Conventionally, those who use their left hand for half or more tasks are defined as left-handers. In the Waltham Forest sample, which is typical of the population as a whole, both in the UK and the West in general, just slightly over ten per cent are left-handed, although, as we shall see later, the proportion is slightly lower in the elderly, and in some other parts of the world.[8]

The graph also shows something else. Although 11.6 per cent of the males are left-handed, only 8.6 per cent of the females are. In other words, left-handedness is somewhat more common in men than women; a result that once again is typical of many studies, which overall find about five left-handed men for every four left-handed women. Not a large difference, but it seems to hold universally and to reflect something important about the biology of left-handedness. Hints of such a sex difference can even be seen in Darwin's immediate family, where three out of eight males were left handed (thirty-eight per cent), compared with one out of six females (seventeen per cent).[9]

The idea that sex and handedness are related goes back a long way in psychology. Freud corresponded with Fliess on the subject in 1897 and 1898, although their main interest was in Fliess's 'Bi-Bi' theory – namely, that bisexuality and bilaterality are related, and that latent left-handedness is associated with latent homosexuality. In the modern era, this idea has had a chequered history. Overall, male homosexuals are slightly more likely to be left-handed than heterosexual males, although interpreting the data is complicated by the fact that, from the 1920s to the 1970s, 'left-handed' was American slang for 'homosexual'. Left-handedness also seems more common in transsexuals; that is, those who have had a surgical sex-change from male to female or female to male. Likewise, American children showing signs of gender identity disorder are more likely to be left-handed. Quite how one interprets such results, particularly when taken with the statistically higher rate of left-handedness in men compared with women, is still far from clear, but there is undoubtedly a pattern that needs explaining.[10]

Figure 7.2 has yet one more thing to reveal. Look carefully at those right-handers scoring zero, and then at those scoring one, two, three or four. Over two-thirds of right-handers are *strong* right-handers. For left-handers, the situation is not as extreme. Compare those scoring ten with those scoring nine, eight, seven or six. Only about one-third of left-handers are *strong* left-handers. To put it another way, right-handers are more strongly handed than left-handers. Partly this is because left-handers live in a 'right-handed world' in which artefacts of all sorts, from microwave cookers to computer keyboards to pianos, are built primarily with the needs of right-handers in mind. As a result, left-handers have adapted to the right-handed world and often use their right hands, even if their natural preference is to use the left. This may explain some of the details of Figure 7.2, although it cannot be the entire explanation, since only one question (the one about scissors) mentions an object specifically designed for use by the right hand.

A fuller explanation is more interesting, and reflects the fact that many left-handers, and for that matter some right-handers, are *inconsistent handers* and carry out some skilled activities with one hand and some with the other. I became aware of this myself as a twelve-year-old at a summer camp. I had been taught to use a hand-axe and, being right-handed, used it in my right hand. I was then taught about the full-length felling axe and, while I was using it, someone asked, 'Are you left-handed?', because I lifted the axe over my left shoulder before bringing it down. It still feels unnatural to do it any other way. The many right- and left-handers in Figure 7.2 who score between one and nine exhibit a similar inconsistency.[11]

Only in the past fifteen years or so have researchers looked seriously at the idea that many individuals are inconsistent in their handedness, their interest mainly due to the work of Michael Peters at the University of Guelph in Ontario. Michael found that about a third of left-handers who write with their left hand prefer to throw a ball with their right hand, and are more accurate with the right hand. Since then, it has also become clear that about two or three per cent of right-handers who write with their right hand also prefer to throw with their left hand. Even if the idea is new in research terms, individuals wrote about the phenomenon long ago. For example, the pioneer sexologist Havelock Ellis said in his autobiography, 'I am right-handed except in the single action of throwing a stone or ball...I have never been able to throw a ball with my right hand, and...I have never written with my left hand.'[12]

Although the discussion so far has been in terms of handedness, there are those who prefer to talk of 'sidedness', because many aspects of behaviour seem to occur preferentially on one side or the other involving not only the two hands, but also the arms, legs, eyes, ears and feet. Some of these asymmetries relate to handedness, but not all.[13]

Footedness is related to handedness, right-handers mostly being right-footed, and left-handers being left-footed. Testing footedness is straightforward – ask someone to kick a football at a goal and the foot they use is the dominant foot. The footedness of professional footballers has been assessed by counting how often they touch the ball with the right or left foot. Most use their preferred foot about eighty-five per cent of the time, and almost none use each foot the same number of times. In other words, even professional footballers are not 'ambi-footed' – that is, ambidextrous with their feet. Like the rest of the population, twenty per cent of footballers are left-footed, a percentage substantially higher than the ten per cent or so of left-*handed* people. Quite a lot of right-handers are therefore 'cross-lateral', writing with their right hand but kicking with their left foot.[14]

Ear dominance is mainly seen when people hold a telephone to one ear, about sixty per cent preferring to listen with the right ear and forty per cent with the left, right-handers tending to favour the former and left-handers the latter. Although ear dominance has been little studied in the past, this may change now that mobile phones are near universal.

In contrast to the neglect of ear dominance, eye dominance has been much more thoroughly studied. To find out if you are right- or left-eye dominant, stretch out your arm and point at some small object in the distance. Now close one eye. If the finger is still aligned with the object then the open eye is your dominant eye. If you now look with the other eye, the non-dominant eye, then the finger should no longer point at the object. That is 'sighting dominance' – an object is preferentially sighted with one eye rather than the other. Eyedness can also be assessed by a simple question about which eye is used for looking through a keyhole or down a microscope. About seventy per cent of people prefer the right eye, and thirty per cent the left. Although left-handers tend to prefer the left and right-handers the right, there are many people who are cross-lateral, writing with one hand but sighting with the eye on the opposite side. Whether that matters is still controversial. In the 1920s, Samuel Orton suggested that cross-dominance was responsible for difficulties in reading, resulting in dyslexia. The theory still has its advocates, but the evidence for it is weak.[15]

People have lots of other lateralities – for instance, more people chew on the right side than the left – but many such lateralities are only of marginal interest. Two, though, have intrigued me for a long while because they are such trivial, silly, unimportant behaviours: hand-clasping and arm-folding. To assess hand-clasping, quickly clasp your hands together with the fingers and thumbs interlaced. Which thumb is on top; the right or the left? Now try doing it the other way round and you'll find you have to pause a little and think about it, and that the fingers do not seem to fit together properly. Your first and instinctive clasp was easier and more natural. Hand-clasping doesn't seem to be learned; indeed, most people aren't even aware of it until it is pointed out to them. It also starts early in life, as shown in Figure 7.4, where my daughter Franziska, then six weeks old, is spontaneously clasping her hands. In Britain, about sixty per cent of people clasp with the left thumb on top, and the proportion is the same in right- and left-handers. Particularly interesting is that as one passes eastwards from Britain, across Europe, Asia and Oceania, so the proportion of left-claspers gradually declines, until in the Solomon Islands, to the east of New Guinea, only thirty per cent of people clasp with the left thumb on top. More mysteriously still, hand-clasping runs in families, albeit rather weakly, so that two left-clasping parents are more likely to have a left-clasping child than are two right-clasping parents.[16]

Arm-folding is another laterality you can assess in yourself. Quickly fold your arms one over the other. Which wrist is on top, right or left? Now try to do it the other way round. When I use this demonstration in lectures there is invariably a delayed burst of laughter as people rotate their arms in front of them, only to realise they have ended up exactly where they started, with the same wrist still on top. Putting the 'wrong'

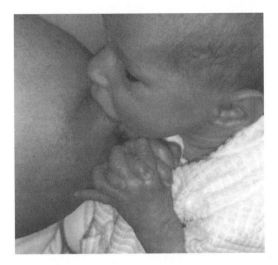

Figure 7.4 Franziska spontaneously hand-clasping at the age of six weeks.

wrist on top is difficult, has to be thought about, and feels very peculiar. In the UK about sixty per cent of people put the left wrist on top, the proportion being the same not only in right- and left-handers, but also in right- and left-hand-claspers. Handedness, hand-clasping and arm-folding must therefore have different causes. The list of minor asymmetries could continue almost indefinitely. For instance, about one in five people can wiggle their ears; and of those who can wiggle just one ear, twice as many can move the left ear as the right ear. However amusing and weird, why asymmetries such as these should exist is almost completely unknown at present.[17]

Handedness, though, really does need explaining, and for a host of reasons it is the behavioural asymmetry into which researchers have put most effort. Firstly, it is more extreme than the other asymmetries, being furthest removed from the fifty-fifty mixture that would result from the chance processes of fluctuating asymmetry. It is also very easy to measure reliably in large numbers of people, either through using questionnaires or simply by observing them, whether in real life or in photographs. So how, when and why is it that most people are right-handed but a sizeable minority are left-handed?

Charles Darwin had particularly wanted to know whether William would be right- or left-handed, yet, despite quite extensive observations, he got it wrong. When, then, is it possible to tell if a child will be right- or left-handed? The handedness of young children is not easy to study, particularly in the first year or two of life. In particular, there seems to be a stage known as the 'chaotic phase' where hand preference swings around from day to day – right today and left tomorrow. The end of the chaotic phase is a bit variable, so handedness is not always clear until eighteen months or even two years of age, but by then the overall direction of handedness will be set for life.[18]

Since the handedness of children typically only becomes apparent in their second year, researchers tended to assume there was little point in looking for handedness before that. Such an assumption, however, was emphatically shown to be wrong by Peter Hepper, of the Queen's University of Belfast, a specialist in the behaviour of human fetuses. Using ultrasound scanning, he found that fetuses, like babies, suck their thumbs, and they do so as early as after twelve weeks of gestation. Hepper looked at several hundred fetuses and found that in over ninety per cent of them it was the right thumb that was sucked. Since the earliest signs of asymmetry in the cortex of the brain can also be found at about the same stage of development, the temptation was to assume that right-thumb-sucking results from early asymmetry in the brain, but subsequently Hepper looked at fetuses of only ten weeks gestation. He found that such young fetuses do not suck their thumbs, but they do move their arms and legs, and so he looked at which arm, the left or the right, was moving more often. In eighty-five per cent of the fetuses, it proved to be the right arm. This could not be taken as an indication that the brain becomes asymmetric earlier than previously thought, because so early in develop-ment the neurones in the brain have not yet become connected to the spinal cord. Such early behavioural asymmetries seem, therefore, to arise in the spinal cord or the limbs themselves, raising the possibility that handedness is not to do with the cortex of the brain but comes from much lower down in the nervous system, although how and where this might be is still not known at the moment.[19]

Darwin had been interested in the development of William's handed-ness because he suspected William could well be left-handed, William's mother being left-handed and the predisposition 'well known to be inherited'. If this was 'well known', it was not because of any systematic scientific studies but simply due to anecdotal and informal observation. Even today there is controversy over the facts, some scientists still arguing that handedness is not inherited. What is not disputed, though, is that handedness runs in families. A while ago, the late Phil Bryden and I studied all the research on this topic that we could find in the scientific literature, which together had looked at more than 70,000 children whose parents were both right-handed, both left-handed, or, like Darwin's family, one of them right- and the other left-handed. The results indicated that where both are right-handed there is a 9.5 per cent chance of having a left-handed child; where one is right- and the other left-handed, there is a 19.5 per cent chance; and where both are left-handed, the chance rises to 26.1 per cent.

Lots of things can be seen in these few numbers. Neither right- nor left-handers 'breed true'; two right-handers having left-handed children and two left-handers having right-handed children. In fact, most children of two left-handed parents are right-handed – three out of four of them. Nonetheless, left-handers are more likely to have left-handed children than right-handers. To put it more precisely, if one parent is left-handed the chance of having left-handed children is increased 2.05 times, and if both are left-handed they are 2.75 times more likely to have a left-handed child than two right-handed parents. Even so, there is still no obvious pattern in the way handedness runs in families. For example, in one large study we carried out, half of the left-handers knew of no other left-hander in the family.[20]

There seems little doubt, then, that left-handedness runs in families. Does that mean it has to be genetic? Not necessarily, for there are many things that run in families that are not inherited through the genes – money, perhaps, being the most obvious example. If your parents are very rich, then you will probably be so too, but there is no gene for wealth, your money instead resulting from cultural inheritance, which is a part of what biologists broadly call 'the environment'. Could the roots of left-handedness similarly lie in 'the environment'? One of the oldest ideas in psychology is that handedness is acquired culturally, through the social pressure of teachers and peers, or through imitation of parents and nursemaids. In the fourth century BC, the philosopher Plato put this position forcefully:

It is due to the folly of nurses and mothers that we have all become limping, so to say, in our hands. For in natural ability the two limbs are almost equally balanced; but we ourselves by habitually using them in a wrong way have made them different.

Certainly there is a case to be made for Plato's argument. After all, we learn so many things by imitating our parents that it would hardly seem surprising if children with a left-handed parent copied that parent and used their left hand; and if both parents were left-handed this would be more likely still. The fact that most children of two left-handed parents become right-handed might simply reflect the strength of the right-handed world, where schools, teachers and society in general conspire to force people to be right-handed. Half a century after Plato, however, the alternative theoretical position was put forward by Aristotle, perhaps the greatest of all biologists, who said handedness occurred 'by nature', and, by implication, was inherited:

For instance if we all constantly practised throwing with our left hands, we should all become ambidextrous; yet the left hand is such by nature, and the right hand is none the less superior to the left, however much we equalize the use of the two. Change of use does not abolish the natural distinction.

For Plato's environmentalist theory, the devil, as always, is in the details – in the details both of the logic and the data. On the logical front, Sir Charles Bell, in the mid-nineteenth century, was one of the first to emphasise that left-handers become left-handed despite the supposed social pressure of living in a right-handed society: 'every thing they see and handle, conduce to make them choose the right hand, yet, will they rather use the left'. But, as Bell then argued, if left-handers are left-handed despite social pressure to be right-handed, it makes no sense to argue that right-handers are right-handed because of social pressure – innate factors may also be important in them. The difficulty in terms of data comes from families in which both parents are right-handed but one grandparent is left-handed. The children in such families are more likely to be left-handed than are children where all the grandparents are right-handed. The left-handedness of the children cannot be due to imitating the parents since both parents are right-handed themselves, and it is doubtful that grandparents have sufficient influence over their grandchildren to teach or force them to be left-handed. It is more likely that the left-handed grandparent carries a gene for left-handedness that has lain dormant in the parent but then manifested itself in the grandchild.[21]

The environment can, however, work in a rich variety of ways, of which the combination of cultural learning, social pressure and teaching is but one facet. In other words, there may be other factors increasing the likelihood of left-handers having left-handed offspring without genes needing to be involved. A theory in vogue when I started my research was that left-handedness results from brain damage due to physical traumas associated with birth. Birth is undoubtedly a dangerous process, even in our age of modern obstetrics, and it must have been far more so in the pre-modern era. The human head has been subject to two conflicting evolutionary pressures; on the one hand to be as large as possible in order to contain the much-expanded brain, and, on the other, to be small enough to fit through the female pelvis. The baby's skull faces a tight journey, slipping, sliding and moulding itself during the journey through the birth canal. As a result, the soft, vulnerable brain sometimes ends up damaged.

How could brain damage during birth make people more likely to be left-handed? Imagine *everyone* is meant to be right-handed, with the left half of the brain producing hand movements in the right hand. What happens if the brain is damaged as it is squeezed through the pelvis? If the right half of the brain is damaged, then there is no effect on handedness, since the normal left half of the brain still controls the right hand. But if the left half is damaged, then the right hand no longer works properly and the right half of the brain takes over for skilled, complex movements. Since the right half of the brain controls the left hand, that person will become left-handed. Obstetric damage (or indeed any form of brain damage early in life) would then produce an increased rate of left-handedness. Could such a theory also explain why left-handedness runs in families? The elegant answer is that it might, because if a woman is left-handed because her mother's small pelvis damaged her brain during birth, then she probably also carries the genes to produce a small pelvis, and thus her children will also run the risk of mild brain damage and hence of being left-handed.[22]

It is a clever theory and, as a young PhD student, I was lucky to be able to test it. In 1958, the British National Child Development Study (NCDS) collected detailed information on the birth of every child born between 3 March and 9 March. Those 16,000 children were followed up when aged seven, eleven and sixteen, and are still being followed through their adult lives, providing one of the great resources of modern psychosocial research. My good luck was that some anonymous but far-sighted researcher included several questions in the survey relating to the handedness of the children. The data, stored in the then ESRC's Research Data Archive at the University of Essex, revealed no hint of association between birth complications and a child subsequently becoming left-handed. Whatever its formal beauty, the theory that left-handedness is due to birth stress was shown to be wrong. I was left, then, once more, with the probability that handedness is associated with genes.[23]

Genetic theories of handedness have always suffered from one obvious and seemingly insuperable problem: twins. Look at Figure 7.5, which shows my daughters Franziska and Anna at about fourteen months of age. Franziska is the one on the left of the photograph, holding the spoon in her left hand, and Anna is on the right, holding the spoon in her right hand. Franziska and Anna are identical twins; they are what is called monozygotic, meaning they started as a single fertilised egg and therefore have exactly the same genes. If handedness is under genetic control, shouldn't Franziska and Anna be identical in their handedness? About one in five pairs of identical twins have discordant handedness, one being right- and the other left-handed. Although it might seem strange, handedness can still be under genetic control, even when identical twins do not have identical handedness. Two things must be considered.

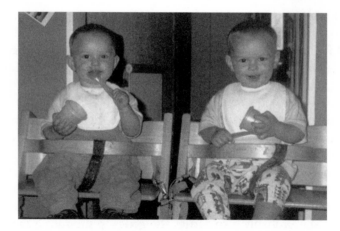

Figure 7.5 Franziska and Anna at the age of fourteen months, Franziska holding the spoon in her left hand and Anna holding the spoon in her right.

Firstly, for something to be genetic it is not the case that identical twins have to be the same in every way with any difference necessarily implying an environmental influence. What *is* true is that identical twins should be more similar than non-identical twins (also known as fraternal or dizygotic twins), who genetically are like ordinary brothers or sisters. A careful analysis of all the data suggests that this is indeed the case, identical twins being more similar in their handedness than non-identical twins.[24]

The second thing to consider with identical twins is quite how handedness might be inherited if it is genetic, and whether there is a mechanism for identical twins having different handedness. Although for most genetic conditions there is a straightforward link between gene and outcome, that isn't always the case, and sometimes chance plays an important role. One such case was described in chapter 5, the *iv* mouse. A double dose of the normal gene results in the mouse's heart being on the left. However, a double dose of the *iv* gene gives a fifty per cent chance that the heart is on the left and a fifty per cent chance that the heart is on the right (*situs inversus*). What would happen to identical twin mice with a double dose of the *iv* gene? Well, one of the mice could have the heart on the right and the other on the left. That indeed seems to happen in those rare cases of human identical twins in which one has the heart on the right (*situs inversus*) and the other has it on the normal side. So, in mice and men, identical twins need not have the same laterality for their heart, despite the side of the heart being under genetic control. A precisely analogous situation occurs in the handedness of twins.[25]

Any acceptable model of the inheritance of handedness must explain not only how identical twins can be different but also why two right-handed parents can have left-handed children, or why three-quarters of the children of two left-handed parents are *right*-handed. Coming to the problem as a PhD student, I felt that any acceptable genetic model of handedness had to learn from other lateralities known to be under genetic control. Few had been well described, but the *iv* mouse, then only recently documented, seemed to be precisely what I was looking for, and suggested that a parallel model for human handedness might fit the data. I therefore played around with a gene (technically, an allele), which I called *C* for *chance*, that in a double dose gives a fifty per cent chance of being right-handed and a fifty per cent chance of being left-handed. This was not only an analogue of the *iv* gene in the mouse, but was also similar to the effects of fluctuating asymmetry, which underlies symmetry and asymmetry in all biological systems (see chapter 5). As well as the *C* gene, my model included a second gene called *D* (for *dextral*), a double dose of which always results in a person being right-handed.[26]

That still left the question of what happened in individuals who had one *D* and one *C*. Although in the mouse a single dose of the *iv* gene is the same as having a double dose of the normal genes – it is what is known as 'recessive' – the same did not seem to be happening for handedness. Some mathematical modelling soon showed that if the *C* gene was recessive, then the model did not fit the large amounts of data available from studies of handedness in families, and it was no better if the model was the other common type; that is, 'dominant'. The model did, however, fit the family and twin data if the two genes, *D* and *C*, were additive or co-dominant, meaning that individuals with one of each gene were midway between those with two *D* genes or two *C* genes. In this model, people with the *DD* pattern of genes (the *DD* genotype) had zero chance of being left-handed, those with the *CC* genotype had a fifty per cent chance, and those with one of each, the *DC* genotype, had a twenty-five per cent chance.

Although very straightforward, such a simple genetic model can explain many crucial features of handedness in families and twins. For example, the handedness of identical twins is easily explained in a way similar to the explanation given for the side of the heart in *iv* mice. The twins with the *DD* genotype are no problem: since *DD* always results in right-handedness, each twin is right-handed and all *DD* twin pairs consist of two right-handers. However, if the twins' genotype is *CC*, then each twin has a fifty per cent chance of left-handedness, that evens probability occurring as a result of fluctuating asymmetry – biological noise. The fifty per cent chance applies separately to each twin, as if each twin separately tossed a coin. On a quarter of occasions, both twins toss a head; on a quarter of occasions, both toss a tail; and in the other half of the cases one tosses a head and the other a tail. It is this latter half who will be discordant in terms of handedness. However, the *CC* genotype forms only a few per cent of twins. For the more numerous *DC* twins, the situation is slightly more complicated mathematically since each has a one in four chance of being left-handed. That means that about one-third of *DC* twin pairs will have one member right-handed and the other left-handed. Taking into account that *DD* twins are more frequent than *DC* twins, who in turn are more frequent than *CC* twins, one finds that about ten to twenty per cent of identical twin pairs should have one right-hander and one left-hander.

The next thing is to see how the model explains the pattern of handedness in families, and, in particular, why two right-handers can have left-handed children, and how the children of two left-handers are mostly right-handed. As far as two right-handed parents are concerned, if they are both *DD*, then, since they only have *D* genes, their children can also only have *D* genes, and so must be *DD*, and hence have to be right-handed. That, though, ignores the seventy-five per cent of *DC* individuals and the fifty per cent of *CC* individuals who are right-handed. Taking right-handers as a whole, most must be *DD* but there are some who are *DC* and a few who are *CC*. Two right-handed parents can therefore be carrying *C* genes, meaning that their children can be *DC* or even sometimes *CC*; and *DC* and *CC* children can be left-handed. Right-handers do not therefore breed true. The converse situation applies to the children of two left-handed parents. The only certainty for left-handed parents is that they definitely are not *DD*; they can, though, be either *DC* or *CC*. The situation is easier to think about when both left-handed parents are *CC*. Each parent carries only *C* genes, so their children only have *C* genes and are the *CC* genotype, giving them a fifty per cent chance of being left-handed. Among left-handed parents, however, *CC* is scarcer than *DC*. Most left-handed parents have the *DC* genotype, and therefore many of their children will have a *D* gene, and if the children are *DC* or even *DD*, then it is likely or even certain that they will be right-handed. That means, overall, that many of the children of two left-handers will be left-handed, but the proportion will be much less than the fifty per cent that would occur if both parents had the *CC* genotype. Working through the numbers systematically, between a quarter and a third of the children of two left-handers could be left-handed. Like right-handers, then, left-handers do not breed true, only a minority of their children being left-handed and the majority right-handed. When one parent is right-handed and the other is left-handed, the situation is midway between that for two right-handed or two left-handed parents.[27]

Although relatively straightforward, this genetic model therefore explains the two otherwise difficult facts about left-handedness: that it does not run very obviously in families and that identical twins often do not have identical handedness. In the next chapter we will also see that it can explain the relationship between handedness and language dominance in the brain.

By now it should be clear that I believe people are right- or left-handed because of the genes they carry. Certainly that seems to be the most parsimonious way of accounting for a mass of data, of which it is otherwise difficult to make coherent sense. Of course, just because a model *fits* does not mean it is necessarily correct. In the modern world, the real proof that handedness is due to a gene will come from the isolation of a sequence of DNA that differs systematically between right- and left-handers. Surprisingly, few people seem to be serious about carrying out such a search. Thankfully, a few are doing so, and it is to be hoped that in the not-too-distant future their search will be successful.[28]

Genetic models have many uses but in the end they only tell us about patterns of DNA. What they cannot do is tell us exactly what it is that *makes* us right- or left-handed. There must be a long chain of causality, which reaches from having one or other sequence of DNA, through to using a pen with only the right or the left hand. Something in that chain produces a difference between the two hands, making them so different that most of us spend our lives preferentially using only one hand for such complicated and highly skilled activities as writing. Such a view could, potentially, be misleading, because there is no doubt that if we absolutely *had* to, then we could learn to use the other hand. The non-dominant hand is connected to a highly serviceable brain which if driven by sufficient motivational forces could learn many skilled activities. Indeed, in everyday life the non-dominant hand is quite capable of learning to type fluently, play the piano and carry out a host of bimanual skills. It can also take over if the dominant hand is impaired by accident or war. The poet Walt Whitman described such cases in a short piece called 'Left-hand writing by soldiers', written just after the end of the American Civil War, in which he had served as a nursing orderly:

April 30, 1866 — Here is a single significant fact, from which one may judge of the character of the American soldiers in this just concluded war: A gentleman in New York City, a while since, took it into his head to collect specimens of writing from soldiers who had lost their right hands in battle, and afterwards learned to use the left...

I have just been looking over some of this writing. A great many of the specimens are written in a beautiful manner. All are good. The writing in nearly all cases slants backwards instead of forward. One piece of writing, from a soldier who lost both arms, was made by holding the pen in his mouth.

Even if, as with Whitman's soldiers, necessity could force us to write with the non-dominant hand, something normally induces us to use the dominant hand. Here we need to distinguish two separate but related aspects of handedness. I am right-handed in two different ways. Put my hands flat on a table and ask me to tap very quickly with just my right index finger or just my left index finger and I am quicker and more regular with my right hand: my right hand has more *skill* than my left hand. Alternatively, put a sweet straight in front of me on the table and I will pick it up with my right hand. However, I could do it just as well with my left hand, which is quite skilled enough for this simple task. Nevertheless, I use my right hand: I have a *preference* for using my right hand. How, though, are skill and preference related? If my right hand is more skilled, it is easy to see that I might prefer to use it, but the reason it is more skilled may be that I prefer to use it and it has therefore had more practice to help make it skilful.[29]

Understanding which comes first, the skill or the preference, is far from easy, because skill and preference are so closely interrelated in most people. However, by looking at children with autism, my colleagues and I had a hint of how skill and preference can be separated. We started by measuring hand preference, asking the children to perform a range of simple, unskilled activities, such as picking up a sweet. Most children with autism preferred the right hand, just as other children do. We then measured the skill of each hand using a standard task in which a row of pegs is moved as quickly as possible from one row of holes to another. As with other children, each child with autism had one hand that was clearly more skilled than the other. The big difference from other children was that half of the children with autism were more skilled with their right hand and the other half with their left. When we looked at hand preference and skill together, there was, in marked contrast to the other children, no relationship between skill and preference. This probably tells us which came first. Only the preference showed a clear dominance of right hand over left, with most children being right-handed. The skill measure showed no overall dominance, half being more skilled with the right and half with the left. The implication is that handedness usually starts as a *preference* for using the right hand, which subsequently results in the right hand becoming more skilled. This link has been broken in the children with autism, so that although there is a clear preference for the right hand, it has not resulted in the right hand becoming more skilled. In other words, preference comes before skill.[30]

The distinction between skill and preference can be seen in people with disabilities affecting the arms. In 1903, Pieroccini described a girl who had a normal left arm but was born with only a stump of a right arm. Efforts were made to teach her to write with the left arm but these were surprisingly unsuccessful. Eventually, an attempt was made to teach her to write by fixing a pencil to the stump of the right arm, and she immediately learned to write without difficulty. A recent example, showing the subtle relationship between skill and inborn preference, was told to me by Michael Peters, himself an expert on lateralisation, who described a client seen by his wife Anne, a social worker:[31]

> The woman has a congenital malformation of the right hand, which Anne describes as flipper-like, without formed separate fingers. She writes with the left hand. However, Anne observed that she used the right hand for gestures during speech, and also used the right hand for adjusting her hair, or smoothing her skirt. When asked, the client stated that she considers herself right-handed because she prefers to use her right hand when she can.

That handedness begins as a preference rather than a skill difference may tell us something about which part of the brain is responsible for handedness. The cerebral cortex carries out so many skilled, complicated tasks that many people have presumed that right-handedness must have something to do with the left cerebral hemisphere. However, this may not be the case. Once one starts to think the impossible and speculate that right- and left-handedness do not originate in the cortex, then one can start looking elsewhere in the brain. For instance, it is possible that handedness arises in one of the most primitive parts of the brain, the brain stem, found at the base of the brain and connecting the brain to the spinal cord. The idea goes back to experiments on laboratory rats which were being given high doses of drugs such as amphetamines that induce stereotypical behaviours – highly repetitive sequences of actions such as paw-washing or whisker-grooming. Some of the rats kept turning in circles, and to stop them getting stuck in the corner of a cage they were tested in a hemispheric bowl, where they could repeatedly turn until the drug wore off. After a while, it became clear that some rats consistently turned clockwise whereas others consistently turned anti-clockwise. A detailed pharmacological analysis found that the amphetamine was affecting the concentration of dopamine, a neuro-transmitter, and that normal rats do not have exactly the same amount of dopamine on the left and the right sides of the brain stem. The amphetamine was amplifying that small normal difference, making one side dominate so that the animal then turned in only one direction.[32]

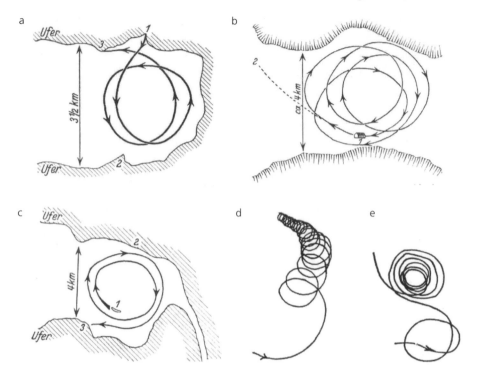

Figure 7.6 Examples of spiral movements (turning tendencies) in humans and animals. **a,** Horse with a sleigh and no driver on a frozen lake, trying to go from 1 to 2. **b,** Three people wandering in fog from the barn at the centre, and trying to go to 2. **c,** Two fishermen on an island 1, trying to row in fog to 2. **d,** A blindfolded swimmer in a lake. **e,** A blindfolded car driver on a Kansas plain.

The tendency of the amphetamine-dosed rats to turn in circles may be related to a phenomenon in humans known as 'turning tendencies', a propensity to circle in one direction. In 1885, Ernst Mach 'learned from a retired army officer that on dark nights or in snow-storms … troops will move approximately in a circle … so that they almost return to their point of departure, though all the time they are under the impression that they are moving forward'. Early experiments blindfolded subjects and set them walking to see if they went in a straight line. Most did not, instead turning in large circles, and eventually returning whence they started – perhaps the explanation for those wintery disasters in which the hero sets out into the blizzard to fetch help, never returns, and after the storm relents is found dead in the snow close to where they started. Figure 7.6 (a, b, c) shows exactly how this can occur. My favourite of the early studies was one carried out in 1928 by Schaeffer, who blindfolded subjects, put them at the wheel of a car, and asked them to drive in a straight line across the near infinite flat plains of Kansas. Actually they went round and round in giant circles (see Figure 7.6e). The same was found for blindfolded swimmers in a lake (Figure 7.6d).

Such turning tendencies are seen in modern research studies by attaching a tiny movement detector to a subject's belt. During the course of a day, most people turn more in one direction than the other, clockwise if they are right-handed and anti-clockwise if they are left-handed. Since people are like laboratory rats in showing differences in dopamine concentrations on the two sides of the brain, it is probable that the mechanism of turning rats and humans is the same. People have other turning tendencies as well. One of the earliest is the tonic neck reflex: lay a newborn baby flat on its back and the head will spontaneously turn to one side or the other, for most babies this being to the right. Could human handedness also be a turning tendency? It could indeed. Sit at a table, place a glass of water straight ahead, just beyond arm's reach, and reach out to pick up the glass. Think carefully about the muscles of the shoulders and upper trunk as you do it, and you'll find that as the arm reaches out so the upper part of the body also turns. Hand preference might well be something to do with primitive turning tendencies, although quite how is not yet clear.[33]

Handedness, whether skill or preference, interests psychologists not only in its own right as an intriguing phenomenon but because it is also related closely to that great discovery of the Victorian age, Dax and Broca's finding that, for most people, language is located in the left half of the brain. It seems inconceivable that Charles Darwin, as well as being interested in handedness, would not also have been fascinated by Broca's discovery, about which in 1866 Walter Moxon, a physician, could write:

> It is, I think, not over venturesome to say, that no observations have for many years excited in the medical world more intense and general interest than those of M. Broca, upon the coincidence of the loss of speech with paralysis of the right side.

Nevertheless, Darwin seems to have written nothing about Broca or cerebral lateralisation, either in print or in his private correspondence. It is a great shame, as there is hardly a person interested in the evolution of language and cerebral dominance who would not give their eye-teeth to know Darwin's thoughts on the topic. The differences between the two halves of the brain occupy the next chapter.[34]

☞   ☜

THE LEFT BRAIN, THE RIGHT BRAIN
AND THE WHOLE BRAIN

We have met both of them already – Louis Pasteur, who made the fundamental discovery that biological molecules are handed, and Ernst Mach,
whose philosophical thought-experiment showed that only asymmetric
systems can distinguish left and right. They are now going to help us
understand more about cerebral asymmetry, not by their theoretical or
experimental work but because each, in mid-career and at the height of
their scientific powers, suffered serious damage to one half of the brain. If
one looks carefully at Figure 8.1 the signs can be seen. Pasteur holds his
stick in the right hand, while the left hand shows the typical posture of
paralysis. The photograph of Mach is more subtle. The left hand may
look strange but it is actually gripping hard on the cane that never left his
hand, even when he was cremated. The right hand is the abnormal one.
Almost totally paralysed, it must have been carefully positioned by a
helper, most likely Mach's devoted wife Ludovica ('Louise'). Another
slight abnormality is also just visible – the right-hand corner of the
mouth droops slightly, probably owing to a partial facial paralysis.[1]

On 19 October 1868, Pasteur, then forty-five years old, was due to talk
at the Académie des Sciences in Paris. He had felt strange that morning,
noticing a tingling sensation over the whole left side of his body.
Nevertheless, at the meeting he read out loud, 'in his usual steady voice',
a paper on silkworms by the Italian scientist Salimbeni. He walked home
with colleagues, had dinner and went to bed at nine o'clock. Soon,
though, he got worse, the earlier symptoms returning, accompanied by
difficulty in speaking. His speech returned sufficiently for him to call for
help, but by now his left arm and leg were becoming paralysed. The
doctors, despite their patient being the doyen of modern scientific
medicine, applied the centuries-old remedy of leeches. Surprisingly, there
was some improvement, with his speech returning, but the paralysis
intensified, and within twenty-four hours his left side was totally

Figure 8.1 Left, Pasteur in the summer of 1892 at the age of sixty-nine. Right, Mach in the years 1913–16 at the age of about seventy-five.

immobile. His speech however improved, with Pasteur wanting to talk about science and also complaining of his left arm: 'It is like lead; if only it could be cut off.' Pasteur had suffered a cerebral haemorrhage, affecting the right side of his brain and hence paralysing the left side of his body. Slowly, over the next few months, some movement returned to the left side, and by January 1869 he could walk again. He never, though, regained proper use of the left hand, and thus needed assistants to help with his experiments. His mind seemed completely unaffected by the incident, there being many creative years to follow, including work on silkworms, beer-making, anthrax, and the vaccine against rabies.[2]

Ernst Mach suffered his stroke in July 1898 aged sixty, fifteen years older than Pasteur. He later described it himself:

> I was in a railway train, when I suddenly observed, with no consciousness of anything else being wrong, that my right arm and leg were completely paralyzed; the paralysis was intermittent, so that from time to time I was able to move again in an apparently quite normal way. After some hours it became continuous and permanent, and there also set in an affection of the right facial muscle, which prevented me from speaking except in a low tone and with some difficulty.

His eldest son, Ludwig, accompanied Mach back to Vienna, where a long convalescence, but little recovery, followed. The disability would always be extensive. Mach's wife Louise helped him bathe, dress and feed; he was unable to leave his house or garden unaided, and he hardly travelled again, yet he did not stop working. He became the 'project co-ordinator', his son Ludwig literally becoming his right-hand man, running experiments in the household laboratory, where the two would sometimes go for days at a time. Mach had always published prolifically, and this continued after the stroke. He had been right-handed and thus could no longer use a pen or pencil with ease, but that didn't stop him writing. Within a few days, he was using a typewriter with his left hand. Over the subsequent thirteen years, he published or substantially revised six books, and wrote a dozen new articles. Lecturing was another matter. The paralysis of the facial muscles meant that speaking in public was a torture both for Mach and his audience, and he soon gave this up. Mach, always the philosopher and careful observer, never stopped observing his own paralysis.[3]

> Very often…I formed the intention to do something with my right hand, and had to think of the impossibility of doing it. To the same source are to be referred the vivid dreams which I had of playing the piano and writing, accompanied by astonishment at the ease with which I wrote and played, and followed by bitter disappointment on awaking. Motor hallucinations also occurred. I often thought that I felt my paralysed hand opening and shutting, and at the same time the total movement seemed to be hampered as if by a loose, but stiff glove. But I only had to look to convince myself that there was not the slightest movement.

Superficially, the damaged brains of these two famous scientists do not tell us much about the two sides of the brain. Broca, a few years before Pasteur had his stroke, convincingly showed that language is usually located in the brain's left hemisphere. However, Mach, despite having right-sided paralysis and left-brain damage had no problems at all with language, merely some difficulties in speaking due to facial paralysis, while Pasteur, who had a left-sided paralysis and right-brain damage, experienced episodes, albeit transient, of loss of speech.

In contrast to these cases, brain damage can also manifest itself in subtle and insidious ways, the symptoms coming and going, as seen in another famous nineteenth-century patient, the novelist Sir Walter Scott. He noticed his first symptoms on 5 January 1826, aged fifty-four. His journal entry reads: 'Much alarmed. I had walked till twelve…and then sat down to my work. To my horror and surprise I could neither write nor spell, but put down one word for another, and wrote nonsense.' By next day he noted, 'My disorder is wearing off.' Problems with reading and writing recurred in 1829, and on 15 February 1830 his speech was affected as well, as he described a week later: 'Anne [Scott] would tell you of an awkward sort of fit I had on Monday last; it lasted about five minutes, during which I lost the power of articulation, or rather of speaking what I wished to say.' The attack left its residue, Scott noting that from then on, 'I seemed to speak with an impediment'. During an attack in April 1831, 'the right side of the face was slightly distorted and the right eye fixed', indicating disease of the left side of the brain. Things reached a crisis at the beginning of 1832, in Naples, 'when the decay of [Scott's] brain had now begun in solemn earnest, and he moved in an interior world of his own'.

In June 1832, nearing Nijmegen in Holland towards the end of a gruelling coach journey home, Scott had his most serious apoplectic attack. The end was near, and he wished to die at home at Abbotsford in Scotland. Almost comatose, he travelled in July 1832 from London to Newhaven by boat, his medical attendant being the young doctor, Thomas Watson, whose case John Reid opened this book. Scott died on 21 September 1832, and two days later a post-mortem examination of the head was carried out. While the right half of the brain was entirely normal, the left half showed severe damage, with three cystic regions as well as areas of softening. Scott's illness with its slow onset and varying pattern of symptoms contrasts markedly with the single catastrophic episodes suffered by both Pasteur and Mach. It showed that left-brain damage does not always and only affect speech but can be confined to affecting writing, reading and spelling. Indeed, patients can show any combination of language deficits, perhaps the most bizarre being 'alexia without agraphia', in which patients can write perfectly but are incapable of reading what they have written.[4]

If the contrasts between Scott, Pasteur and Mach seem confusing, then that is entirely typical of neurology and neuropsychology. Some patients have enormous areas of brain damage with no apparent psychological impairment, whereas others have tiny, tightly localised areas of damage with devastating symptoms. It is an important lesson. A problem for any historical analysis is looking back with the benefit of hindsight, making our predecessors seem so stupid in not seeing what to us is so obvious. Broca and Dax were undoubtedly correct when they observed that patients exhibiting severe, lasting problems with language tended to have damage to the left rather than the right half of the brain, but this does not mean that all patients with left-sided brain damage have language problems, or that such problems, particularly transient difficulties such as those Pasteur suffered, cannot occur with right-sided brain damage.

A question often asked is why no astute physician before Dax and Broca saw the link between language loss and left brain damage. One intriguing theory is that the advent of universal literacy in the mid-nineteenth century *made* our brains asymmetric, and hence there was nothing to be discovered before then. Though ingenious, the idea is almost certainly wrong, since illiterate patients show the same pattern of language loss after left hemisphere brain damage as literate patients.

The ground had been prepared for associating language with only one half of the brain long before Dax and Broca. The Greek physician Hippocrates in the fifth century BC had noticed instances of temporary speechlessness following an epileptic fit 'with paralysis of the…arm and right side of the body', but the link was not followed up. This wasn't because physicians ignored language problems or the side of brain damage in their patients. Sir Thomas Watson, in the fifth edition of his textbook of medicine in 1871, accepted from his own clinical notes the link between language loss and left-sided brain damage (he would, of course, have noticed this in relation to Sir Walter Scott's last illness). Indeed, the association between right-sided paralysis and language loss was noticed even by novelists. In *Wilhelm Meister's Apprenticeship*, published in 1796, forty years before Dax gave his paper in Montpellier, Goethe, wrote: 'altogether unexpectedly my father had a shock of palsy; it lamed his right side and deprived him of the proper use of speech'. If one wants a much more recent example, Jeanette Winterson, in her novel *Gut Symmetries*, describes how: 'In the night David had a stroke. In the morning he was paralysed on his right side. He could not call out. He could not speak.' If Goethe and Winterson could notice the association between speech loss and the side of hemiplegia, how was it that Marc Dax was the first to notice that patients with aphasia were more likely to have right-sided paralysis and hence left-sided brain damage?[5]

To find out, Arthur Benton checked the published case histories in three treatises on brain disease, published in 1761, 1825 and 1829–40 by Morgagni, Bouillaud and Andral. Altogether, forty-six patients had damage to only one half of the brain, but only some showed 'speechlessness'. The table below shows the various combinations.

Table 8.1: Brain Damage and speech loss			
	Speech loss	No speech loss	% with speech loss
Left-sided brain damage	15	6	71% (15/21)
Right-sided brain damage	5	20	20% (5/25)
% with left-sided brain damage	75% (15/20)	29% (6/26)	

The pattern seems clear enough. Although most patients with left-sided brain damage had speech loss (seventy-one per cent), few patients with right-sided brain damage had speech loss (twenty per cent). Putting it another way, three-quarters of the patients with speech loss had left-sided brain damage, compared to the half we would expect were this simply down to chance.[6]

Unfortunately, it wasn't so obvious to Morgagni, Bouillaud and Andral, because they analysed sub-groups of ten, twenty-five and eleven patients respectively. Using these smaller samples, rather than the whole group together, the results yield nothing of statistical significance. The problem of statistical significance is worth taking a little further, calculation showing that a sample of at least forty is needed to be ninety per cent certain of discovering a significant statistical difference. Dax was the first systematically to analyse a large number of cases.[7]

Part of the problem, even today, is that 'speech loss' covers an ill-defined mishmash of conditions, many of which are unrelated to the fact that language originates in the left hemisphere of the brain. After all, if Pasteur and Mach had been included, then Mach might well have been put down as 'speech loss', because he couldn't speak properly, whereas Pasteur might not have counted, because the problem was only transitory. Neurology and neuropsychology have therefore concentrated on defining and classifying the precise problems experienced by individual patients with brain damage. Even the term '*speech*lessness' is problematic, failing to distinguish whether the problem relates simply to the vocal tract and the production of spoken words, or whether there is a more general problem of *language* and its components. Detailed analysis has identified a number of separate syndromes, some of which we will come to in a moment.

The situation for the neurologist in Broca's day was complicated by the fact that not only was 'speechlessness' vague and ill-defined, but so also was the associated brain damage. Patients with precise and restricted sets of symptoms and equally precise and restricted areas of damage to the brain were needed. In other words, patients such as the seventy-five-year-old woman admitted in 1875 to the cantonal hospital in Geneva, Switzerland, with paralysis of the right arm and leg, and near total loss of speech: 'Incapable of speaking, she pronounced only isolated syllables without any meaning such as *Eh, eh*: Ah, oi; –– eh, *baba* – ah! *ba, ba, za-za-ya*. One day she uttered *maman*.' The latter was the sole proper word she spoke. Eventually, the woman died of pneumonia, and at post-mortem her brain showed a surprisingly tiny area of damage, restricted to the frontal lobe of the left hemisphere (Figure 8.2). The symptoms and the location of the damage were remarkably similar to those Broca reported

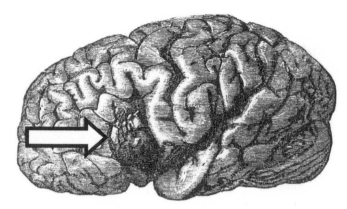

Figure 8.2 The left hemisphere of the brain of a seventy-five-year-old Swiss woman with a right-sided hemiplegia and a complete loss of language. The front of the brain is to the left and the area of damage, which was described as 'a slightly yellowish softening', is clearly seen about a third of the way across in the rear part of what is called the frontal lobe.

in his patients. The syndrome is now known as Broca's aphasia, aphasia meaning a loss of language ability.[8]

It was soon recognised that not all aphasia is Broca's aphasia, other symptoms also occurring after left-hemisphere damage. As so often occurs, careful clinical observation described these syndromes before names were created for them. For instance, in 1834, in Dublin, Dr Jonathan Osborne described a twenty-six-year-old scholar of Trinity College suffering an apoplectic fit that affected the left half of his brain:

> He spoke, but what he uttered was quite unintelligible, although he laboured under no paralytic affection, and uttered a variety of syllables, with the greatest apparent ease. When he came to Dublin, his extraordinary jargon caused him to be treated as a foreigner, in the hotel where he stopped, and when he went to the college to see a friend, he was unable to express his wish to the gate porter, and succeeded only by pointing to the apartments which his friend had occupied.

Osborne tested him systematically with a specific passage. The remarkable pot-pourri of language-like sounds he produced was very similar to that produced by patients affected by what we now call Wernicke's aphasia or 'jargon aphasia':

> In order to ascertain and place on record the peculiar imperfection of language which he exhibited, I selected the following sentence from the By-laws of the College of Physicians, viz.: *'It shall be in the power of the*

College to examine or not examine any Licentiate, previously to his admission to a Fellowship, as they shall think fit.' Having set him to read this aloud, he read as follows: 'An the be what in the temother of the trothotodoo, to majorum or that emidrate, ein einkrastrai, mestreit to ketra totombreidei, to ra fromtreido asthat kekritest.'

However bizarre, the language has a peculiar, poetic, beauty to it, reminiscent of James Joyce's *Finnegans Wake*. Many patients with language problems can show some recovery – a further complication for neuropsychologists trying to understand the deficits – and Osborne's patient had done just that when tested eight months later:[9]

On repeating the same by-law of the College of Physicians before mentioned, after me, he spoke as follows: *'It may be in the power of the College to evhavine or not, ariatin any Licentiate seviously to his amission to a spolowship, as they shall think fit.'* More lately he has repeated the same by-law after me perfectly well with the exception of the word power, which he constantly pronounced *prier*.

Another variant of language problems due to left-sided brain damage was seen in another Dublin patient, admitted to Steven's Hospital. On 17 March 1832, James Fagan, a twenty-three-year-old pipe-maker had a drunken brawl with a dragoon, who struck Fagan's head with his sword, knocking him unconscious, fracturing the bone and exposing the brain. After several operations, Fagan was discharged on 15 April, having largely recovered. His doctor records:

Fagan was…able…to resume his work as a pipe-maker. I saw him this 20th of July, 1832; his health is excellent, but his memory of *words* but not of *things*, is greatly impaired; he told me 'he knew every thing as well as he ever did, but he could not put a name on any thing'. I showed him a button, he laughed, and said, 'I know what it is very well, it is a ba, ba, ba, — Och! I can't say it, but there it is', pointing to a button on his own coat.

Fagan continued to live an irregular life after leaving hospital. He was frequently drunk, and after a debauch on 22 August, 'he nearly lost all power in the right arm and hand, and the right side of the face was obviously affected by paralysis'. Two days later, an abscess at the site of the original wound was drained and again he recovered, although weakness of the right arm and leg showed the left hemisphere of the brain had been damaged. He also still had severe problems finding words; a condition we now call 'nominal aphasia' or 'anomia':[10]

he cannot repeat proper names, but miscals almost every thing; although he can perfectly describe the use of it, he calls, for instance, a watch, a gate; a book, a pipe, &c; ... it is remarkable, however, that the moment he employs a wrong word he is conscious of his mistake, and is most anxious to correct it. ... He counted 5 on his fingers; but could not say the word 'finger', though he made many attempts to do so. He called, his thumb, 'friend'. When desired to say 'stirabout', he said, and invariably says, 'buttermilk'; but was immediately conscious of his error, and said, 'I know that's not the name of it.'

Although there are various types of language problem and aphasia, common to them all is left- rather than right-hemisphere damage. A key problem for neuropsychology has been precisely to define what it is that the left hemisphere does which the right hemisphere does not. Dozens of differences have been identified but, in most, the advantage of the left hemisphere is only relative rather than absolute – that is, the left hemisphere is *better* than the right but the right hemisphere nonetheless has some capability in a given function. If the difference is only relative, it makes little sense that damage to Broca's area (on the left side only) can produce near total loss of language ability. Researchers have therefore used brain scanners to find tasks that are *only* carried out in the left hemisphere. A part of language that seems to depend solely on the left hemisphere is syntax or grammar. Look at the following two sentences: which of them is wrong?

The boogles are blundling the bladget.

The boogles is blundling the bladget.

Even though these sentences are meaningless, containing only nonsense words, it is still obvious that the second is grammatically incorrect, the singular 'is' not agreeing with the plural 'boogles'. In the experiment carried out by Peter Indefrey and his colleagues in Nijmegen, subjects decided whether sentences were grammatically correct while lying in a brain scanner. Activity occurred entirely on the left-hand side of the brain in an area that significantly overlaps with that described as Broca's area (see Figure 8.3).[11]

Damage to the left hemisphere can also produce difficulties in activities unrelated to language. In the condition called 'apraxia' (more strictly, in ideational and ideomotor apraxias), patients have difficulty producing skilled, complicated movements. This is not due to paralysis or weakness, but because of impaired control of the limbs. Consider an everyday action such as striking a match. The subtleties of the movements

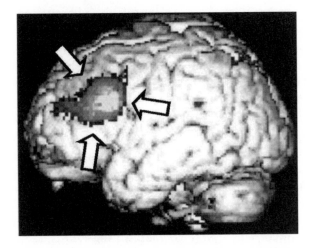

Figure 8.3 The arrows show the area of the left hemisphere, which is active in processing grammar (syntax).

involved can only fully be understood by attempting to programme a robot to perform the task. The box, held in one hand, must point in the right direction. The match, held in the other hand and at the correct end, must push against the box at a certain pressure: not too hard or the fragile match will break, and not too gently or there will be insufficient friction to make the match strike. Next, the match must be accelerated along the box: too slowly and again it will not strike; too fast and it will snap. Finally, the match must be lifted away as it strikes so that the box does not ignite in turn. Patients with apraxia have trouble with such tasks, either because they cannot conceptualise the *idea* of the movement, or because they cannot put the idea into practice. Kinnier Wilson summarised it beautifully in 1908, discussing a French patient:

> [She] was asked to lift her right arm, but after crossing it over her body, putting her hand in her left axilla [armpit], and making various energetic but hopeless efforts, she said plaintively, 'Je comprends bien ce que vous voulez, mais je ne parviens pas à le faire' [I understand well what you want me to do, but I can't manage to do it], and there lies the whole situation in a nutshell.

Although apraxia is due to damage to the left half of the brain, problems usually occur with both the right and the left hand. The left hemisphere must therefore process instructions for carrying out complex movements with either hand. Since speech and writing also involve fast, intricate, carefully co-ordinated movements, this may indicate a fundamental link between movement and language, both originating in the left hemi-sphere because both have 'grammatical' rules. The 'sentence' of striking a match is made up of separate movements, the 'words' of motor action that, like words in sentences, only make sense in the proper order. Order meaningless the them the garbage wrong put and is in result (Put them in the wrong order and the result is meaningless garbage).[12]

So far, this chapter has concentrated on the left hemisphere, which is undoubtedly concerned with language and action. If you are wondering what happens in the right hemisphere, you are in good company, numerous neurologists having asked the same. One of the first was Hughlings Jackson, the greatest of English neurologists, who worked at the National Hospital for Neurology and Neurosurgery at Queen Square in London. In 1874, he suggested that 'the right side is the chief seat of the revival of images in the *recognition* of objects, places, persons, etc.'; what we would now call 'non-verbal' processing.[13]

Writers and readers, obsessed as they are with words, forget that thought doesn't always involve language but is often instead concerned with processing pictures, images and three-dimensional space. Think about a cube. How many sides has it got? How many corners? If one side is painted black and all the others white, how many white sides will the black side touch? If an ant starts at one corner and crawls along the edges to the diagonally furthermost corner, how many edges will it crawl along? Such problems are not solved with words but by generating in the mind a picture of a cube, a 'mental image', which is then turned around to 'see' the answers. Words are only involved in asking the question and giving the answer. These are the sorts of task for which the right hemisphere is specialised.[14]

The right hemisphere is involved in the process called perception – 'making sense' of the sensory world. It involves not only sight, but also touch, sound, taste, and so on. Perceiving visual images is so natural and immediate that its working only becomes apparent when it goes wrong. Right hemisphere damage can cause visual agnosia, in which patients cannot understand what they are seeing. Their eyes work normally, as do the lower levels of visual processing in the brain, so that they are aware of light and shade, lines and blobs, but they cannot combine these compo-nents into something coherent and meaningful. Figure 8.4 may help to give a little insight into the experience of being agnosic. See what you can 'see' in it, and notice how the English word *see* has two very different meanings. You should have no trouble seeing areas of black and white but can you see anything else in this picture? This has to be a picture of something, but what? Even now, you probably cannot see this picture of a very famous person, but that is because the picture is purposely printed

Figure 8.4 A demonstration of what it feels like to be agnosic. See text for further details.

upside down. Turn the book upside down and try again. The picture is purposely difficult to see. A few clues may help. The person is a man...an *old* man...with white hair...looking straight at the camera...a scientist ...the most famous scientist of the twentieth century...a physicist...The word 'relativity' will probably give you the answer if you haven't got it already. At some point, the meaningless blobs suddenly become organised, 'make sense', and you 'see' a picture of Einstein. If you are still having problems look at the picture from a distance and blur your eyes.[15]

In the moments before you recognised the picture as Einstein you understood the experience and the frustration of patients with agnosia. Knowing something is there, knowing other people know what it is, even knowing about the constituent parts. All that is missing is the big picture. Oliver Sacks described it beautifully in relation to his patient Dr P, a talented musician with agnosia:

'What is this?' I asked, holding up a glove.
'May I examine it?' he asked, and, taking it from me, he proceeded to examine it...
'A continuous surface,' he announced at last, 'infolded on itself. It appears to have' – he hesitated – 'five outpouchings, if this is the word.'
'Yes,' I said cautiously. 'You have given me a description. Now tell me what it is.'
'A container of some sort?'
'Yes,' I said cautiously, 'and what would it contain?'
'It would contain its contents!' said Dr P., with a laugh. 'There are many possibilities. It could be a change purse, for example, for coins of five sizes. It could...'

I interrupted the barmy flow. 'Does it look familiar? Do you think it might contain, might fit, a part of your body?'
No light of recognition dawned on his face.

Agnosic patients do sometimes recognise objects, but they manage it in subtle ways, for instance, by mentally tracing the outline of the shape. Confuse the outline, as for instance in Figure 8.5, where the images overlie one another, and agnosic patients find it almost impossible to recognise what is in the picture. Agnosic patients also find it difficult if objects are seen from unusual, non-canonical views. The saw in Figure 8.6a is easy to recognise even for patients with right-hemisphere damage, but that in Figure 8.6b is far more difficult.[16]

Although pictures are two-dimensional, the real world is three-dimensional, which gives particular problems to patients with right-hemisphere damage. Life is one big three-dimensional problem. You lie in bed and there is a cup of coffee on the bedside table. You have to reach out, pick it up, lift it without spilling, and tilt it as it goes between your lips. You have to get dressed. The shirt is an interesting problem. It will fit over your arms and chest, two long tubes on a cylinder. But the shirt looks nothing like that. It lies flat over the chair, no tubes are visible, and the front seems to be cut in half, with tiny circular things down one side which have to match a series of little slits down the other side. How does one possibly wear that? Difficulty in dressing – dressing apraxia – is terribly disabling, virtually precluding an independent existence, but it is neglected as a condition. It results from right-hemisphere damage, and can be seen as part of the wider, more general, problem of constructional apraxia. Although dressing apraxia and constructional apraxia are both apraxias, they differ dramatically from the ideational and ideomotor apraxias mentioned earlier, which resulted from left-hemisphere damage. All are apraxias because they involve a problem in movement. The differences between the right- and the left-hemisphere apraxias are best seen in split-brain patients where the right and left hemispheres are completely separated, the connections between the two hemispheres having been cut.[17]

It is easy to talk as if the left hemisphere and right hemisphere are entirely separate organs, each even with their own personality, but of course they are not. Both work together to create a single individual. The two hemispheres are connected by a large bundle of fibres, the corpus callosum, by which the hemispheres communicate and co-operate. That process can be seen even in the production of language, which is not entirely the left-hemisphere task that it is made out to be. If it were, then patients with right-hemisphere damage only should have unimpaired language skills. Certainly they can talk normally, with a wide vocabulary

Figure 8.5 Poppelreuter in 1917 showed that patients with agnosia have great difficulty in identifying objects such as these in which the outlines have been confused because the objects overlap.

a   b

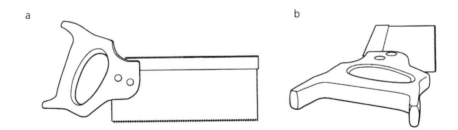

Figure 8.6 Two views of a saw (**a**) in a canonical view and (**b**) in a non-canonical view in which there is a lot of foreshortening due to perspective. Agnosic patients have far more difficulty with the foreshortened view.

and good grammar. Their language, though, is not normal, lacking the musical quality of speech, prosody, whereby the tone goes up and down, and the words accelerate and decelerate or get louder and softer, providing emotion and emphasis. Speech without prosody is like those computer-synthesised voices one hears on telephones. Prosody is not the only part of language dependent on the right hemisphere: metaphor, sarcasm and humour also come from the right hemisphere. In short, the rich communicative system we call language depends on both the right and the left hemisphere working together, each making its own distinctive contribution.[18]

Left Hand
Pre-Op

Right Hand
Pre-Op

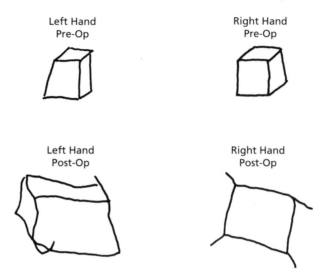

Left Hand
Post-Op

Right Hand
Post-Op

Figure 8.7 The split-brain patient PS has been asked to draw, from memory, a picture of a cube, using the left hand and then the right hand. The top row shows the drawings done before the operation, and the bottom row the pictures done after the split-brain operation.

If normal functioning requires the cerebral hemispheres to work together, what happens if the hemispheres are disconnected from each other? Occasionally neurosurgeons do just this to treat severe epilepsy, standard drugs not stopping the convulsions that start in one hemisphere and then spread through the corpus callosum to the opposite hemisphere, resulting in a generalised convulsion. Cutting the corpus callosum – 'splitting the brain' – prevents the convulsions spreading from one hemisphere to the other, so the patient stays conscious during an attack, a great practical advantage. Patients with a 'split brain' are also of great interest to neuroscientists trying to understand how the brain works, and one of the first people to study them, Roger Sperry, was awarded the Nobel Prize in 1981.

PS was fifteen when, in 1976, he had his brain split to treat the epileptic fits he'd suffered since he was two years old; an operation that proved highly effective in ameliorating his condition. As part of an experiment, however, he was asked both before and after his operation to draw two simple cubes, one with his left hand and one with his right. The cubes in the top half of Figure 8.7 are those he drew before his operation, and both are quite well drawn, that done with his left hand being slightly shakier, as is to be expected in a right-hander. Look now at the drawings shown at the bottom of Figure 8.7. Not only are neither of these drawings as good as those drawn before the operation but they are also very different from each other. That done with the left hand is very shaky, the lines not being as straight as before the operation, nor meeting properly at the corners. Nevertheless, the drawing is clearly still of a cube. The drawing done with the right hand is another matter altogether. The lines may be straighter, slicker, better formed, and meeting properly; but the picture looks nothing like a cube, all sense of three-dimensions being utterly lost. Neither a right nor a left hemisphere on its own can draw a cube properly.[19]

The corpus callosum makes one full brain out of two half-brains. When the corpus callosum is cut, each hemisphere does what it can, but neither has a complete cognitive toolkit. The right hemisphere, driving the left hand, understands three-dimensional space, and can represent a three-dimensional object using lines on a flat piece of paper. Its problem is that it can't draw very well, and doesn't know how to draw straight lines that join where they should join. It is like a patient with an ideational apraxia – the instructions for moving the hand properly are missing. Exactly the opposite applies to the left hemisphere, which drives the right hand. It knows how to make the right hand produce straight lines that meet at what should be the cube's corners, but it has no understanding of three-dimensional space or perspective, or of how lines look when projected on to a flat surface, so the cube is unrecognisable. The left hemisphere is like a patient with constructional apraxia. PS not only has a split brain but is like two different patients with brain lesions, one with right-hemisphere damage and the other with left-hemisphere damage. However tempting it is to talk of right and left hemispheres in isolation, they are actually two half-brains, designed to work together as a smooth, single, integrated whole in one entire, complete brain.[20]

The normally seamless amalgamation of the different approaches of the two hemispheres can be seen when solving a simple problem. For example, answer the following: 'Every state has a flag. Zambia is a state. Does Zambia have a flag?' It is very straightforward as brain-teasers go: there is no trick, and, as you'll already have worked out, the answer is 'Yes'. What is not obvious, though, is that questions such as this have two separate components, each teasing a different half of the brain, as experiments have made clear. In these experiments, the functioning of one half of the brain is temporarily stopped by the electroconvulsive shock therapy used to treat some severe mental illnesses. Deglin and Kinsbourne studied patients in St Petersburg who were having shock treatment, and asked them questions similar to the one above about Zambia. Before the shock, none had any trouble with the questions, but afterwards the answer they gave depended on which half of the brain was shocked, and which was still working. After a right-sided shock, leaving just the left hemisphere to solve the problem, the patients were almost chillingly logical in the way they solved the problems: 'It is written here that each state has a flag, and that Zambia is a state. Therefore Zambia has a flag.' There is something Spock-like about such answers, as if they come from a computer or an automaton. An entirely different sort of response occurred when the shock was left-sided, leaving just the right hemisphere to solve the problem: 'I've never been to Zambia and know nothing about its flag.' However true that may be, it has nothing to do with solving what is essentially a logical problem. Given the premises, the answer *has* to be 'Yes'. The right hemisphere seems to lack proper logic and therefore tries to solve the problem using everyday knowledge about the world. In this case, since it knows nothing about Zambia, it can give no answer.

Consider a slightly different question: 'All monkeys climb trees. The porcupine is a monkey. Does the porcupine climb trees?' Here one of the premises is false, since porcupines are not monkeys. That, though, doesn't affect the logical structure of the question. If all monkeys do climb trees and the porcupine were a monkey then it would be true that porcupines climb trees. This sort of problem is called a counterfactual. How do the right and left hemispheres cope with it? After a left-sided shock, the functioning right hemisphere of one patient commented, with great indignation, 'Porcupine? How can it climb trees? It is not a monkey. It is prickly like a hedgehog. It's wrong here!' The right hemisphere has knowledge of porcupines and knows what they can and can't do. With a right-sided shock the functioning left hemisphere of the patient replied completely differently: 'Since the porcupine is a monkey, then it climbs trees.' When the experimenter said, 'But you do know that a porcupine is not a monkey?', back came the reply, 'It is written so on the card.' The left hemisphere knows how to handle logic and the right hemisphere knows about the world. Put the two together and one gets a powerful thinking machine. Use either on its own and the result can be bizarre or absurd.[21]

The wrong answer to the porcupine question strikes us as both humorous and odd. Humour often involves an absurdity, a bizarreness, a failure to recognise two different and incompatible perspectives, or the merging together of separate meanings. The 'sense of humour' of the two hemispheres seems to be different. Consider the following exercise, used in a study of right- and left-brain-damaged patients. The patients were given the following three lines, and then asked to choose a punchline from several alternatives.

A new housekeeper was accused of helping herself to her master's liquor. She told him, 'I'll have you know, sir, I come from honest English parents.'

Patients with right-hemisphere damage, and hence a functioning left hemisphere, preferred the following punchline,

He said, 'All the same, the next time the liquor disappears you're fired.'

Logical, but definitely unfunny. Patients with left-hemisphere damage, and hence a functioning right hemisphere, chose an entirely different answer that might seem funny in a pantomime but is hardly an appropriate punchline:

Then the housekeeper saw a mouse and jumped into her master's lap.

The real punchline succeeds in being funny by appealing both to the left hemisphere's sense of logic and the right hemisphere's knowledge of hidden meanings – in this case, of the word 'extraction':[22]

He said, 'I'm not concerned with your English parents. What's worrying me is your Scotch extraction.'

As well as being involved in understanding three-dimensional space, having knowledge about the world, and being involved in humour, the right hemisphere has other functions as well. It is particularly involved in attention, a process central to cognition yet often ignored, forgotten or unnoticed. Without the ability to focus attention, the mind would be flooded with irrelevant, pointless, useless information that would over-whelm it. Much going on in the world around us is boring and does not need high-level processing, this being reserved for the interesting, impor-tant and potentially life-threatening things that happen. Attention allows us to ignore what is irrelevant to the task or activity we are involved in. We are completely unaware of many things until someone mentions them to us. For instance, the seat on which you are sitting is pressing on your buttocks. There! Suddenly it floods into consciousness and for a few moments you will not be able to forget it. Your buttocks are the focus of your thoughts and yet a few moments ago you were unaware of them. That is attention.

When attention goes wrong, then very strange symptoms can occur. An example can be seen in the novelist Charles Dickens, who was unwell for the last five years of his life. In February 1866, when he was aged fifty-four, four years before he died, he had clear signs of heart disease. The most serious problem, though, appeared three years later, on 23 April 1869. While in Chester on a gruelling tour of public readings, Dickens noticed strange symptoms. He wrote to his doctor, Frank Beard, who rushed him to London, where he was seen by Sir Thomas Watson. Watson's report was clear and specific:

> After unusual irritability, [Dickens] found himself, last Saturday or Sunday, giddy, with a tendency to go backwards, and to turn round...He had some odd feeling of insecurity about his left leg, as if there was something unnatural about his heel; but he could lift and did not drag his leg. Also he spoke of some strangeness of his left hand and arm; missed the spot on which he wished to lay that hand, unless he looked carefully at it; felt an unreadiness to lift his hands towards his head, especially his left hand – when for instance, he was brushing his hair.

Something was seriously wrong with the left side of Dickens' body and, by implication, the right side of his brain. Watson was in no doubt about the situation: '[Dickens] had been on the brink of an attack of paralysis of his left side, and possibly of apoplexy.' The damage to the right half of his brain was fatally confirmed on 8 June 1870 when, at dinner, he stood and then collapsed heavily to his left side, falling unconscious. He died just after six o'clock the next evening.[23]

Dickens' illness does not seem particularly remarkable, but there was a strange symptom, not yet mentioned, which at the time (1868) had not been described in the medical literature. While walking to the house of John Forster, his friend and future biographer, he noticed that 'he could read only the halves of the letters of the shop doors that were on his right as he looked'. On 21 March 1870 a similar thing occurred:

> he told us that as he came along, walking up the length of Oxford Street, the same incident had recurred as on the day of a former dinner with us, and he had not been able to read, all the way, more than the right-hand half of the names over the shops.

Dickens was suffering from what is now called 'neglect dyslexia'. When reading, patients usually do one of two things. They may read only the right-hand half of a word, particularly if it is a legitimate word: so DATE may be read as ATE, FRIGHT as RIGHT, TRAIN as RAIN, or, embarrassingly for a man looking for a public toilet, WOMEN as MEN. Alternatively, a word is read of the correct length, but with the left half being guessed at: for instance, SAWMILL as WINDMILL, CAKE as MAKE, or TOGETHER as WHETHER.[24]

Neglect dyslexia is a variant of a far more common condition usually called by the simpler term neglect, but also known as visual neglect, hemi-neglect, spatial neglect, unilateral neglect and hemi-inattention. A famous case was Federico Fellini, the Italian film director, who as well as winning Oscars for his films *La strada* [*The Road*] (1954), *Le notte de Cabiria* [*Nights of Cabiria*] (1957), *Otto e mezzo* [*8½*] *(1963)* and *Amarcord* [*I Remember*] (1973), was a talented painter and cartoonist. In March 1993, he flew to Los Angeles to receive a fifth Oscar, this time for lifetime achievement in film-making. He knew he was a sick man, and in June of the same year he had cardiac bypass surgery in Zurich. He was not a good patient, fretting to leave Switzerland and return to Italy. Eventually, the doctors let him return not to Rome but to Rimini, his birthplace. He had a suite at the Grand Hotel and, in August, while his wife was in Rome, he collapsed in his room, conscious but unable to move or reach a phone. He lay helpless on the floor for three-quarters of an hour, until found by a maid. At the hospital in Rimini, a CT scan (Figure 8.8) showed a thrombosis in the rear part of his right cerebral hemisphere, this being the cause of the paralysis in his left arm and leg that confined him to a wheelchair for the rest of his life.

Fellini was transferred to the hospital in Ferrara for rehabilitation, a process he found slow and frustrating, and despite reassurances, he 'saw in the doctors' eyes disbelief that he was going to make … [a] total recovery'. Fellini's wife, Giulietta, was also seriously ill at the same time, having been diagnosed with an inoperable brain tumour, which she tried unsuccessfully to keep from Fellini. Their fiftieth wedding anniversary was due on 30 October and Fellini persuaded the doctors to let him go to Rome, where he hoped to celebrate the anniversary. Two weeks before it, he and Giulietta went out for Sunday lunch on their own. That evening at the hospital in Rome Fellini had a second, more massive stroke, and went into a coma from which he never awoke. He was pronounced brain-dead in the intensive care unit, but was ventilated until 31 October, the day after the fiftieth wedding anniversary. Giulietta died six months later on 23 March 1994.[25]

As well as the immediate and obvious problem of paralysis, Fellini showed clear signs of visual neglect, a problem that occurs frequently after right-hemisphere strokes and is surprisingly disabling; indeed, the amount of neglect is the best predictor of recovery from such strokes. Neglect is precisely what it says it is. The patient ignores, neglects, half of the world; typically, the left half. Although left-sided neglect may at first seem merely to reflect the fact that the right hemisphere is responsible for the left-hand side of the body and associated perception, the situation is not that simple, because right-sided neglect after a left-hemisphere stroke is much rarer. This indicates that it is the right hemisphere that is responsible for attention.[26]

Neglect is easily tested by giving patients a horizontal line such as this

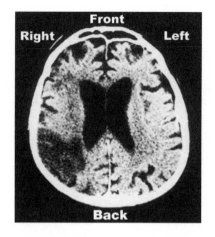

Figure 8.8 CT scan of the brain of the film director Federico Fellini, one week after his stroke. There is a wedge-shaped, dark area in the rear part of the right hemisphere, which is best seen by comparing it with the normal left-hand side. It is in the temporo-parietal region.

and asking them to put a pencil mark exactly in the middle. You might like to try it.

Patients with neglect put the mark way over to the right-hand side, apparently ignoring much of the left-hand part of the line. Fellini did precisely this, but his extravert personality, florid imagination and near compulsive desire to draw, meant that he also added a few extra details, as seen in Figure 8.9.[27]

Patients also neglect many other aspects of the left half of space. Examples include eating only the food on the right-hand half of a plate, reading only the right-hand half of a page, remembering only the buildings on the right-hand half of public places, washing only the right half of the face, dressing only the right arm and right leg, and so on. Not all patients have every symptom – Fellini, for instance, did not show personal neglect in washing or dressing.[28]

Some patients with neglect draw only the right half of objects. The left-hand drawing of Figure 8.10, for instance, depicts only the right-hand half of a clock face, and shows only half the numbers. In contrast, although in the right-hand drawing the numbers are mostly confined to the right-hand side of the clock, all twelve numbers are present, and hence are in the wrong places.[29]

Neglect is not just an inability to see half the world, as though one is

Figure 8.9 Two of the line-bisection tasks carried out by Fellini. Fellini was unable to resist the temptation to carry on drawing after he had made the mark indicating where he saw the centre. In the lower example, the right-hand character says, 'Go to the half-way point!' whereas the left-hand figure replies, 'Forget it!'.

watching only half a television screen. The difficulty is with attention, and since attention concentrates on *objects*, what ultimately is neglected is the left half of objects. That though raises the question of what we mean by a visual object. In Figure 8.11, a fifty-nine-year-old woman with a right-hemisphere stroke has copied a drawing of two flowers in a single pot. Half of each flower is drawn, as is half of the pot. The fact that she can draw half of each object surely implies that she has some knowledge of the entire object in each case. Herein is the reason why neglect so interests neuroscientists.[30]

Although it is often said that neglect patients lack insight into their condition, that was certainly not so for Fellini, who even drew a picture in which the character asks, 'Where is the left?' (Figure 8.12).

It might be thought that a lifetime working in the visual arts might counter neglect, but that is far from the case. Several artists with neglect have continued painting and drawing, but the entire left half of what they were painting is left out of their pictures. Figure 8.13 shows a portrait by the English artist Tom Greenshields – even the careful scrutiny required in the act of painting does not prevent one neglecting half of visual space.[31]

Damage to the right half of the brain can also result in other unusual

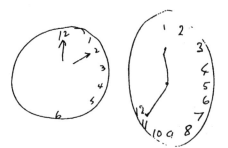

Figure 8.10 A patient with neglect has been asked to copy a drawing of a clock. The left-hand drawing, done soon after the stroke, shows omission of the numbers on the left-hand side of the clock face, whereas the drawing on the right, after some recovery, shows transposition of numbers from the left-hand side to the right-hand side of the clock face.

Figure 8.11 A patient with neglect was asked to copy the drawing of flowers on the left. Their copy is shown on the right.

Figure 8.12 A spontaneous drawing made by Fellini three to four weeks after his stroke. The character, who is clearly Fellini himself, asks, 'Dov'é la sinistra??' – 'Where is the left?'.

Figure 8.13 Portrait by the English artist Tom Greenshields, carried out after he had suffered a stroke affecting the right hemisphere of his brain. The drawing, which is entirely on the right half of the paper, shows neglect, only half of the figure being drawn.

syndromes. In a condition known as misoplegia patients suffering paralysis can show a virulent hatred of the affected limb, a problem Fellini showed to some extent, describing his paralysed left arm as 'a bloated damp bunch of asparagus'. In extreme cases, patients may violently hit the affected limb and shout abuse at it. In another condition known as anosognosia, there is denial that physical disease is present, a condition that Charles Dickens may have shown. In extreme cases, a patient may not even admit that paralysed limbs are their own, as in a seventy-three-year-old-woman with paralysis of the left half of her body who, in 1956, was admitted to hospital in Bucharest. She had suffered a stroke the day before, but when asked to show her left arm or leg she instead indicated the right arm or leg. Eventually, the doctor pointed at her left hand, asking, 'Whose hand is this?' 'It's the hand of the patient in the next bed,' came the reply. Sensing further explanation was needed, the patient added, 'I asked her to put her hand on my tummy because I felt cold.' When the doctor pinched the patient's left hand, she merely said, 'I feel you're pinching the hand of the patient next to me.' The patient died a month later, and the post-mortem showed damage in the parietal lobe of the right hemisphere, the same location usually said to be associated with neglect and other defects of attention.[32]

A curious feature of brain lateralisation is the abilities or functions that are lateralised, often with no obvious reason. Swallowing is one of them. Dysphagia, the medical name for difficulty in swallowing, occurs in over a third of patients with strokes in one half of the brain. Fellini had dysphagia, as evidenced at the lunch in a Roman restaurant with his wife Giulietta, which turned out to be their final meal together.

> Eating with abandon and talking animatedly, Fellini suddenly began choking. The cause was a piece of mozzarella. The stroke had affected his ability to swallow, but in the pleasure of the moment, he had forgotten.

Most people have a dominant hemisphere for swallowing and, rather peculiarly, it is the right hemisphere for about half the population, and the left hemisphere for the rest.[33]

Many psychological processes are lateralised and, as Joseph Hellige has put it, 'there are now so many...tasks that show asymmetry that there is little hope of ever compiling a complete list'. All I can give here is a brief flavour of some of them. Ironically, flavour is itself lateralised. Smells and flavours are rated more highly when inhaled into the right nostril and subsequently processed by the right rather than the left hemisphere. The left hemisphere, however, is more accurate at identifying and naming smells. Right-hemisphere damage can lead to the 'gourmand syndrome', in which there is a sudden onset of an obsessional interest in fine foods. Amongst other fine things of life, the right hemisphere is particularly important for music, recognising and remembering melodies, and singing. Rhythm, on the other hand, is a left-hemisphere speciality, as is absolute pitch. Music, therefore, like other complex human activities, requires the integration of right and left hemispheres. Musical hallucinations can occur after right-hemisphere damage, as can musicogenic epilepsy – convulsive fits produced by particular pieces of music. Even sex is lateralised. An experiment in Paris analysed brain activity in heterosexual men while they lay in a PET scanner watching sexually explicit films. To check that the films were having the desired effect, the subjects had devices attached to the penis to measure the size of their erections. Sexual arousal activated two areas in the right hemisphere associated with motivation, and an area in the left hemisphere involved in involuntary

Figure 8.14 Does the bar at the top or the bottom look darker?

responses. Like other activities, sex depends on two sides, as it were, coming together.[34]

Differences between the two sides of the brain – functional asymmetries – are present in everyone, and can sometimes be seen in simple demonstrations. Look at Figure 8.14 and decide if the top or bottom bar looks darker. The two are actually identical. Having said that, about three-quarters of people given the test say the top bar looks darker. The reason is that just as patients with right-hemisphere damage ignore the left half of space, so normal individuals over-exaggerate it; a process called pseudoneglect. The top figure looks darker because its dark patch is on the left and has more psychological impact. Pseudoneglect occurs even in the simple task of putting a mark exactly in the middle of a line, as in the test earlier in this chapter. If you had a go at that, go back and use a ruler to measure whether your mark is precisely in the middle of the line. It probably is not, most people putting the mark a few percentage points to the left of centre because the left half of space attracts more attention than the right. This also makes people more likely to bump into objects on their right side, because they pay less attention there. Pseudoneglect may also explain why the 'balance point' in pictures is shifted towards the left side, why the left foreground of a picture seems closer than the right, why names for pictures more often refer to items in the left foreground, and why when actors want to come on stage unnoticed they enter from the right. Figure 8.15 shows another variant on this theme. Which of the two faces looks happier? The faces are called chimeric, being made from two separate parts. For most people, the bottom face, with its happy half on the left, looks happier. In part this is due to greater attention being paid to the left, and in part to the fact that the right hemisphere is particularly involved in recognising emotions.[35]

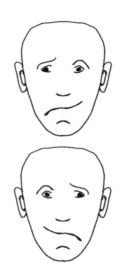

Figure 8.15 Look directly at the nose of each face and say which individual looks happier, the one at the top or the one at the bottom.

Demonstrating that language is normally located in the left hemisphere of people without brain damage was first carried out in the 1960s, using what is called dichotic listening. The subject wears a pair of stereo headphones, but instead of each ear hearing a slightly different version of the same thing (as when listening to music in stereo), the two ears hear completely different things; perhaps one set of words in the right ear and a totally different set of words in the left. The words heard by the right ear are processed more accurately, because the right ear mainly connects to the left hemisphere.[36]

The story so far seems pleasingly straightforward: the left hemisphere processes language (including speech, reading, writing and spelling) perhaps because it is better at handling the fast, sequential events found in these tasks, whereas the right hemisphere carries out a whole string of 'non-verbal' tasks, typified by the highly parallel, holistic analyses needed to understand visual images and make sense of three-dimensional space. However, straightforward though the story seems, it is only a story; and like all stories it tells us as much about ourselves and our need to hear such tales as it tells about the way the brain actually works. Certainly it is a story told in dozens of introductory psychology and neuroscience textbooks, and one that pops up regularly in popular science books. Indeed, Lauren Harris points out it even appears as a 'core fact' in a book entitled *What Literate Americans Know*. The big mistake in the idea is its casual presumption that everyone's brain is the same. This is simply not the case. As we have already seen earlier in this book, ten per cent of people are left-handed, their dominant hand controlled by the right rather than the left hemisphere. The brain is more complicated than simplistic descriptions, as can be seen in relation to language.[37]

Within a year or two of Broca's announcement that language was located in the left hemisphere of the brain, it became obvious to him and others that this wasn't always the case. In some people, perhaps five per cent or so, language was located in the *right* hemisphere. For example, in August 1866, Hughlings Jackson described the following case:

> On Friday last I saw at the Hospital for Epilepsy and Paralysis a patient who, with hemiplegia of the *left* side, had considerable defect of speech. …This poor fellow lamented that the paralysis should have seized him on the left side, because, he said, he was 'a left-handed man'.

If left-handers were the reverse of right-handers, having language in the right hemisphere rather than the left, then that might be seen as restoring some of the symmetry so conspicuously lost when Broca described language as being located solely in the left hemisphere.[38]

It was a nice idea but it was wrong, although it lasted many years and still appears today, perhaps because the concept is so psychologically seductive. Within a few years of Broca, the major medical and psychology textbooks emphatically gave it as a rule. William James, a founding father of psychology, in his *Principles of Psychology*, put it succinctly: '[language] in right-handed people is found on the left hemisphere, and in left-handed people on the right hemisphere'. Such a widely promulgated 'rule' can rapidly distort the scientific literature because the only cases published are those that fit the rule, every new patient seemingly verifying it yet further. Although some physicians did notice patients who seemed to be exceptions to the rule, particularly right-handers, the theory was propped up for many years by a variety of theoretical devices that made it extremely hard to refute. The most powerful of these suggested that the patients were not 'really' right- or left-handed. Most right-handers can do some tasks with their left hand and most left-handers do some tasks with their right hand, so it was a persuasive argument. Left-handers in whom language was located in the left hemisphere were reclassified by claiming that they had been forced to write with their right hand and that this had shifted the hemisphere for language to the other side. Perhaps the most difficult group were right-handers in whom language was apparently located in the right hemisphere. The explanation given here was that there must be a left-hander in their family and that they were thus 'honorary left-handers'.[39]

Perhaps least satisfactory of all was the concept of 'latent left-handedness'. If a person in whom language was located in the left hemisphere wrote with their right hand, did everything with their right hand and had no left-handed relatives, the rule insisted that they must still be left-handed, and so it was said that they had 'latent left-handedness'. The concept is seen in its most refined form in the description by Alexandria Luria, the eminent Soviet neuropsychologist:

> ... it is possible to determine *subtle signs of left-handedness* ... [A]mong the *morphological* signs of latent left-handedness are a large left hand, a well developed venous system on the back of the left hand, a wide finger nail on the fifth finger of the left hand, and highly developed expressive musculature on the right side of the face... The number of *functional signs* of latent left-handedness is considerably greater... Under condition[s] of high affect... latent left-handedness may manifest itself, and the individual may switch to his left hand... A detailed interrogation often reveals irregularities in the dominance of the right hand...

The interrogation, the search for the subtle signs, the tiny slips that reveal the truth – it is all suggestive of a medieval witch hunt, seeking out the stigmata, the witch-marks catalogued in the *Malleus Maleficarum*. Perhaps that finger nail is a little too large? Are those veins too developed? Such 'irregularities' can always be found, and thus anyone could be classified as a latent left-hander. Think only of the fact that Luria included in his list of distinguishing features such criteria as left-hand clasping and left-eye dominance. Since sixty per cent of the British population are left-hand claspers, and thirty per cent are left-eye dominant, most of the population will be latent left-handers on these criteria alone. Clearly, though, if everyone is a latent left-hander, then the scientific utility of the concept is precisely zero.[40]

If left-handers are not simply the mirror-image of right-handers, what is the real relationship between handedness and language dominance? The proper relationship was largely worked out in the years after the Second World War, and shows an intriguing mathematical pattern that is not at all easy to explain. It can be seen in a simple table:

Table 8.2: The relationship between handedness and language dominance (%)		
	Left-sided language	Right-sided language
Right-handers	~95	~5
Left-handers	~70	~30

Several things must be pointed out. Firstly, handedness and language dominance are undoubtedly associated, right-handers being more likely than left-handers to have language in the left hemisphere of the brain. Equally clearly, though, left-handers are not the mirror image of right-handers. If they were, ninety-five per cent of left-handers would have language in the *right* hemisphere, whereas only about thirty per cent do so. So the majority of left-handers, like the majority of right-handers, have language located in the left hemisphere. Nevertheless, a left-hander is still about five or six times more likely than a right-hander to have language located in the right hemisphere. An interesting twist to the statistics is that, since there are eight or nine times as many right-handers as left-handers in the population, the majority of people with right-sided language are *right*-handed. These numbers provide a real conundrum, the explanation of which is essential if we are to understand brain lateralisation.[41]

If the situation for handedness and language dominance seems complicated, one also needs to think about the so-called right-hemisphere processes. Are they always in the right hemisphere? Again, the answer is no. Altogether there are eight possible combinations of handedness, language dominance and lateralisation of non-verbal processes, each of which occurs in the population. Once more, though, this oversimplifies the situation, because I have talked as if 'language dominance' is a single thing, responsible for speech, reading, writing, spelling and so on. However, the evidence from patients with brain damage suggests that some of these abilities can be located in one hemisphere and the rest in the other. For an example we can again turn to Sir Thomas Watson, only this time with himself as the patient. On 22 October 1882, at the age of ninety, he had a stroke. He fell to the left side, when he put out his tongue it went to the left, and his face was flattened on the left side, so his doctors quite properly diagnosed a thrombosis in the right cerebral hemisphere. Interest in the illness of this grand old man of British medicine was such that the *British Medical Journal* published fortnightly bulletins. He did not lose his speech, and he maintained a professional interest in his own case, although 'He said he could not explain his symptoms'. Most puzzling is that, despite having a right-sided stroke, 'In trying to write a letter, he could not remember how to spell the simplest words.' Spelling should surely, like other language functions, be in the left hemisphere, so it was strange to see difficulties occurring after right-hemisphere damage. To make matters worse, there were no other problems with language. The implication is that some components of language can be located in one hemisphere whilst others are in the other.[42]

These complicated patterns of brain organisation can be explained using the same genetic model that was used to explain handedness in the previous chapter. Remember that there are two genes, D and C, and that those individuals who have a double dose of D, the DD genotype, are all right-handed. Something early in their development pushes their brain one way, and puts the motor control centre in the left hemisphere. The model for language also assumes that exactly the same thing happens to the developing language centre, so that it also is pushed across into the left hemisphere. The *DD* individuals therefore look much like the text-book paradigms, all being right-handed with language in the left hemi-sphere. What, though, about the *DC* and *CC* individuals, many of whom are left-handed? It is easiest to deal first with the *CC* individuals.

The *CC* genes effectively do nothing to determine whether the motor control centre is in the right or the left half of the brain, and thus exactly half of *CC* individuals become left-handed and the other half right-handed. A biological coin is tossed, and it falls one way or the other, due to chance and chance alone. If, though, the *CC* gene has no influence upon the side of the motor control centre, the natural thing is to assume it also has no influence upon the side of the language centre, which will also have a fifty-fifty chance of being on the left or the right. So far so good. The neat thing at this point is to realise that if it is chance that pushes the motor centre one way or the other, and chance also that pushes the language centre one way or the other, then those two chance processes are probably independent. It is as if two separate coins are being tossed. The result is that a quarter of *CC* individuals are left-handed with language in the left side, a quarter are left-handed with language in the right side, a quarter are right-handed with language in the left side, and a quarter are right-handed with language in the right side. In other words, all possible combinations occur.

For *DC* individuals, much the same thing happens except that since three-quarters are right-handed and one-quarter left-handed, then it is also the case that three-quarters have language in the left side and one-quarter in the right side. Again, it is as if two separate but biased coins are tossed. We can now work through the calculations for all the three geno-types, *DD*, *DC* and *CC*, remembering that *CC* individuals are quite rare, *DC* individuals more common, and *DD* individuals the most common. The final result is clear. If ten per cent of people are left-handed, then 7.8 per cent of right-handers and thirty per cent of left-handers will have lan-guage in the right hemisphere; definitely a good approximation to the numbers in the table earlier. It is when numbers like this pop out of the calculations that scientists believe mathematical models say something useful. When I first saw those numbers back in the late 1970s, I became very excited, and since then I still haven't found any serious reason for doubting the basic model. The details can be tweaked, but the central idea still seems pretty good.[43]

Having said that, statistics can be misleading or unreliable. This is particularly so in relation to the proportion of right-handers in whom language is located in the right hemisphere. Although in Table 8.2 I gave a figure of about five per cent, it would have been easy to find some other number, varying from the one per cent or so found in clinical data based on long-term follow-up of patients, through to eight per cent for patients with acute strokes, and up to fifteen or even twenty per cent if the figures are derived from using the somewhat unreliable dichotic listening tests in normal subjects. That proportion, though, is central to the whole process, and if it were much outside the range of five to ten per cent, then the model could indeed be wanting. It was more than twenty years after the theoretical model was developed that elegant new technology finally gave an accurate estimate of this important number, and it is to this that we finally turn.[44]

Neurosurgeons often perform difficult and dangerous operations deep inside the brain. It was realised early on that if they knew in which side of the brain language was located they could then operate from the other side. As James Gardner, a neurosurgeon in the 1940s, put it, 'removal of a tumour at the cost of a patient's speech is scarcely an accomplishment on which to congratulate oneself'. However, it is difficult to know with certainty in which hemisphere language is located, and for half a century the only reliable method was a test developed by a young Japanese neurosurgeon, Juhn Wada. He was working in Hokkaido just after the end of the Second World War, and conditions were grim:

> The old order had crumbled, and signs of inflation, poverty, and undernutrition were everywhere. My dinner as a young duty doctor at our University Hospital was a lone piece of unskinned potato and on the rare occasion a bowl of rice glue.

Despite doing brain surgery, Wada had no training and no teachers, and he learned his techniques from a textbook of neurosurgery sent by his brother, a surgeon in Boston.[45]

Wada was concerned originally with a different problem. Patients sometimes develop severe unstoppable epilepsy, *status epilepticus*, in which each attack triggers another. Wada wondered whether the vicious circle could be broken by anaesthetising just half of the brain by injecting a general anaesthetic into the carotid artery that supplies blood to one or other half of the brain. One day, a young Japanese boy who worked as a cook at the local US Army base was admitted to the hospital. A drunken GI had told the boy that he could shoot his hat off his head. The GI's aim was good but not good enough, removing not only the boy's hat but the scalp, bone and top layer of the brain on the left side. An operation was needed but the boy was in *status epilepticus*, and no operation could be done until the convulsions had stopped. Wada persuaded the boy and his family to try the new treatment. Sodium amytal, a short-acting barbiturate anaesthetic, was injected directly into the left carotid artery using the crude but surprisingly effective method of thrusting a long needle straight into the neck. The boy's fits stopped immediately, and he became temporarily paralysed on his right side, confirming that the left half of the brain was anaesthetised. In addition, he became completely mute. After ten minutes or so, as the drug washed out of the brain, so the paralysis and mutism wore off. Wada's treatment had worked and, at the same time, given clear evidence that this boy's language centres were in the left hemisphere of his brain.[46]

Over the next year or so, Wada tested fifteen more patients, all right-handed, and this time he was more systematic, finding that, in every case, speech was stopped when the drug was injected into the left carotid artery and was unimpaired when injected into the right carotid artery. Five years later Wada was granted sabbatical leave to go to the Montreal Neurological Institute, where he demonstrated his amytal technique. Within a couple of years it was a routine procedure prior to undertaking neurosurgery.[47]

However unsubtle biologically, there is no question that the Wada technique works. It does have occasional complications, but these have been reduced by modern methods. To a psychologist, though, its problems are twofold. Firstly, it is mainly carried out on people who, by definition, have abnormal brains, a fact that makes generalising to the rest of the population rather difficult. Secondly, it is simply too invasive to use on large numbers of normal people, which is what scientific research needed to do. Although, over the years, other techniques for assessing language dominance were developed, none could precisely identify the side of language in an individual. The Wada test, therefore, remained the gold standard. That changed, though, at the end of the 1990s, with some very new technology: transcranial Doppler ultrasonography.[48]

Doppler ultrasonography is similar to the ultrasonic scanning routinely used in obstetric units for looking at the developing fetus, but it has an additional subtlety. A beam of ultrasound is sent into an artery, and the corpuscles in the blood bounce back some of the sound waves so that they can be detected by a probe. The label 'Doppler' refers to the Doppler shift, the phenomenon in which sounds, such as the whistle of a speeding train, rise as the source comes toward you, and fall as it goes away. The faster the train the larger the difference in the whistle's pitch as it approaches and goes away. The corpuscles in the blood do the same thing. The faster they are moving towards the scanner, the more the pitch of the ultrasound is raised, meaning that the speed of the blood in the artery can be measured.

Transcranial Doppler ultrasonography puts the probes on the side of the head, where they measure the blood flow in the arteries of the brain: it has been described as 'a stethoscope for the brain'. The method is non-invasive, easy to administer, and appears to be completely safe. Stefan Knecht and his colleagues at the University of Münster in Germany have adapted it to assess language-processing in the brain. Subjects sit in front of a computer screen on which appears a randomly chosen letter of the alphabet, say 'T'. The subjects then have fifteen seconds in which to think silently of as many words as possible beginning with that letter, and they then have five seconds to tell the experimenters those words. Word generation is mainly carried out by the temporal lobe of the brain, which is located just above the ear on each side. It is hard work for the brain, and the increased blood supply is detected by transcranial Doppler ultrasonography in the middle cerebral artery that supplies the temporal lobe. There is a temporal lobe on each side of the brain, but if only one side is generating the words, then blood flow should correspondingly only increase on that side. By comparing blood flow in the right and left arteries it is then possible to determine whether a person is right- or left-hemisphere dominant for language, and because the technique is quick, easy, relatively cheap and non-invasive, it can be performed on a large number of subjects.

In 2000, Knecht and his colleagues published two important papers concerning tests they had performed on large numbers of subjects, using this method. In the vast majority of right-handers, the blood flow increased on the left side, just as Broca would have expected. What was particularly interesting, though, was the small group of right-handers for whom the flow of blood increased on the right side, meaning they had right-sided language. Twelve of the 204 right-handers showed evidence of language on the right side of their brain, very close to the proportion of 5.9 per cent predicted by the genetic model. In a second paper, Knecht and his colleagues also looked at a large group of 122 left-handers, of whom twenty-nine, just under twenty-four per cent, had right-sided language, a figure comfortably close to the predicted value of thirty per cent.[49]

This chapter has shown that the right and left hemispheres of the brain differ in a host of ways. But why do they differ? How long have humans predominantly been right-handed, why are some left-handed, and what made cerebral specialisation evolve? Those are questions for the next chapter.

EHUD, SON OF GERA

Ehud, son of Gera the Benjamite, is perhaps the first left-hander whose name we know. We also know what he did, for, around 1200 BC,

> the Israelites sent him to pay their tribute to Eglon king of Moab. Ehud made himself a two-edged sword, only fifteen inches long, which he fastened on his right side under his clothes and he brought the tribute to Eglon the king of Moab.

After paying the tribute, Ehud asked Eglon for a word in private:

> Eglon then came up to [Ehud]…and Ehud reached with his left hand, drew the sword from his right side and drove it into [Eglon's] belly. The hilt went in after the blade and the fat closed over the blade.

Ehud, though, was not the only left-hander. The book of Judges also describes how the Benjamites were in a bitter civil war with their fellow Israelites:

> They flocked from their cities to Gibeah to go to war with the Israelites, and that day they mustered out of their cities twenty-six thousand men armed with swords. There were also seven hundred picked men from Gibeah, left-handed men, who could sling a stone and not miss by a hair's breadth.

Interpreting the passage is not easy, but usually it is taken as meaning that the seven hundred picked men were a part of the twenty-six thousand, making 2.7 per cent of the population left-handed; or to be more precise, a *minimum* of 2.7 per cent left-handers.[1]

Judges provides what is almost certainly the first documentary evidence of how common left-handedness was in the historical past. However, from then until late in the nineteenth century, no-one seems to have counted how many people were left-handed. That most people were right-handed never seems to have been in any doubt, for otherwise laws such as that enacted in fifth century BC Athens would have had little point:

> The Athenians were commanded by a number of generals, among whom was Philocles, who had recently persuaded the people to pass a decree that all prisoners of war should have their right thumbs cut off to prevent them holding a spear, although they could still handle an oar.

Nor is there much doubt that there was a noticeable but nevertheless small minority who were left-handed. Homer, Aristotle, Plato and other Greeks clearly met left-handers on a daily basis, but what percentage of ancient peoples were left-handed is not stated in any written evidence. If, however, we are to understand handedness, we need some idea of how many people have been left-handed in the past, so that we can work out an evolutionary timescale in relation to right- and left-handedness. Hopefully this will enable us to construct some sort of explanation of how and why human right-handedness evolved, and how this relates to human language. Such is the business of this chapter.[2]

One of the first statistically sound modern studies of the incidence of left-handedness was undertaken in 1871 by William Ogle, a physician at St George's Hospital in London, who, 'unable to find any reliable statistics' on handedness, asked the next two thousand patients that he saw in his clinics whether they were right- or left-handed. Eighty-five, which is 4.25 per cent, said they were left-handed, a proportion very similar to that found in a half-dozen or so other studies carried out just before the First World War, but that is less than a half of the ten per cent described in chapter 7, which is typical of modern studies. A difference that large needs explaining.[3]

The largest ever study of handedness took place in 1986, and, somewhat surprisingly, this had its origins in a study of how well people could smell. *National Geographic* magazine was publishing an article on smell and, to illustrate the article, a 'scratch-and-sniff' card was included, which could be returned by post. Readers indicated what they thought the smells were and then ticked a couple of boxes to indicate their age, sex, and the hand used for writing. The cards were returned by huge numbers of people – over 1,100,000 Americans – which allowed a wonderfully fine-grained statistical analysis. Figure 9.1 charts the percentage of left-handed men and women in relation to the year of birth. As in

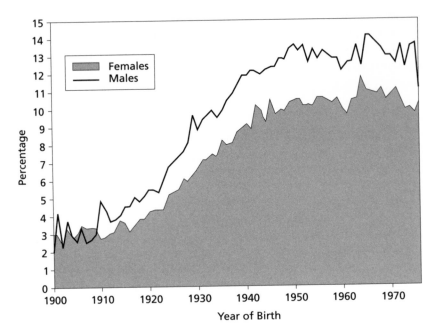

Figure 9.1 Left-handedness during the twentieth century. The percentage of people who are left-handed is plotted vertically, with men indicated by the solid black line, and women by the grey shaded area.

most other large studies, there are somewhat more left-handed men than women. However, the most dramatic result is that the proportion of left-handers is seen to have increased substantially during the twentieth century. For those born before 1910, about three per cent are left-handed, a proportion similar to that found by Ogle thirty years earlier. However, as the years pass so the proportion of left-handers steadily increases until by the end of the Second World War the proportion of left-handers seems to have reached a peak at between ten and eleven per cent of women and about thirteen per cent of men. Dramatic increases clearly took place within half a century. Something must have caused them, but what?[4]

The historical data in America and elsewhere in the West clearly show that left-handedness is more common now than it was a hundred years ago. Different countries also show different rates of left-handedness. Studying such differences has not been easy because researchers have often used different methods to assess handedness, meaning that apparent differences could be artefactual. Phil Bryden, at the University of Waterloo in Ontario, systematically collected data to resolve this problem. In a collaboration with Maharaj Singh, from India, and Yukihide Ida, from Japan, a single questionnaire was designed for use in Canada, India and Japan. Subsequently Taha Amir and I have also used the same questionnaire in the United Arab Emirates. The results are striking. In Canada and the UK, the proportion of left-handers is about 11.5 per cent. However, as one moves across Asia the proportion falls, becoming 7.5 per cent in the Emirates, 5.8 per cent in India and four per cent in Japan. Such differences are not restricted to Asia. A study in Africa, using broadly comparable methods, found that 7.9 per cent of people in the Ivory Coast and 5.1 per cent of people in the Sudan were left-handed. Just as we can ask why there are historical differences in the rate of left-handedness in Europe and North America, so also we can ask why, even at the end of the twentieth century, there were large differences in the incidence of left-handedness in Europe, America, Asia and Africa.[5]

In biology differences between groups can only be explained by differences in genes or environment. Since right- and left-handedness are under genetic control, it is possible that the historical and geographical differences described above are due to differences in the incidence of genes associated with left-handedness – the more C genes there are in a population, then the more left-handers there will be. Although this is possible, most social scientists would probably vote in favour of environmental factors, such as differences in the social environment, as affording a better explanation. There are many social problems involved with being left-handed in a right-handed world, problems that have been greater in the past than they are now, and which may well be greater in some non-Western countries, where a more traditional approach to the symbolic use of the right hand for certain activities persists. Broadly speaking, all such environmental effects can be called 'social pressure'. The result is that people may be born left-handed but are effectively forced to become right-handed for most activities. In surveys and studies they are therefore counted as right-handers.

Distinguishing the role of genetic factors from that of social pressure in causing geographical and historical differences in handedness is not easy – be there only a few C genes or lots of social pressure the outcome is exactly the same: not many left-handers. The challenge, then, has been to find a way of distinguishing genetic from social factors. This was a problem that Phil Bryden and I chewed over many times while visiting each other. In July 1996, while visiting Waterloo, just a month before Phil died suddenly while at a conference in Montreal, Phil and I realised how to get around the impasse. Social and genetic explanations can be distinguished so long as one knows how handedness runs in families; and we knew precisely that for Phil's Canadian and Indian data. In order, though, to understand those results, let me first explain a little of the theory, using the three imaginary countries of Euchiria, Lowgenia and Hipressia.[6]

Euchiria is a country where a left-hander can be a left-hander, social pressure to be right-handed being unknown; and it so happens that exactly ten per cent of Euchirians are left-handed. The genetics of hand-edness are the same as those described in chapter 7, and so handedness runs in families to a moderate extent. As genetic theory predicts, 7.8 per cent of the children of two right-handed parents are left-handed, com-pared with 19.5 per cent when one or both parents are left-handed. In other words, left-handed children are 2.5 times more common when a parent is left-handed than when both parents are right-handed. Because of the way handedness is inherited twenty per cent of the genes in the gene pool are C genes (that is, twice ten per cent) and eighty per cent are D genes.

Lowgenia, like Euchiria, does not pressurise left-handers to be right-handed. However, left-handers are somewhat rarer because the propor-tion of C genes is lower, being half that in Euchiria. Because only ten per cent of the gene pool consists of C genes, the proportion of left-handers in Lowgenia is five per cent, rather than the ten per cent found in Euchiria. That may seem obvious enough, but what is not so intuitive is the way handedness runs in families in Lowgenia. The basic principle is the same but it has a surprising consequence. When both parents are right-handed, 3.8 per cent of the children are left-handed; about half the proportion found in Euchiria. However, if a parent is left-handed, sixteen per cent of the children are left-handed, a figure hardly different from Euchiria. The outcome is that children in Lowgenia are 4.2 times more likely to be left-handed if a parent is left-handed than if both parents are right-handed. Left-handedness, therefore, runs more strongly in families in Lowgenia than it does in Euchiria, where the comparable figure was 2.5 times. As the number of C genes falls, so handedness runs *more strong-ly* in families. Although not actually a paradoxical result, it does at first seem counterintuitive.

Hipressia is not a good place for left-handers to live. Hipressians don't like left-handers and exert strong social pressure on them from birth to behave exactly like right-handers. Despite this pressure, only half of left-handers end up as right-handers. Hipressia has the same proportion of C genes as Euchiria, twenty per cent of the gene pool, so that, without social pressure, ten per cent of Hipressians would be left-handed. However, social pressure forces half of these to be right-handed, meaning that social surveys report that five per cent of the population is left-handed, precisely the same proportion as in Lowgenia. The way handed-ness runs in families in Hipressia is, however, rather different from that in Lowgenia. Amongst the children of two right-handed Hipressian parents, 4.5 per cent are left-handed, but if a parent is left-handed, then

9.9 per cent of the children are left-handed. A Hipressian child is there-fore 2.2 times more likely to be left-handed if they have a left-handed parent.

Left-handedness therefore runs *less strongly* in families in Hipressia, with its ratio of 2.2, than in Lowgenia with its ratio of 4.5, or in Euchiria with its ratio of 2.5. The best way to see what is happening is to realise that those Hipressian left-handers forced to behave like right-handers by social pressure are indistinguishable socially from right-handers, *but they still carry the genes of left-handers.* Many children who apparently have two right-handed parents actually have a parent who was originally left-handed and who continues to carry *C* genes that can be passed on to their offspring. Even taking into account that social pressure will make half of all genetically left-handed children become right-handed, in Hipressia there are still rather more left-handers among the children of two right-handed parents than in Lowgenia (4.5 per cent *vs* 3.8 per cent). Social pressure in Hipressia also means families with a left-handed parent have relatively fewer left-handed children than in Lowgenia (9.9 per cent *vs* sixteen per cent). Put the two effects together and Hipressia and Lowgenia are very different. We now have a way of knowing whether a low rate of incidence of left-handedness, as in Lowgenia or Hipressia, is due to fewer genes or to social pressure. If handedness runs more strongly in families, then fewer genes is the answer, as in Lowgenia, whereas if handedness runs weakly in families, then social pressure is the answer, as in Hipressia.[7]

Having found a way to distinguish genetic factors from social pressures, Phil and I could then decide whether real countries with lower rates of left-handedness, such as India and Japan, were more like Lowgenia or Hipressia. The answer for Phil's data from India and Canada was exceed-ingly clear. Left-handedness was not only rarer in India but it also *ran more strongly in families.* India, therefore, was like Lowgenia. The same result was subsequently found for the Japanese data, for data from the United Arab Emirates, and for other researchers' data from the Ivory Coast and the Sudan. All these countries behaved like Lowgenia, and nowhere could we find evidence of countries behaving like Hipressia. Even though social scientists might have predicted the converse, the con-clusion was clear enough – left-handers are less common in India and other countries because there are fewer left-handed genes, not because of social pressure. This doesn't mean that social pressure is unimportant, only that it might work in more subtle ways than the crude manner sug-gested in our hypothetical Hipressia. To see in what ways such pressure may have been exerted, we need to look at the historical data from Western Europe and North America.[8]

If the past is another country, is that country Hipressia or Lowgenia? Studies of handedness in families in the West date back to the early twentieth century. One, in 1927 by Chamberlain, asked a large group of students entering Ohio State University about their own handedness and that of their siblings, most of whom would have been born between 1900 and 1920. Of 7714 children, 368 were left-handed, a proportion of 4.8 per cent, similar to both Ogle's 1871 value and that of the *National Geographic* respondents born before 1920. We need, though, to look more carefully at the family data before we can say whether this low rate is due to genes or social pressure. The results were as follows. For two right-handed parents, 4.3 per cent of children were left-handed, compared with 12.3 per cent, if a parent was left-handed. A child was therefore 2.9 times more likely to be left-handed if he or she had a left-handed parent. Chamberlain's result was not exceptional. In five family studies of children born before 1939, 7.3 per cent of whom were left-handed, the child of a left-handed parent was 3.29 times more likely to be left-handed. By contrast, in family studies of children born after 1955, of whom 13.3 per cent were left-handed, the child of a left-handed parent was only 1.64 times more likely to be left-handed. If that modern figure of 1.64 represents Euchiria, then the earlier figure of 3.29 is like Lowgenia rather than Hipressia. The lower rate of left-handedness at the beginning of the twentieth century was due to fewer *C* genes and not due to social pressure. There is an exact parallel between modern India and Japan, and the West a century ago.[9]

Western populations at the end of the twentieth century seem to have more than twice as many *C* genes as they had at the beginning. Something must have caused that change. Genes increase in frequency if they are more successful and decrease if they are less successful. Success means only one thing: having more offspring. We need to know, then, whether left- and right-handers have the same numbers of children, and fortunately we can work that out from historical studies. In families where children were born after 1955, two right-handed parents had an average of 2.49 children, a right- and a left-handed parent averaged 2.6 children, and two left-handers averaged 2.57 children. The differences are small, left-handed parents having just slightly more offspring. In the first half of the twentieth century, however, the picture is very different. Families were larger overall, and two right-handed parents averaged 3.1 children. However, if one parent was left-handed there were only 2.69 children, and if both parents were left-handed, then there was an average of only 2.32 children in the family. In other words, two right-handers had thirty-four per cent more children than two left-handers. That is a surprisingly large difference, which might well have affected the proportion of genes in the population associated with left-handedness.[10]

Why did left-handers have fewer children, and how is this finding compatible with our earlier conclusion that low incidences of left-handedness, geographically or historically, are not due to social pressure? The answer is that the social pressure described in Hipressia was *direct* social pressure – an individual left-hander was forced to be right-handed, perhaps by being beaten or otherwise abused at school, home or elsewhere in society. Pressure, in other words, was directly placed on individual left-handers. Changing handedness is not easy though, so, unsurprisingly, relatively few left-handers ever change as a result of direct pressure. Much more subtle is *indirect* social pressure.

Indirect social pressure is a far more insidious force than direct pressure. A little gentle gossip, a few carefully aimed barbs, a sneer or two at the right moment, and people find themselves ostracised, isolated, spurned and, as a result, possibly even no longer considered as a possible sexual partner. Within small, pre-modern, pre-technological societies, the effects of such pressure could be quite large, particularly in a world where most people married partners who had grown up within the neighbouring ten or twenty miles. Indirect social pressure does not stop people altogether from finding a partner or having children, but it might delay the process, and the later one starts to have children the fewer one is likely to have. Such effects would have been greater in the past, when families were larger and birth control less effective. Such are the subtle ways in which indirect social pressure might act. Presumably, if there is evidence that this process was at work at the beginning of the twentieth century in the West, it may also be acting in the less-developed world at the beginning of the twenty-first century.

Although it is obvious from Figure 9.1 that there were large changes during the twentieth century in the incidence of left-handedness, it is also clear that, by the end of the century, the incidence of left-handedness *had stopped rising*. There seems to be a natural or 'proper' rate of left-handedness in the population. The indirect social pressure exerted by our Victorian forebears may have prevented left-handers having children, but when those pressures were removed the rate of left-handedness rose and returned to what looks like its 'proper' level. There seems to be some sort of balance between the proportions of right- and left-handers, with ten to twelve per cent being a stable level. That, though, begs many questions concerning why left-handedness occurs and how the incidence is determined; and from these arise bigger questions about how and why right-handedness and left-handedness evolved over much longer time-scales than a single century.

Working out why something happened requires one to first know how and when it happened, which isn't always easy to discover when it comes to the history of handedness. Apart from that one unreliable estimate concerning Old Testament Israel, there are no written data before the late nineteenth century. There are, however, other ingenious sources of evidence. One such is art, in which people have been portrayed carrying out a variety of actions. Stan Coren and Clare Porac, psychologists in Vancouver, looked at paintings, drawings and sculptures in which a person used one hand or the other to carry out some skilled action, such as throwing a spear. They looked at over a thousand such pictures from a range of cultures, the earliest dating from before 3000 BC. Figure 9.2 shows that the vast majority of people in such pictures used their right hand. However, about eight per cent used the *left* hand, a proportion remarkably similar to modern estimates of the rate of left-handedness. Furthermore, that figure is remarkably constant across five thousand years, with perhaps just a slight decrease in the nineteenth century.[11]

Handedness can also leave its traces in various artefacts. The silver spoon in Figure 9.3 is part of the Mildenhall Treasure, dating from fourth-century Roman Britain; one of thirty-four objects discovered in 1942 when a Suffolk farmer ploughed his field four inches deeper than usual. The spoon's handle has been produced by holding one end of the metal while twisting the other. A right-hander would do this by using the left hand to hold one end firm while twisting the other end clockwise, using the powerful supinator muscles of the right forearm. The result, confusingly, is a left-handed twist in the metal. The spoon is kept in room 49 of the British Museum in London, along with other metalwork from Roman Britain, and Celtic and Bronze Age Europe. In a selection of torques, necklaces, bracelets, and domestic items such as bucket handles, forks and the like, dated to between 1600 BC and 400 AD I found seventy-nine items with a left-handed spiral. Eight, however, had a right-handed spiral, slightly over ten per cent, the ancient left-handers thus having left a record of their existence.[12]

Evidence of an earlier left-hander still comes from Ötzi; the 'man in the ice' who lived about 3200 BC and died high in the Italian Alps. He was buried in the ice of a glacier, where he lay for over five thousand years until, on 19 September 1991, the ice melted sufficiently for his well-preserved body to be found. Amongst the artefacts he was carrying were several twisted cords produced by a right-hander, most probably Ötzi himself (Figure 3.3). Ötzi also carried a bow, along with arrows in a quiver, two of which were finished with a stone arrow-head. The feathers at the tail of the arrow were held on by fletching, a thin cord wound around the shaft of the arrow. When right-handers fletch, they invariably wind in a left-handed spiral, as was the case with one of these arrows, but the other has a right-handed spiral, suggesting that it was made by a left-handed fletcher.[13]

Stone tools can also indicate left-handedness. Various tools at a site in

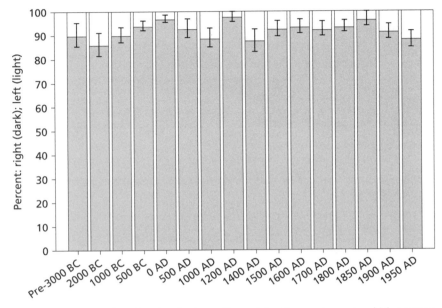

Figure 9.2 The proportion of works of art over the past five thousand years in which a person is using the right hand (grey bars) or the left hand (white bars) to carry out an action. The error bars indicate ninety-five per cent confidence intervals.

Belgium, dating back about 9000 years, were 'refitted', that is, they were put back together to discover the single piece of stone from which they originated. One such stone was used to produce several tools, two of which were fine-pointed borers. Detailed microscopic analysis of the marks left on the tools showed they had been turned anti-clockwise, as a left-hander would do. Taken overall, about five per cent of the tools at the site were probably used by left-handers.[14]

At this point, the trail of left-handers starts to run cold, although there is one further piece of evidence from Boxgrove in England, just south of the Sussex Downs. Five hundred thousand years ago, this would have been an area of freshwater marshes, inhabited by rhinoceros and giant deer. About 150 stone hand-axes, used for butchering such animals, have been found here. So well preserved is the site that, in three cases, one can see the chippings left behind as the axe was knapped. In one case, the 1715 separate chips show a denser pattern on the right-hand side, suggesting the knapper was left-handed.[16]

Even if left-handers have been around for a long while, the majority of humans are right-handed and have been so for at least two million years; back, that is, to the days of *Homo habilis*, the precursor of modern *Homo sapiens*. Several things suggest that *Homo habilis* was right-handed. Like modern man, *Homo habilis* had troubles with food getting stuck between

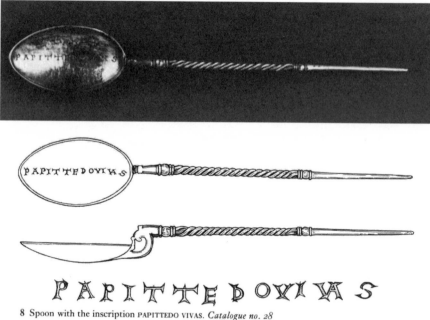

8 Spoon with the inscription PAPITTEDO VIVAS. *Catalogue no. 28*

Figure 9.3 Silver spoon from the Mildenhall Treasure in the British Museum.

the teeth, so an early invention was the toothpick, which, from the wear patterns found on teeth one and a half million years old, seem to have been mostly held in the right hand. Occasionally, the bones of a skeleton can also suggest handedness, although a problem here is that bones other than the skull are rarely fossilised, particularly in the pairs needed to compare right and left. One skeleton, found in Kenya, at Nariokotome and dated to 1.6 million years ago, did have both clavicles (collar bones) preserved, and these showed where the powerful deltoid muscle from the shoulder was attached. The muscle on the right was larger than that on the left, suggesting a stronger right arm and hence right-handedness.[17]

Much better preserved than bones are stone tools, the near indestructible relics of the first beginnings of human culture. W. H. Auden recognised our debt to their unhonoured begetters:

> …what a prodigious step to have taken.
> There should be monuments, there should be odes,
> to the nameless heroes who took it first,
> to the first flaker of flints,
> who forgot his dinner…
> where should we be but for them?

Those tools also tell the handedness of the flakers of flint. Most right-handers use the left hand to hold the stone core while pieces are removed by a hammer stone held in the right hand. The hammering imparts a slight twisting movement to the core, and the chips that fly off have a slight but definite twist, allowing one to tell the handedness of the knapper. The anthropologist Nicholas Toth looked at chips from Koobi Fora, a site in Kenya dated to 1.8 million years ago, and found the chips were typical of those produced by right-handers, suggesting right-handedness was predominant.[15]

Our earliest ancestors, then, like us, were mostly right-handed. What of their ancestors? That is a key question for understanding the evolution of handedness, and one that is keenly debated at present. To answer it, we need to ask whether other animals besides humans exhibit 'handedness'. Broadly speaking, the answer is yes, although only humans show right-handedness. Take a cat, for example. Most cats hook food from the bottom of a tin using one paw rather than the other. Like humans, there-fore, they show a preference for one limb. Where cats differ from people is that half prefer the right paw and half the left paw. In contrast, ninety per cent of humans would use the right hand. Humans, in other words, show right-handedness whereas cats do not.

Human handedness runs in families, whereas that does not seem to be the case in animals, two left-pawed cats being as likely to have a left-pawed kitten as two right-pawed cats. Put in terms of the genetic model described earlier, it is as if all animals had a double dose of the C gene. I say 'as if' because, strictly, the C gene is a null gene that does precisely nothing, its effect being to let fluctuating asymmetry determine handed-ness. It is easier, however, to talk as though some sequence of DNA has a causal effect. In such a scenario, right-handedness, and the D gene that causes it, are specifically human characteristics.[18]

That, then, is the simple story. However, in the past twenty years researchers have found a host of subtle asymmetries in the way animals use their right and left arms/legs/paws/wings/horns/claws/flippers or whatever. These undoubtedly are cases of handedness, but the controver-sial question is whether they are right-handedness. Take, for example, work among apes, our nearest living relatives. The handedness of gorillas is well documented, thanks to the work of Dick Byrne, who spent many hours observing the eating habits of mountain gorillas in Rwanda. These obtain their favourite plant foods via a complex sequence of actions. For instance, when eating the leaves of a nettle, they hold the base of the plant tightly with the right hand, put the left hand loosely around the base, and then slide the left hand up the stem to form a bunch of leaves, which they then grip with the right hand (see the flow diagram in Figure

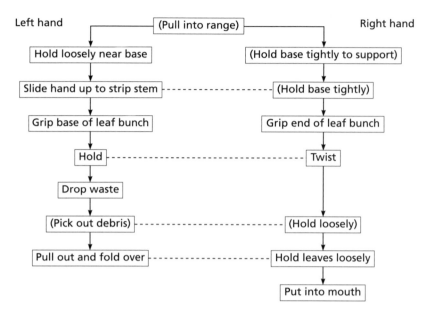

Figure 9.4 How a right-handed mountain gorilla uses the right and left hands to harvest the leaves of a nettle. For a left-handed gorilla right and left are reversed.

9.4). For a left-handed gorilla, the entire sequence is back to front. Just as with cats, half of gorillas are right-handed and half are left-handed. Again, they show handedness, but not right-handedness. Chimpanzees, however, are different. Studies in the wild suggest that about half of chimpanzees prefer to use their right hand and half their left. However, given more exacting tasks in the laboratory, there is a small but significant tendency for more chimpanzees to use the right rather than the left hand, particularly in bimanual tasks where one hand supports and the other acts upon an object. Chimpanzees may, therefore, show right-handedness. However, if this is so, the excess of right-handers over left is still fairly small, only about sixty per cent being right-handed; not nearly as extreme as the nine right-handed humans for every one left-hander.[19]

The genetically close relationship between chimpanzees and humans makes it tempting to assume that any tendency towards right-handedness in chimpanzees must be due to the same factors that cause right-handedness in humans. It may also lead us to speculate about the existence of some common ancestor, eight or ten million years ago, who was also right-handed. Neither hypothesis, however, is reasonable. There may have been a common right-handed ancestor, but it is equally possible that right-handedness evolved in chimpanzees half a million years ago, long after separating from the human line. Right-handedness in chimps and in humans may be entirely separate, having no common evolutionary origin. Just because two things appear to be similar does not mean they evolved in the same way and at the same time. A lizard living 290 million years ago may have been the first animal to have run on two legs, but that does not mean that the human ability to walk on two legs is somehow related to it. The lizard's walking ability is an *analogue* of human walking, not a *homologue*.[20]

If chimpanzees were the only animals showing asymmetries at the population level, then it would be easier to believe that their right-handedness had something to do with human right-handedness. However, further population-level behavioural asymmetries have been found in other animals. The laboratory mouse is a good example, since measuring handedness in mice is fairly straightforward. Some attractive food is placed in a narrow perspex tube mounted in the wall of the cage, the animal reaches into the tube and takes the food with one paw or the other, and the researcher counts how many times out of fifty the right paw is used. Most mice have an individual preference but, for a long while, it was thought half were right-handed and half left-handed. As more strains of mice were studied, however, so it became clear that some strains do show right-handedness, although, as with chimpanzees, the effect is small, fifty-five to sixty per cent preferring the right paw. Similar effects can be found in a wide range of animals. For instance, humpback whales like to slap one of their two flippers on the surface of the sea, and for three-quarters of them it is the right one that they choose. Chimpanzees, mice and whales are, obviously, all mammals. As we shall see, though, asymmetries can be found in a much wider range of animals.[21]

In the early twelfth century, the Chinese emperor Hui Tsung commissioned his artists to paint a portrait of the peacocks in his garden. The emperor, however, shook his head when in the finished work he saw a peacock lifting its right foot to step on to a flower bed. 'A peacock always raises its left foot first to climb,' he said. Parrots similarly show left-footedness, and even domestic chicks prefer to scratch the ground with the right foot. Some toads can show consistent asymmetries, the common toad preferentially using its right leg to remove an object on its snout, although the green toad and cane toad do not. The mosquito fish prefers to detour to the left around objects, as do some other species of fish. However, although some species of toads and fish show these asymmetries, other quite closely related species show no asymmetry at all, suggesting that such asymmetries are quite idiosyncratic. Asymmetries are not restricted to living species. In the floodplains of the South African Karoo, 240 million years ago, lived a mammal-like reptile, *Diictodon*, which looked something like a dachshund with a lizard's skin. It dug unusual helical burrows, some of which have been fossilised to produce a perfect cast known as a Devil's Corkscrew. As with corkscrews in general, the burrows are right-handed, suggesting that *Diictodon* was also asymmetric. The record for the oldest-known behavioural asymmetry goes to the trilobites, some of which lived over 500 million years ago. Many fossil specimens exhibit signs of damage that look as though they may be due to attacks by would-be predators. The injuries are two to three times more likely to be on the trilobite's right side. Whether they reflect asymmetry in the behaviour of the trilobite or its predator is not yet clear.[22]

If interpreting behavioural asymmetries is problematic, it is no more so than interpreting the far commoner anatomical asymmetries that pervade so many organisms. Sometimes these are very obvious, forming the basis for speciation, as in the flatfish, discussed already in chapter 5, which, as Darwin said, 'are remarkable for their asymmetrical bodies'. There are two different ways in which a flatfish might lie on the sea-bed; either on its left or right side. The most primitive flatfish show fluctuating asymmetry, half of the fish in a species lying on their right side and half on their left. Soles and flounders are almost entirely consistent in the side they lie on, with some species all lying on the right and others all lying on the left. Flatfish are not the only species in which asymmetry forms the basis for speciation. Foraminifera ('forams') are tiny, ubiquitous oceanic protozoans, which, when fossilised on the sea-bed, are used by biologists and geologists to date strata. There are hundreds of different species, which differ principally in the shape of their shells. In one case, *Neogloboquadrina pachyderma*, there were known to be right- and left-handed forms, these being mirror-images of one another. Conventional wisdom said that the species took one form when the water was warm, and the other when it was cold. Only with the advent of molecular genetics was this interpretation shown to be totally wrong. Just like right- and left-handed flatfish, the right- and left-hand forms of *N. pachyderma* are in fact different species with different DNA.[23]

Examples of anatomical asymmetries in animals could go on for ever, for they are varied and fascinating. Snail shells mostly spiral clockwise, but occasionally spiral anti-clockwise; the male fiddler crab and snapping shrimp have one front claw that is massively enlarged in relation to the other, probably a result of sexual selection; hermit crabs are specialised for living in right- or left-handed shells, the two forms having diverged about eighty million years ago; some butterflies have a left forewing that is larger than the right, which makes the animal fly round in circles, which increases its chance of attracting a mate; the asymmetrically winged cricket 'sings' with the left wing on top of the right; owls have their ears at different heights so that they can pinpoint sounds without moving the head; and last, but far from least, the garter snake, instead of having one penis, has two hemipenises, the right being larger than the left.[24]

Lord Rutherford, a physicist, famously said that 'All science is either physics or stamp collecting.' Collecting lists of asymmetries may be fun, but the advance of science depends on proving or disproving theories. Why has handedness developed, both in humans and in animals? The answer, at least in part, lies in what the pioneer eighteenth-century economist Adam Smith called 'the division of labour'. He used the example of pin-making, a task then done by hand and involving about eighteen distinct operations. Working alone, a person could make about twenty pins a day. With specialisation, each person concentrating on one or two of the actions, Smith found that ten persons could make forty-eight thousand pins in a day – nearly five thousand each. The increased productivity resulted in part from the fact that '[T]he improvement of the dexterity of the workman necessarily increases the quantity of the work he can perform; and the division of labour...necessarily increases very much the dexterity of the workman.' Practising just one task lets one become far better than one can be by practising multiple tasks. We can apply this reasoning to the 'division of labour' between the two hands. The chimpanzees in the Gombe National Park, Tanzania, beautifully illustrate the point. They 'fish' for termites by delicately inserting a thin twig or piece of grass into the entrance of a termite mound. The termites attack the invading object, biting on to it with their mandibles. If the stick is slowly withdrawn, the soldier termites will come with it, and can then be eaten. In November, the peak time of year, chimpanzees 'fish' in this way for two hours a day, devoting over ten per cent of their waking hours to collecting this important food source. Some chimpanzees fish with either hand. Others, however, are more specialised, always fishing with the same hand, and, as Adam Smith would have guessed, they do far better, collecting thirty-six per cent more termites. Specialisation improves efficiency, even between the two hands, and so handedness will always confer an advantage. Notice, though, that it does not matter which hand is used – and, in fact, half the Tanzanian chimps specialise in using the right hand, and half in using the left. Explaining right-handedness involves something different from explaining handedness.[25]

The most obvious characteristic of right-handedness is the term that Adam Smith used: 'dexterity'. The right hand is particularly skilful, carrying out fine, rapid and complex manipulations of objects – as in making pins. In large part, this is due entirely to the brain that controls its movements, but there is another part of the equation that we have almost totally ignored – the hand itself. It is a wondrous organ, without which human handedness would be a paltry thing of little consequence, akin perhaps to the behaviour of those whales that flap their right flipper against the surface of the sea. It is time to put the hand back into handedness.

Aristotle described the human hand as 'an instrument that represents many instruments'. Whereas other animals have limbs specifically shaped by evolution for climbing, running, flying, swimming, or even just hanging around, as in the sloth, humans have no such specialisation, the human hand being developed for a range of activities. It is the perfect all-purpose tool, the Swiss Army knife of limbs, so adaptable that it solves problems such as hunting, running, flying, swimming, and so on, by making machines that do *all* those things. It needs, of course, to be connected to a brain that can control such a tool, but the brain would also be far less useful without such a hand connected to it. Brains and hands co-evolved.[26]

Sir Philip Sidney, the sixteenth-century poet and soldier, wrote a whimsical poem about a group of pre-lapsarian animals who, despite living in perfect harmony, nevertheless begged Jove for a king to rule over them. Each animal contributed its best part to the new ruler, and so the lion gave its heart, the elephant its memory, the parrot its tongue, the cow her eyes, the fox its craftiness, the eagle its vision, and finally, the ape gave 'the instrument of instruments, the hand'. The result, predictably, was a disaster, for the newly created king turned out to be Man, who indeed ruled over the beasts of the earth, but not to their benefit. The fable puts the human hand in its proper biological place, for if one were to look anywhere for it, the ape would be the best place. Figure 9.5 shows a wide range of ape and primate hands, including the gibbon (*Hylobates*), the orang-utan (*Pongo*), the chimpanzee (*Pan*), and the gorilla. All the hands are serviceable in their particular ways, but none is functionally as good as the human hand. What is so special about the human hand?

Figure 9.5 doesn't quite show how the hands of primates differ from each other. Our own hand, *Homo*, looks like all but two of them – *Ateles*, the spider monkey, and *Colobus*, the colobus monkey – in having four fingers and a thumb. The basic layout of the human hand is not therefore in itself exceptional, being yet another variation on the highly adaptable vertebrate forelimb. The features that really distinguish it are found in its functions. Put your hand flat on the table. You should be able to see all five fingernails. Now touch the pad of your thumb against the pad of your index, middle, ring and little fingers. That action is unique to humans, and is called *opposition*. Look carefully at your fingers and thumbs while doing it and you will see that the nail of the thumb and of

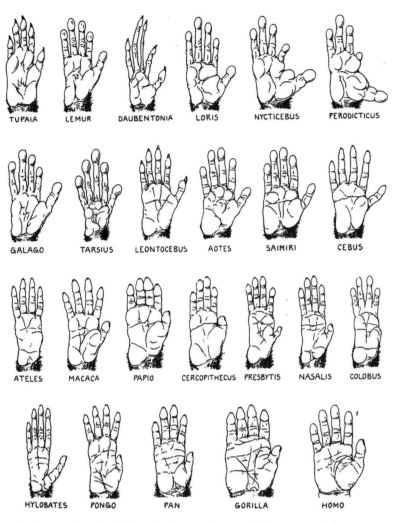

Figure 9.5 The right hands of different primates. *Lemur* to *Tarsius* are prosimians; *Leontocebus* to *Ateles* are New World monkeys; *Macaca* to *Colobus* are Old World monkeys; *Hylobates* to *Gorilla* are apes. *Tupaia* is not now recognised as a primate by most authorities.

the finger are now on opposite sides – if you can see one you can't see the other. Opposition occurs because the thumb rotates, which is unique to humans; and the little finger also rotates to some extent – so-called ulnar opposition. An opposable thumb and an opposable little finger together produce the two important grips of the human hand – the *precision grip* and the *power grip*. Picking up a needle or an egg, wielding a hammer, unscrewing the tight cap of a jar – all require opposition of the thumb or the little finger to make the grip work. Apes can pick up objects but not in the precise or forceful way that humans can.[27]

Most primates have *prehensile* hands, meaning that the fingers wrap around objects and can hold them. This ability is vital if one's life mainly involves moving around in trees from branch to branch – and, indeed, many primates also have prehensile toes, or even, in New World monkeys, a prehensile tail, which can also grip branches. Such finger-grasping is not, however, sufficient for skilled tool usage, as patients with an amputated thumb will affirm. Skilled movements need precision grips and power grips. A pencil is held between the pads of the thumb, first finger and second finger – the 'three-finger chuck' form of the precision grip. A variant used for precise tasks with larger objects, such as putting a screw-top on a large jar, has the pads of all five fingers arranged around the object – the 'five-finger chuck'. Hammering or digging uses the power or squeeze grip, in which the fingers curl around the handle from one side and the thumb curls around from the other side, reinforcing the hold. Particularly crucial is the way the entire hand tips to the side of the little finger, so forearm and hammer are aligned. In a fencer, the arm and the sword together form a single straight line from shoulder to sword tip. A chimpanzee cannot adopt that posture, so always wields a stick at an angle to the forearm and cannot use it with precision.[28]

The human hand has evolved in several stages. The hand of *Australopithecus*, dating back three to four million years, differs from modern humans and chimpanzees, and yet the thumb is surprisingly modern, suggesting that the three-finger chuck may have been possible. However, the little finger is very unhuman, being unable to rotate towards the palm, so these early hominids were not able to perform a five-finger chuck or power grip. Using large stones for hammering other stones to make tools would have been impossible (confirmed by the fact that stone tools have never been found with *Australopithecus*). Nonetheless, the three-finger chuck would have been advantageous for some skilled tasks, perhaps even for throwing small stones. Evidence that *Australopithecus* could throw is also found in the pelvis. A key part of throwing involves placing one foot forward and rapidly rotating the pelvis, so that the weight of the whole body is behind the throw. Chimpanzees can't do that, because the *gluteus maximus*, the large muscle comprising the buttock, is attached to the pelvis in a different way. The *gluteus maximus* in *Australopithecus* was arranged in a way that would have made overarm throwing possible.[29]

When the power grip, so essential for handling tools and heavy objects, finally emerged is unclear. Fossil remains of complete hands are exceptionally rare. The single specimen of a *Homo erectus* hand found (at Olduvai Gorge in Tanzania, dating back around 1.75 million years) might have been able to form a power grip, but the evidence is not completely compelling. There is, then, a frustratingly long wait until the next fossil hand, which is Neanderthal and only 50,000 years old. However, even though fossil hands are wanting, stone tools made by human hands are common and well preserved. Something must have dramatically changed to make the creation of such tools possible, and we must assume that the five-finger chuck and power grip must have existed two or even two and a half million years ago. The hand became a sophisticated tool, capable of adapting to the needs of a technological society.[30]

A sophisticated tool needs a sophisticated controller, for as the celebrated anatomist Frederic Wood Jones put it, 'What we are admiring in the multitude of actions of the useful human hand is the human cerebral perfection, not the bones, muscles and joints.' Brain and hand each, in fact, need the other for, in the words of the novelist Robertson Davies, 'the hand speaks to the brain as surely as the brain speaks to the hand'. The evolving hand and brain were in continual dialogue. Three million or so years ago, the brain of *Australopithecus* weighed in at about 450 grams, whereas by one and a half million years ago, *Homo erectus* had a brain weighing twice as much. By half a million years ago, early *Homo sapiens* had a brain weighing 1300 to 1400 grams, similar to modern humans. This evolution of the brain is central to understanding handedness, for the evolution of the hand alone cannot account for the existence of *right*-handedness, since never in human evolution does the anatomy of the right hand significantly differ from that of the left. Right-handedness must have come from the evolution of the brain.[31]

Between three and two million years ago the human brain must have become asymmetric. It was, of course, already in two parts, the right- and left-cerebral hemispheres, connected by the giant bundle of nerve fibres we call the corpus callosum. Although fairly large and fast, the corpus callosum is slow and of limited capacity given the vast amount of nervous traffic within each hemisphere. Imagine a large global corporation with headquarters in the northern and southern hemispheres, each with its own powerful computer but the two connected by old-fashioned telephone lines. The two computers would be able to work together but they would struggle to work *as one*. Communication delays and the limited capacity of the phone lines would not allow proper integration. The best strategy would be for each computer to specialise, each dealing with matters of interest in its half of the world, with high-level tasks divided between the systems, one handling, say, production, and the other dealing with retailing. Each system would need to communicate only the data specifically needed by the other, thus requiring far less network capacity between the two. A risk for any such system would be the problem known as 'lock-out'. There are stories of early rockets with two computers on board in which the countdown reached zero but nothing happened because each computer was waiting for a signal from the other. The solution, here, is to nominate one computer as 'master' and the other as 'slave'; for any task, one computer has to be dominant, preventing lock-out. This may provide an explanation of why as brains get larger so they become lateralised and one side becomes dominant. As we have seen before, however, it does not explain why the *left* hemisphere controls the right hand and is dominant for language.

Ultimately cerebral lateralisation must result from some difference in the way that the left and right halves of the brain function or are put together. Introductory books on psychology often include a table comparing the processing styles of the two hemispheres, which looks something like this:

Table 9.1 Processing styles of the two hemispheres of the brain	
Left hemisphere	Right hemisphere
Verbal	Visuo-spatial
Symbols	Images
Analytic	Holistic
Intellect	Intuition
Successive	Simultaneous
Serial	Parallel
Convergent	Divergent
Realistic	Impulsive
Abstract	Concrete
Objective	Subjective
Rational	Metaphorical

Although these categories offer a general summary of the difference between the two hemispheres, it is nevertheless at a high level of abstraction. Brains, though, consist of nothing but billions of neurones interconnected by billions of synapses. During development, the only thing that genes can do is specify how many and what type of neurones there will be in a particular area, how they will interconnect with other neurones near them, and which distant areas they will connect to. Since the left and right halves of the brain differ in their functions and there is a genetic basis for that difference, then in the final analysis, this must be due to some low-level difference associated with neurones and their connections. Differences of that sort might eventually make one hemisphere better for the serial processing of symbols and hence for analytic and logical high-level analysis, and the other best suited for the parallel processing of images, but such specialisations cannot themselves be programmed by genes.[32]

The two halves of the brain can only differ genetically in simple properties of nerves and the way they work. Nerves work principally in terms of time, impulses arriving at various points on their surfaces and either dissipating, if too spread out in time, or summating, if close enough together for the nerve to generate an impulse in turn, which then stimulates other nerves. How those impulses spread depends on the network of nervous connections, how close they are together, how many nerves connect with every other nerve, and so on. Anything that affects the timing or the layout of nerve cells could therefore affect the sort of processing carried out, in just the same way as the speed, architecture and memory of a computer determine how it interacts in real time with people or other computers. It is possible that timing differs between the two halves of the brain. This idea has been particularly explored by Mike Nicholls at the University of Melbourne in a series of almost minimalist experiments. A person was asked to listen to a burst of simple white noise, rather like the hiss of an untuned radio, lasting about a quarter of a second. On half the occasions there was a short gap of silence in the middle of the burst of noise, and this gap could be as short as two-hundredths of a second and still be noticed. A variation on the experiment compared the two hemispheres by presenting the sounds to one ear only, and found that the gap can be ten to fifteen per cent shorter when it is heard by the right ear (and hence the left half of the brain). The left half of the brain seems to work slightly faster than the right half, similar results being found not only for gaps in sounds, but also for a flashing light or a gentle touch to a finger.[33]

Timing is critical for the speech and language which so typify the functions of the left hemisphere. For instance, the sounds /b/ and /p/ in the words 'big' and 'pig' are different because in 'big' the larynx starts vibrating about a twentieth of a second earlier than it does in 'pig'. Such small timing differences must be heard by the brain listening to the speech and must be produced by the brain generating the speech. Lorin Elias, at the University of Saskatoon in Canada, has shown that people who are better at hearing the difference between /b/ and /d/ with their right ear are also better at Mike Nicholls' gap-detection tasks. Language may therefore be in the left hemisphere primarily because that hemisphere works faster. Timing not only affects simple phonetic distinctions, but is also crucial for understanding the grammatical sense in a sentence. Read the following two sentences out loud:

The sheriff saw the Indian and the cowboy noticed the horses in the bushes.

The sheriff saw the Indian, and the cowboy noticed the horses in the bushes.

That tiny little comma in the middle makes a huge difference to the meaning. The first sentence can be broken down as '(The sheriff) saw ((the Indian and the cowboy) noticed (the horses in the bushes))'; it is the Indian and the cowboy who notice the horses. The second sentence breaks down in an entirely different way: '(The sheriff saw the Indian), and (the cowboy noticed the horses in the bushes)'; now it is only the cowboy who notices the horses. The sentences have a different grammatical structure, which when spoken aloud, or indeed even when read silently, depends entirely on the slight pause after 'Indian', which is indicated in writing by the comma. A faster mental processor benefits language because a set of sounds or words can have many possible meanings. The more quickly one can sift through these to find the proper meaning, the more efficient will be one's grasp of language. Again, this might explain why the left hemisphere is the one that handles language.[34]

Language is not the only task requiring precision timing; another such is throwing. The role of timing for throwing is seen in an elegant study in which experienced sportsmen threw tennis balls at a target while the position of each joint in their arm was accurately monitored. Figure 9.6 shows typical throws of a right-handed subject using the right and the left arm. The right arm is altogether smoother and more co-ordinated, whereas the left arm shows larger and more sudden changes of direction. Although the tips of the fingers of both arms manage to move in a smooth trajectory, there is a big difference in the precise moment at which the ball is released by the fingers. For a target three metres away, releasing the ball four thousandths of a second too early or too late means hitting a quarter of a metre too high or too low. For primitive man, using stones to hunt, that makes the difference between dinner and hunger. Using the right hand, ninety-five per cent of throws were within a ten millisecond release window, and forty per cent hit the target. Using the left, however, less than sixty per cent were within that time window, meaning the target was only hit about fifteen per cent of the time. Timing differences between the hemispheres could well therefore explain why the left hemisphere is dominant for language and motor skills.[35]

The differences between the right and left hemisphere, despite being best described in terms of high-level behaviours such as language, may

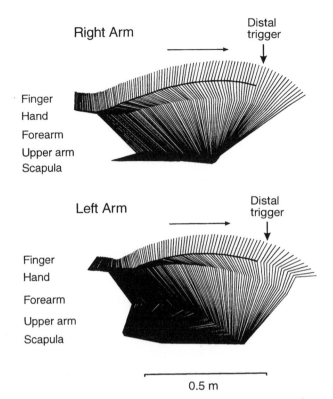

Figure 9.6 Movements of the right and left arm while throwing a ball, recorded from a single subject. The ball is released at the point marked 'Distal trigger'.

ultimately be dependent on the relatively low-level organisation of neuronal interconnections. Several studies have looked at the detailed interconnections of nerve cells in the two hemispheres and found that in humans the left hemisphere is more complex than the right, with less or no difference being found in chimpanzees and monkeys. Whether this complexity provides the neural underpinning for differences in processing speed is too early to tell, but it must be a possibility. A key evolutionary question, therefore, is what happened two or three million years ago that made the right and left hemispheres different, resulting in the development of right-handedness and language. What novel gene could be involved in this, where could it have come from, and how could it be working?[36]

Genes don't come from nowhere. New genes are sequences of DNA that usually start out as genes producing something else. Copies of sequences of chromosomes are made, usually by accident, and so for a while there are two identical sequences doing identical things. Mutation, however, means that this situation doesn't usually last long, one sequence becoming slightly different from the other. If it becomes different in a harmful way, then the individuals carrying the new gene disappear from the population. Sometimes, though, a mutation produces a slight change in the way that one copy of the gene works – perhaps turning on slightly later in development, perhaps recognising slightly different cell types, or perhaps acting in a different organ or tissue. Genes that affect left–right asymmetry are surprisingly rare, with very few well-described cases in vertebrates. There may be hundreds of asymmetries known, but many and perhaps even the vast majority are probably secondary to other asymmetries, of which the heart is the main one. Think about that curious asymmetry of the garter snake mentioned earlier, the right penis of which is larger than the left. It is part of a more extensive asymmetry because the right testis is also larger, as is the case in other vertebrates including humans. However, in those humans who have their heart on the other side, *situs inversus*, it is the left testis that is larger, showing this asymmetry to be secondary to the heart side in humans. No doubt the same will also be found in snakes.[37]

The most important asymmetry in vertebrates is undoubtedly of the heart, and it occurs because of a complex cascade of proteins that trigger each other and eventually make the cells on one side of the early heart tube grow at a slightly greater rate than the other side. From then onwards, the heart will be on the left rather than the right in that individual. The timing of several genes and the location of the tissues in which they find themselves, as well as their specificity for making heart cells, all work together to produce an apparently simple outcome. Imagine that a copy of one of the genes involved in that process mutated slightly so that instead of affecting cells on one side of the early heart tube it instead made cells in the developing brain grow slightly more quickly on the left than the right. That is perhaps all the gene for right-handedness and left-language lateralisation needed in order to change the brain's micro-architecture, making it work that bit faster, with perhaps slightly more useful connections. The rest, as they say, would be history – our history; and a history we are able to tell because we have language to articulate it.[38]

Evolutionary theories are easily spawned and notoriously difficult to refute. Indeed, Steve Jones once said that 'Evolution is to allegory as statues are to birdshit. It is a convenient platform upon which to deposit badly digested ideas.' To be useful, any evolutionary theory has to be testable. Fortunately, this one is – or, at least, will be. If it is correct, then the genes for handedness, once they are found, should be very similar indeed to the genes that determine the side of the heart in vertebrates.

For the moment, we must be content to wait for those genes to be identified.

There is still an important piece missing from the jigsaw of the evolution of handedness and cerebral lateralisation – how do left-handers fit into the picture? Since humans apparently became right-handed because of the evolution of the D gene, and since modern left-handers carry one or more copies of the C gene, it might seem reasonable to suppose that left-handedness is simply due to that ancient gene, the ur- or C gene. That, though, cannot be the case. If the D gene is responsible for speeding up one of the cerebral hemispheres, thus facilitating spoken language, grammar and precision control of tools, then its absence would mean left-handers do not have spoken language, grammar or precise control of tools – an assertion that is patently false. A more subtle view of the evolution of human left-handedness is required. To begin with, we must distinguish the modern C gene responsible for human left-handedness from the ancient C gene, which can be called C^*. The C^* gene was actually a nothingness as far as handedness is concerned; merely an empty cipher. When the D gene mutated into existence, its massive advantage over the C^* gene in the field of language and manual dexterity meant that the C^* gene was rapidly eliminated from the gene pool. The C^* gene became extinct. At that time in our past, therefore, everyone would have carried two copies of the D gene, and would, in consequence, have been right-handed.[39]

If the modern C is not the ancient C^* gene, then the C gene must have come from somewhere else. There is no direct evidence at the moment as to where this might have been, but the most likely explanation is that sometime in the past two million years, the C gene mutated from the D gene. While the C gene maintained the timing advantages of the D gene (or it would itself have become extinct), it managed to do so without language and manual dexterity always being forced into the left hemisphere. However, from then on, as in the modern world, the D and C genes were both present in the population. That, though, raises another important challenge; to explain how two different genes, D and C, can exist in the population at the same time.

Genes continually compete with one another. Theory shows that if one of two genes has even the smallest of advantages over the other, then inevitably and inexorably the fitter gene eliminates the less fit gene, albeit over hundreds or even thousands of generations. That must have occurred when the C^* gene was driven to extinction by the new D gene, with its advantages in terms of language and dexterity. In fact, one gene does not even need to be fitter than the other for one to be eliminated. Even with two genes of identical fitness, one or the other will always ultimately be eliminated from the gene pool owing to random drift. It may take several thousand generations, but in evolutionary time that is a mere instant. Polymorphisms – two different forms caused by two different genes – are therefore inherently unstable. Herein lies a conundrum, since handedness is a polymorphism, yet the proportion of left-handers appears to have been fairly stable for five thousand years, and probably ten or even a hundred times longer. Since random drift or a small advantage for either the D or the C gene would have eliminated one of them, some force must maintain both genes in the gene pool.[40]

Population genetics knows several ways of making polymorphisms stable or 'balanced'. One requires continual new mutations – as occurs with the condition haemophilia – but the C gene is too common for that mechanism to be feasible. The best-understood way in which polymorphisms are balanced is by what is called 'heterozygote advantage'. Individuals with one copy of each gene, heterozygotes (the DC individuals in the case of handedness), are fitter than homozygotes, individuals with either two D or two C genes. The classic example is sickle-cell anaemia: the people with one sickle gene and one standard gene are fitter overall because they are more resistant to malaria than individuals without the sickle gene, and they do not suffer the severe complications of having two copies of the sickle gene. The challenge is to understand what balances the D and C genes. If heterozygote advantage is the causal mechanism, that means DC individuals will have an advantage over DD or CC individuals.[41]

We are now solidly into the realms of speculation, and although empirical data cannot help us much, that doesn't mean anything goes, for a theory has to make scientific sense and be plausible in the light of what we already know about the workings of D and C genes. This means, for instance, that we can reject any idea that left-handers are advantaged over right-handers – we've already seen that modern right- and left-handers have similar numbers of offspring. Another potential error is to confuse the gene that makes left-handedness possible, the C gene, with left-handedness itself. Although it sounds counterintuitive, DC individuals can be fitter than DD or CC individuals without, on average, left-handers being fitter than right-handers. One has to remember that a majority of DC individuals are right-handed, and that CC left-handers will have a lower fitness.

In looking for an advantage for the C gene – and specifically for the DC genotype – a good starting place is the most striking feature of the C gene: its ability to confer randomness on the organisation of the brain, not only for manual dexterity and language (as described in detail in chapter 8) but almost certainly for a host of other cerebral asymmetries, such as those for reading, writing, visuo-spatial processing and emotion. Although it might seem paradoxical, randomness, at least in small amounts, can benefit complex systems. The idea I will present here, which I call the *theory of random cerebral variation*, suggests that human brains are not all alike, some having fundamentally different organisations, and that such variation sometimes produces brains better able to carry out particular complex tasks. It is an unusual theory, since it is usually assumed that all brains are essentially similar, and that differences only occur owing to pathology or aberration. However, this theory celebrates variation and difference, and provides an explanation for the lay belief that some people literally 'think differently' or have their brain 'wired differently'.[42]

If randomness seems an unlikely source of benefit, remember that biology often harnesses random variation in a positive way. The heart of Darwin's theory of natural selection is that mutations occur at random, and that, despite many being disadvantageous, selection picks out the few that are beneficial and maintains them. Likewise, in the immune system, immunoglobulins are produced by randomly combining gene types resulting in small amounts of huge numbers of antibodies, the few that meet antigens then being produced in large amounts by clonal selection in order to fight infection. Random behaviour can also be useful, a rabbit running from a fox turning at random to make the fox less likely to catch it. Complex systems of any sort show an interesting tension between organisation and randomness. Complexity theorist Stuart Kauffman has shown how the right amount of randomness or chaos facilitates many complex processes. Contrast the frozen organisational uniformity of ice, with every water molecule locked to its neighbours so that no movement or change can occur, with the chaotic randomness of water vapour, the molecules flying around so fast that structural organisation is impossible. Between these extremes is liquid water, which allows movement and change, yet also aggregation and organisation. Life itself occurs on the edge of chaos, tiny alterations precipitating larger changes that are maintained without being destroyed.[43]

Central to the possibility of randomness being beneficial is that there is not too much randomness, for otherwise it will destroy the very thing it tries to create. Think of those little puzzles you sometimes see, containing ball-bearings that have to be rolled into shallow holes. Too large a movement and the balls roll everywhere, displacing those already in holes, yet unplaced balls still need something to move them to the right place. A surprisingly good strategy is gently to nudge the puzzle, to shake it slightly, so that the free balls shoot to more useful areas without dislodging those balls already in place. The right amount of randomness is similarly needed in the process of annealing – hardening a piece of metal by heating, cooling and re-heating (injecting random heat energy, in other words) – slow cooling allowing the crystals to form properly.

If *DC* and *CC* individuals have randomness built into their brain development, *DD* individuals in contrast have the cold certainty of an ice crystal. Every *DD* individual has a brain organised in the same way as every other *DD* individual – language (be it speech, writing, or reading), along with fine motor control and praxis being located in the left hemisphere; and visuo-spatial analysis, the recognition of colours and faces, attention and emotion located in the right hemisphere. The standard textbook description, in other words. There is no doubt that this way of building a human brain is effective; probably two thirds of people have such a brain. If *DD* individuals are like ice, then *CC* individuals are like water vapour, their various brain functions flying at random into right and left hemispheres, each with an evens chance of being on the left or right. That sort of radical re-organisation may well not be beneficial, and perhaps explains why anomalies of handedness and brain lateralisation are found in a wide range of conditions, including dyslexia, stuttering, autism and schizophrenia. The different modular components of language probably work better close to one another in the same hemisphere, rather than scattered all over the two hemispheres and having to communicate through the slow and inefficient corpus callosum.[44]

If the few per cent of people who are *CC* are disadvantaged by their chaotic lack of cerebral organisation – relative, that is, to the standard, off-the-peg, one-size-fits-all brain of *DD* individuals – how might *DC* individuals be advantaged over both *DD* and *CC* individuals? The key is that in the *DC* individual there is only a one in four chance of any modular function being in a different place from usual; that is, only a twenty-five per cent chance of handedness or writing being located in the right hemisphere, or of facial and emotional processing being located in the left. We do not know how many modular processes there are, but suppose it is a dozen. In every *DD* individual, all twelve will be properly lined up to the right and left. The average *CC* individual is very different, having six modules in their usual place and six on the opposite side to normal, giving a high probability of inefficient or even impossible connections being needed. The *DC* individual is not quite so haphazard or so atypical. Typically nine of the twelve modules will be in their usual place, and for many *DC* individuals there will be just two or even one module in a different place from usual. For most modules it will therefore be business as usual. The question is whether an occasional atypical module can benefit rather than disrupt.[45]

A recurrent theme in the scientific literature on left-handedness is the claim that talented or gifted individuals in such disciplines as music, mathematics or the visual arts are more likely to be left-handed. Many such claims are undoubtedly spurious, being based on small sample sizes, statistical aberrations, and so on. That, though, does not mean that the idea is entirely without foundation, especially since, at present, we have little theoretical understanding of how such talents may arise. The idea that brains sometimes are literally wired differently may therefore have some potential, particularly if the difference is sufficiently small to avoid disrupting overall performance. Take the simplest case of someone with just one module in a different place from usual. Might that be beneficial?

The answer, of course, depends on which module is randomly located and where. Imagine that a module for understanding three-dimensional space is in the left hemisphere rather than the right, so that it is now located alongside modules involved in fast, accurate, precise control of the hand; that might well benefit drawing or the visual arts, or perhaps ball control in sport. There might also be disadvantages, perhaps a tendency to lose one's way or a clumsiness in certain activities, but as long as the advantages for one task outweigh the costs involved with others, then a rearrangement of brain organisation may benefit that individual. Alternatively, should a module specialised for understanding emotions be located in the left hemisphere rather than the right, so that it now sits alongside left-hemisphere modules involved in the production of spoken or written language, that might be beneficial for writing poetry or being an actor. To ring the changes once more, a typically left-hemisphere module involved in the symbolic processing of language, if located in the right hemisphere alongside modules concerned with processing three-dimensional space, might perhaps make it easier to carry out some forms of mathematics, such as topology.

Although beneficial combinations of modules may occur more commonly in left-handers, and more precisely in *DC* individuals, be they right- or left-handed, there is no need to assume that left-handers overall will be better at a particular skill than right-handers. Imagine that, say, ability at writing songs, is enhanced by having a 'right-hemisphere' module normally involved in prosodics in the left hemisphere, alongside other modules for processing the meaning of words. Such a combination will occur more often in left-handers than right-handers. However, many other cerebral combinations will also occur more often in left-handers, most of which will provide no benefit for writing songs, and some of which may impair that ability. The average song-writing ability of left-handers will therefore show little or no difference from that of right-handers. Among talented song-writers, there may well be an excess of left-handers, but there will also be an excess of left-handers with *poor* song-writing ability. The theory of random cerebral variation is a theory not about the average or the mean ability of right- and left-handers, but about an increased *variability* amongst left-handers; chapter 8 showed that cerebral variability is one of the few undoubted facts about left-handedness, particularly in relation to language.

So far this book has concentrated on individuals who are right- or left-handed. In human societies, however, individuals do not live in splendid isolation, but instead interact socially. When there are choices to be made about left and right, then the ways people interact are important. That is the subject of the next chapter.

☞   ☜

THREE MEN WENT TO MOW

In 1871, Thomas Carlyle (Figure 10.1) was seventy-six years old and, to many, a grand old man of English literature, who had dominated English critical writing and thought for forty years. At the height of his powers, in February 1840, Emma Darwin, the wife of Charles Darwin, wrote of him to her Aunt Jessie:

> I have been reading Carlyle, like all the rest of the world. He fascinates one and puts one out of patience. He has been writing a sort of pamphlet on the state of England called 'Chartism'. It is full of compassion and good feeling but utterly unreasonable. Charles keeps on reading and abusing him. He is very pleasant to talk to anyhow, he is so very natural, and I don't think his writings at all so.

Carlyle's vast and varied output covered biography, history, translation, lectures, social criticism and an enormous correspondence. His views were often controversial, and in the case of his article on slavery, extreme to the point of alienating even his staunchest supporters. In 1866, his beloved but often maltreated wife, Jane Welsh Carlyle, died after a stormy, sometimes successful, marriage of forty years, and in 1871 he was writing his *Reminiscences*, the book with its paean to Jane that, in 1881, his publishers would rush out just three weeks after Carlyle himself died. He was far from happy. 'Very sad, sunless, is the hue of this now almost empty world to me,' is how he described his feelings. Such melancholy was partly due to the loss of all those dear to him, and partly to his own growing infirmity. For a writer, the ultimate tragedy had occurred – he was no longer able to write.[1]

Carlyle's right hand had become 'mutinous', allowing him to write only with 'much puddle and confused bother'. To quote his own words, from his essay on Chartism, 'Were it not a cruel thing to see … [a] strong

Figure 10.1 Thomas Carlyle writing in his sound-proofed room in Cheyne Walk, 1857. It is clear that he is right-handed.

man with his right arm lamed?' The shaking became evident in the spring of 1863, and in March 1869 Queen Victoria commented on it when Carlyle met her. Froude, Carlyle's biographer, described the symptoms:

> a tremulous motion began to show itself in his right hand, which made writing difficult and threatened to make it impossible. It was a twitching of the muscles, an involuntary ateral jerk of the arm when he tried to use it.

Although diagnosis is difficult across a hundred and thirty years, the most likely explanation is Parkinson's disease. Carlyle described the situation well to his brother John, in May 1870, in a letter written in blue pencil, since he could no longer use a pen:

> Gloomy, mournful, musing, silent, looking back on the unalterable, and forward to the inevitable and inexorable...since I lost the power of penmanship, and have properly no means of working at my own trade, the only one I ever learned to work at. A great loss this of my right hand.

By October 1870 this 'worst evil of all' meant he had great difficulty even in drinking a cup of tea. At the beginning of June 1871, he referred again to the 'terrible loss' of the right hand. 'Alas! alas! for I might still work if I had my hand, and the night cometh wherein no man can work.' Later that month, he took Froude a lárge parcel of papers on which he could no longer work; it was the *Reminiscences*.[2]

On 15 June 1871, Carlyle went out for a morning walk. His appearance to his neighbours 'was unmistakable ... walking slowly along the Chelsea embankment towards the town, eyes down cast upon the ground before his feet – a meditative walk'. The night before had been bad, as his Journal records: 'this morning ... out walking, unslept, and dreary enough in the windy sunshine'. On that walk his thoughts, perhaps prompted by his mutinous right hand, turned to the wider implications of right-handedness. Nor was it for the first time: 'I have often thought of all that.'[3] His Journal puts down the thoughts which he 'never saw ... so clearly as this morning':

> He that has seen three mowers, one of whom is left-handed, trying to work together, and how impossible it is, has witnessed the simplest form of an impossibility, which but for the distinction of a 'right hand' would have pervaded all human beings.

It is an image remote to us, in these days of highly mechanised agriculture. Words of Tolstoy, in bucolic scenes from *Anna Karenina*, perhaps help us picture it better.

> ... the peasants came into sight, some with coats on, some in their shirts, following one behind another in a long string, each swinging his scythe in his own manner. [Levin] counted forty-two of them ... He heard nothing save the swish of the knives, saw ... the crescent curve of the cut grass, the grass and flower-heads slowly and rhythmically falling about the blade of [the] scythe ... On the short rows the mowers bunched together ... their scythes ringing when they touched.

We must remind ourselves of the fearsome instrument that was a full-length scythe: ''tis sharp as a razor', as Levin is told. A curved piece of steel with a carefully honed edge, several feet in length, swung in a huge arc around the mower. To get one's feet in the way was to risk serious injury. With a team of men scything their way across a field it was vital that they all be synchronised. To have one man doing everything back to front – left-handed, in other words – would be to risk disaster.[4]

Carlyle's comment about the mowers raises a host of issues concerning laterality in a world of interaction. For a solitary mower, there is no problem but when several mowers work together, then difficulties arise. So what are the rules governing social interaction and what problems arise from these? Certainly, left-handers encounter problems at the dinner table. The simple act of drinking soup can be a messy business as the left elbow of the left-hander bumps into the right elbow of the right-hander supping on his or her left.[5]

Whenever two people carry out a lateralised task together, the way one of them carries it out will inevitably affect the other's performance. Society, therefore, has to develop rules for such situations, be they laws, codes, rule-books or forms of etiquette.

Often the simplest solution to problems of handedness is a mere convention that everyone obeys. 'When you meet someone, shake their right hand.' That is easy. Whenever, wherever, whomsoever you meet, the rule works, and it hardly disadvantages anyone. Likewise, 'Lay the table, so the knife is on the right and the fork is on the left.' Again, this is easy, and everyone knows where they are. Left-handers may feel mildly disadvantaged, but the inconvenience is so minimal that few complain. Indeed, one often only becomes aware that a rule exists, when the accepted order of a task is reversed, as might occur at a bridge table if a player dealt the cards anti-clockwise rather than the more usual clockwise.[6]

What, though, about more complicated interactions, such as mowing? In the rest of this chapter we will look at questions such as whether people write from left to right, or vice versa; whether they drive on the right or the left; and whether left-handed sportsmen are advantaged and left-handed surgeons disadvantaged.

Writing

Writing is highly asymmetric. As you read this, so the eye starts at the left of the page, scans across to the right, and then flips back to the left to begin the next line. In time, this process becomes second nature, to the point that those of us who live in Europe or North America might be tempted to conclude there is something 'natural' or 'correct' about it. That, though, would represent a very Western view of the world. Across vast areas of the globe, writing is not from left to right ('rightwards'), but from right to left ('leftwards'), as in Arabic, Hebrew or Urdu. Why, then, is English, the language of this book, a rightward language?

Writing seems to have evolved on only a very few occasions, with only two major foci – in Egypt and the Euphrates basin in the fourth millennium BC; and in China during the second millennium BC, or perhaps somewhat earlier. Early writing was initially *pictographic*, each symbol being a picture of the object it represented (as in early Egyptian hieroglyphs). Later, it became *ideographic*, each symbol still representing a word but the symbol being arbitrary. Such systems can be very inefficient, since every symbol has to be learned, as with Chinese characters and Japanese *kanji*. Schoolchildren spend many years learning the 1000 sinograms necessary for reading ninety per cent of typical texts; to read ninety-nine per cent of texts 2400 sinograms are needed.[7]

Writing has been called 'visible language', and to be efficient it has to bear some relation to words *as they are spoken*. Pictographs and ideographs do not do that. The first system that explicitly set down sounds was a *syllabic* script. Here, each character stood for a syllable (e.g. /ba/, /be/, /da/, /de/, and so on). Since languages have many syllables, they also have to have many signs to represent them, so that Elamite, for instance, had 111 syllabic signs, and the Sumerian/Akkadian system had about 600. Such languages were still unwieldy and inefficient, but they were a good deal better than pictographs and ideographs. The most important step in developing most modern scripts occurred with the development of a new type of writing, the earliest examples of which date to about 1700 BC and were discovered in the turquoise mines at Serabit al-Khadim in Sinai, although others have since been found over quite a wide area. The important thing in this writing was that the number of signs was very small – less than thirty. This meant that they could not represent syllables, since there weren't enough of them, but instead had to be alphabetic, each sign representing a single consonant or a vowel. This is the proto-Sinaitic/proto-Canaanite script, from which a vast number of the world's scripts are derived – including the one in which this book is written.[8]

A problem that developed with syllabic scripts, and then even more so with alphabetic scripts, was that they had to be written in a clear order. Pictographs, on the other hand, can be put down in any old order (and often seem to have been). The difference between 'DOG' and 'GOD' in an alphabetic script depends crucially on whether the letters are read from right to left or left to right. Clearly some system seems necessary, although proto-Sinaitic/proto-Canaanite apparently did not have one, some inscriptions reading from right to left and others from left to right. It doesn't matter if one is only reading a few short words **siht sa hcus**. It would, though, be annoying **etisoppo eht ni sretho dna noitcerid eno ni secnetnes emos gnivah detrats ylmodnar eno fi ro segassap regnol rof**. There would rapidly be a drive to standardise the direction of writing – the 'ductus', to give its technical name – which, indeed, is exactly what happened. By about 1050 BC Early Phoenician was written entirely from right to left, anticipating Modern Arabic and Hebrew, which likewise are read from right to left.

Cultural inertia makes changes in a language more difficult, initial choices being maintained for a long while – a modern analogy being the QWERTY typewriter keyboard, which, however illogical and inefficient, will probably be with us for a long while to come. However, English script is directly descended from Early Phoenician. So what happened? Why is this book read from left to right, rather than from right to left? Like so much of Western civilisation, it goes back to the Greeks. Something happened, so that by the sixth century BC, Greek was written from left to right. Since the Roman script came from Greek, and almost all modern Western European scripts came from the Roman script, English is also written from left to right. Similarly, modern Slavic scripts, Russian in particular, are also written left to right, being descended from the Cyrillic script, developed by Saints Cyril and Methodius in the ninth century AD from Greek.

What happened in the five centuries between Early Phoenician and Classical Greek is not at all clear, there being a gaping lack of evidence. The standard theory is that archaic Greek was written from right to left like Phoenician, that there was then an intermediate stage, and that, eventually, it was written from left to right. There was certainly a reversal of some sort, because by the end of the fourth century BC the Greek comedian Theognetos could make fun of someone 'who wrote backwards'. The problem is determining the form taken by this intermediate stage, which is known as *boustrophedon*, a Greek word describing the way an ox ploughs a field, walking one way, turning at the end, and walking back (Figure 10.2) '[T]o cassay the earthcrust at all of hours, furrowards, bagawards, like yoxen at the turnpaht', as James Joyce put it in *Finnegans Wake*.[9]

Boustrophedon can take two forms, as can be seen in Figure 10.3. The one on the left is *boustrophedon* proper, and the one on the right is *false boustrophedon*. In most written scripts, it is not only the entire direction of writing that is asymmetric but also the individual letters. When writing *boustrophedon*, how does one write the letters that go back the other way? Is it as they are normally written, as in false *boustrophedon*, or in mirror-image, as in *boustrophedon* proper? Both have their problems. Many lower-case letters in English are asymmetric, with only about half a dozen being symmetric. If the 'retrograde' (right to left) part of *boustrophedon* is written in mirror characters, then there are forty-six letters in the alphabet instead of twenty-six. Although reading the mirrored letters seems surprisingly easy, writing them is another matter, and *boustrophedon* would almost double the work load for children learning to write. Since children anyway have problems in distinguishing the mirror-pairs of 'b-d' 'p-q', it will hardly help them if they have to remember that 'b' is pronounced /b/ when going one way and /d/ when going the other.[10]

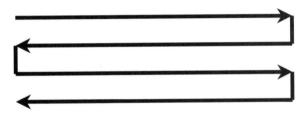

Figure 10.2 *Boustrophedon* – the path of an oxen ploughing a field.

Boustrophedon

uobǝɥdoɹʇsnoꓭ

Boustrophedon

uobǝɥdoɹʇsnoꓭ

Boustrophedon

nodehportsuoꓭ

Boustrophedon

nodehportsuoꓭ

Figure 10.3 True *boustrophedon* and false *boustrophedon*.

What about the other strategy, of using just the standard twenty-six letters and simply putting them in the reverse direction during the retrograde part of the writing, as in false *boustrophedon*? Surprisingly, this is more difficult to read. The problem is seen with letter pairs such as 'ph', which typically represents a single /f/ sound, and is thus read as a single character. Reading the reverse, 'hp', is problematic because the entire appearance of the pair of letters has changed. In reading, there are two separate processes, 'phonemic', looking at individual letters for their sounds, and 'graphemic', treating an entire word as a single visual input. False *boustrophedon* disrupts graphemic reading, and hence is more difficult to read. That is probably why early Greek *boustrophedon* used mirror-characters in the retrograde part of the writing.[11]

Boustrophedon, the writing of the ox, is, as it were, on the horns of a dilemma; either it is easier to read and more difficult to write or vice versa. It is not surprising that it rapidly died out in ancient writing. Perhaps more surprising are moves to reintroduce it. Computers can be programmed so that only the standard twenty-six letters have to be typed on the keyboard, but the screen display or printout has normal or mirror-reversed letters according to the direction of the script. Enthusiasts claim *boustrophedon* is easier and quicker to read because the eye does not have to find its way back to the beginning of the next line, just move down slightly. Whether it will catch on is debatable. Something in its favour is that individuals can use it on their own computers and yet transmit documents which other users can display conventionally. *Boustrophedon* was also used in the Frere and Moon systems of printing for the blind (early predecessors of Braille) because it solved the problem of finding the beginning of the next line, which previously required large areas of blank paper between lines.[12]

A risk when offering any historical description is what has been called 'the Whig interpretation of history'; the easy presumption that everything leads straightforwardly and inexorably to the highest state of humankind. Such an interpretation fails to look at the entire historical picture, ignoring the losers – in our case, the writing systems that became extinct. Figure 10.4 therefore provides a tentative map of the historical development of scripts, showing which are rightwards, which leftwards, and which indeterminate.[13]

Many early scripts are quite indeterminate in their direction, so scribes must have had practice writing in both directions. One might have expected that the easier of the two methods would ultimately be adopted and some have argued that, for right-handers, the natural direction would be left to right. Yet, as the chart makes clear, Early Phoenician was written from right to left. Part of the problem here is that we automatically equate what we are familiar with to what is easiest and most natural. Even Herodotus, the great Greek historian (c. 484 BC–c. 424 BC), seemed incredulous that the Egyptians attempted to justify their method of writing: 'In writing or calculating, instead of going like the Greeks, from left to right, the Egyptians go from right to left – and obstinately maintain that theirs is the dexterous method, ours being left-handed and awkward.'

One factor that needs to be considered is the nature of writing itself. Although today we use pen and paper, that is far from typical historically. Many inscriptions are carved in stone, and so have tended to survive the vicissitudes of history. Inscriptions on papyrus are more fragile, but were probably commoner for everyday writing. Writing rightwards on papyrus, as with paper, might seem easier for a right-hander, since there is less chance of smudging the wet ink. Ink, however, dries quickly, and modern left-handers cope surprisingly well, as do right-handers in leftward scripts such as Arabic. Many early scripts were in cuneiform, in which a triangular stylus is used to make a series of marks on the surface of a wet clay tablet, which is then hardened by baking in the sun or an oven. Again, it might seem that rightwards writing should be easier for right-handed cuneiform writers, since they are less likely to smudge what they have already written. It is an appealing theory, but although most cuneiform scripts are written from left to right, some, such as Hittite and Akkadian, could be written either way, and proto-Elamite seems generally to have been written right to left. The *coup de grâce* for any theory that uses smudging to explain the direction of cuneiform comes from those very few modern people who have tried it: 'practical experience in writing cuneiform on clay shows ... that with good quality clay very little smudging takes place and a conscious effort is needed to erase signs'.[14]

Although it is tempting to conclude that writing from left to right is somehow better, more natural, more efficient and more readily attuned to a brain in which language is typically located on the left side, there is little or no hard evidence for such effects – which is probably why so many right to left scripts exist in the world. While looking at the question in 1949, Gordon Hewes pointed out that most theorists did not even review the whole range of scripts, often simply talking of 'Asiatic' going from right to left, and ignoring the large numbers of rightward scripts on the Indian subcontinent. Scripts that were leftward were described as 'abnormal', and global personality traits were attributed to the peoples who used them, suggesting they were 'introspective, contemplative, etc., in contrast to the situation where writing is directed outward, away from the median line, "towards ends and objects, away from the 'I'".' Such speculations are palpable nonsense. What they do tell is that, yet again, Hertz's left–right symbolism is at work: rightward good, leftward bad sums it up.

Where then do differences in writing direction come from? Hewes' own conclusion seems quite acceptable nowadays: that it was merely a combination of historical, economic and religious factors that meant some scripts survived and others did not, with no relationship at all to their laterality. The situation is analogous to that in evolution, described by Stephen Jay Gould in his book *Wonderful Life*. Organisms such as ourselves have survived while millions of species have become extinct. The temptation is to suggest that we have survived because we are intrinsically fitter and more adapted. Often, though, we do not need such explanations, requiring only what Gould calls *contingency*. Things happen; consequences follow. That is the warp and the weft of history, and one does not need to invoke biological fitness, adaptation or naturalness, any more than historians talk about the relative fitness of Gavrilo Princip and Archduke Franz Ferdinand of Bosnia, whom he assassinated, thereby causing the First World War. Contingency (long chains of events, each dependent upon the one before, but without any inherent advantage) probably best explains the direction of scripts. At some point, for instance, Greek influences on the South Arabian script meant that the latter switched direction from leftwards to rightwards, leading to Ethiopic and modern Amharic scripts, this development having nothing to do with either form being easier to write.[15]

Many aspects of laterality in everyday life are, quite simply, contingent.

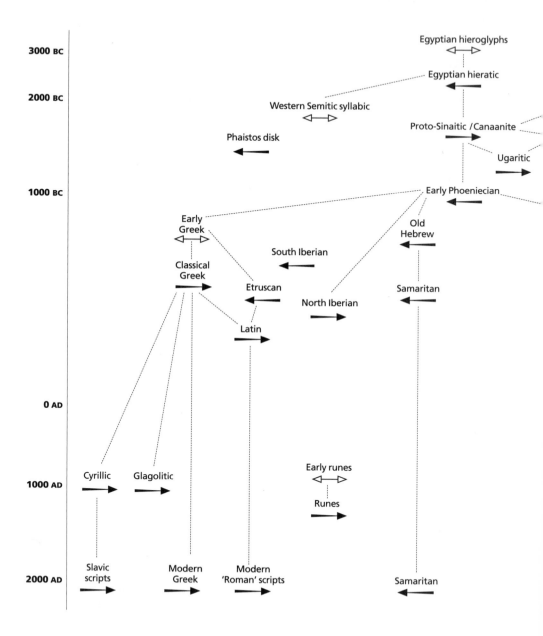

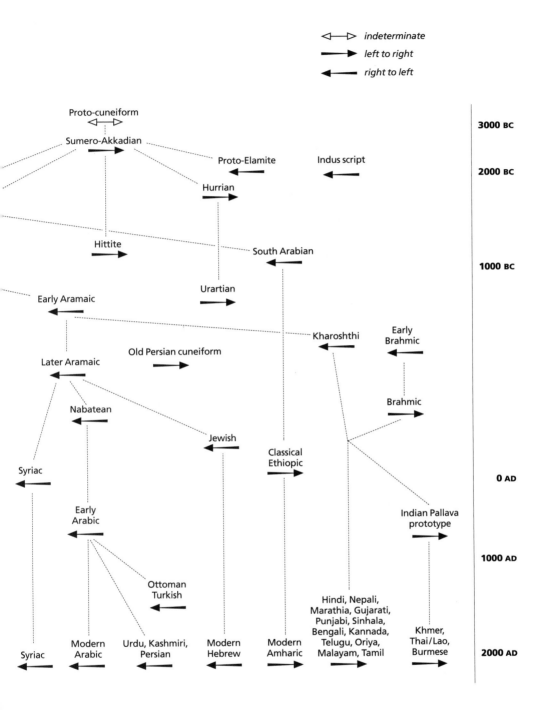

Figure 10.4 The historical development of the direction of writing in different scripts.

That doesn't make them any less interesting. And not *all* are contingent. That is the challenge for science: namely, to work out when there are deeper organising principles that apply across people, groups and cultures, which are implicit in our biology and neurobiology. The bottom line, however, has to be contingency. To go beyond that requires hard evidence; not a mere presupposition of advantage or selection.

Having looked at writing, and concluded that contingency alone is probably all that is needed to account for differences between cultural groups, let us look at some other lateralities, which reflect other underlying principles of organisation relating to social interactions between people and interactions between societies.

Driving

In 1806, John Lambert, a British visitor to North America, described how, 'In Canada, as well as in some parts of the United States, it is a custom among the people to drive to the right side of the road, which to the eye of an Englishman has a very awkward appearance.' That observation holds equally well today. If we are brought up driving on one side of the road, then it never seems quite correct to be driving on the other side. Lambert was certainly correct to talk about this as *custom*, but is that all it is, or is driving on the right or left more natural in some sense? If it is only custom, how did it get that way? It is worth noting that phrase, 'in *some parts* of the United States'. Americans may today drive almost uniformly to the right (although the US Virgin Islands still drive on the left), but it has not always been that way. In other parts of the world, people have not even driven consistently on one side of the road or the other. In his *Italian Journey* Goethe thought it sufficiently unusual to note that in Rome in 1788, during Carnival,

> An hour and a half before sunset, the more eminent and wealthy Romans set out in their carriages in one long, unbroken line and drive for an hour or more. The carriages start from the Palazzo Venezia, keeping to the left of the street…The returning carriages keep to *their* left, so that the two-way traffic remains orderly.

The situation though was clearly not typical for,

> as soon as the evening bells have rung, all semblance of order disappears. Looking for the quickest way home, each driver turns wherever he likes, frequently blocking the way and holding up other carriages…

Similarly in eighteenth-century Paris,

> The streets…often clogged with Gridlock under the Old Regime, because carriages drove on either side of the road and got stuck in face-offs, unable to back up, owing to the vehicles behind them and the difficulty of putting horses into reverse.[16]

Driving on one particular side, the 'Rule of the Road', is only necessary when traffic density reaches a certain critical level. With little traffic, vehicles, like pedestrians on a pavement, can weave around each other, particularly if not travelling too fast. However, once the density gets higher, traffic tends to organise itself spontaneously into streams, as do crowds of pedestrians entering or leaving a football stadium, it being easier to follow in someone's 'slipstream' than to push against the crowd unaided. The rule of the road is largely a compromise between an individual's desire to go wherever he or she pleases, and the benefits of doing the same as other people: benefits of speed and reduced accidents. If, however, the benefits of doing one's own thing are high enough, then the rule of the road will be ignored. W. G. Sebald comments, for example, on how '…the Maltese, with a death-defying insouciance quite beyond comprehension, drove neither on the left nor on the right, but always on the shady side of the road'. Ultimately the rule of the road is only a rule, although often it seems far more immutable than that. At the end of Nabokov's *Lolita*, Humbert Humbert, driving away after murdering Clare Quilty, says:

> The road now stretched across open country, and it occurred to me – not by way of protest, not as a symbol, or anything like that, but merely as a novel experience – that since I had disregarded all laws of humanity, I might as well disregard the rule of traffic. So I crossed to the left side of the highway and checked the feeling, and the feeling was good… nothing could be nearer to the elimination of basic physical laws than deliberately driving on the wrong side of the road.[17]

Given that a rule of the road is needed, there seem to be only two choices, *right* or *left*. In fact, as Geoffrey Miller points out, there are *three* stable equilibria: 'drive on the right', 'drive on the left' and 'drive randomly on the right or left' – and he suggests that the latter is the rule used by the taxi drivers of Bangalore. 'Stable' here means that there is no incentive for any individual to change their behaviour given how everybody else is behaving. It may be advantageous for everyone to change, but that cannot be implemented by single individuals changing their

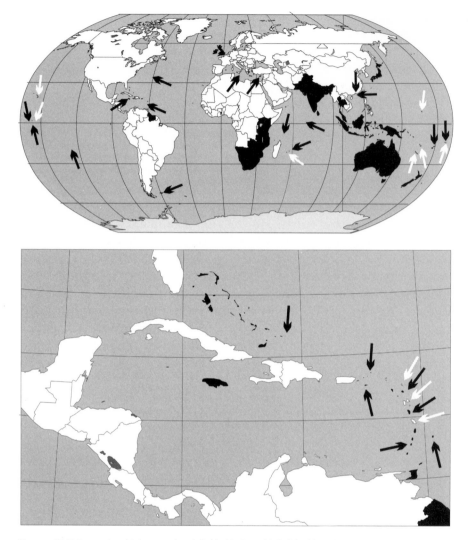

Figure 10.5 Countries driving on the right (white) and left (black) in the year 2000. White and black arrows point to islands that are too small to be seen at this global resolution. The separate map shows the Caribbean, where the situation is particularly complicated.

behaviour. Governments and police forces are needed to implement such rules on behalf of society.[18]

So which countries choose to drive on the left, and which on the right, and how did they get like that? The first thing to do is correct a very common misapprehension, particularly prevalent in North America, that only one country in the world drives on the left: that small island in the North Sea, Britain. It is far from the truth, as the maps in Figure 10.5 show.

Driving on the left is the rule in quite a lot of the world. Geographically, this amounts to only seventeen per cent of the world's area, largely because the vast and sparsely populated areas of Russia and Canada drive on the right, but when one looks at the world's population, about thirty-two per cent drive on the left, principally reflecting the densely packed populations of India, Indonesia, Pakistan, Japan and Bangladesh.[19]

How did countries decide on which side to drive? Canada is a perfect example of the complex evolution of the rule of the road, driving to a large extent reflecting Canada's colonial history. Quebec (Lower Canada) was originally French and hence drove on the right, and Upper Canada (Ontario) followed suit (although it is claimed Toronto once drove on the left). The maritime provinces of Nova Scotia, New Brunswick, Prince Edward Island and Newfoundland followed Britain in driving on the left, as did Vancouver Island and New Caledonia, which became British Columbia. As the country expanded westward into the prairies, Manitoba, Saskatchewan and Alberta followed Ontario in driving on the right, and so it continued until just after the First World War. American tourists, however, posed a problem, causing accidents when they drove from Washington State (driving on the right) into British Columbia (where driving was on the left). The change to uniformly driving on the right took place in two stages: in the interior on 1 July 1920, and in Victoria and Vancouver on 1 January 1922. Most of the maritime provinces quickly followed: New Brunswick in 1922, Nova Scotia in 1923, and Prince Edward Island in 1924. Newfoundland and Labrador was the last, changing its rule of the road in 1947, before joining the Confederation in 1949. A unified rule of the road had taken a long while, and change was inevitably accompanied by concerns about accidents. The fact that few occurred probably reflected the low traffic density, the care taken, and special measures such as in Nova Scotia where, for two months after changeover, drivers had a sign on the windscreen saying 'Keep to the Right'.[20]

The example of Canada shows that the apparent coherence in modern maps belies a confused history. The 1903 Baedeker guide to Italy said, for instance, 'The rule of the road varies in different parts of Italy. In Rome and its vicinity the rule is the same as in England i.e. keep to the left in meeting, to the right in overtaking vehicles. In most other districts, however, this rule is reversed.' Again, notice that word 'most'. In Rome, driving was on the left, as Goethe had described it, as it was also in Milan, Turin, Florence and Naples, but elsewhere it was on the right.

Myths abound concerning reasons for driving to right or left. Rome is typical, it being claimed that in 1300 Pope Boniface VIII decreed pilgrims should keep to the left. It is a nice theory, except that some say he proclaimed pilgrims should keep to the *right*, and searches have not yet found any documentary evidence either way. Mythology is everywhere on this subject. Many explanations centre around the need for pedestrians to keep their sword in their right hand, for horse-riders to keep the whip in their right hand, or for horses to be mounted from the left. Alternatively, the perversities of the French revolutionaries or the *diktat* of Napoleon are often invoked. The problem of all such explanations is that they do not account for the sheer variability that exists, even within countries.[21]

Italy was not alone in having some regions opt for driving on the right and others for driving on the left. Austria had traditionally driven on the left, but during the Napoleonic conquests (as a result of which countries such as the Low Countries, Switzerland, Germany, Italy, Poland and Spain now drive on the right), the French only invaded the western part of Austria, the Tyrol, and hence it alone drove on the right. The rest of Austria drove on the left, and continued to do so until the *Anschluss* of 1938, when, by declaration, the whole of Austria changed to driving on the right. The situation was confused, with motorists being unable to see traffic signs, and in Vienna the trams continued to drive on the left until track and signals could be changed. Hitler also made Czechoslovakia and Hungary drive on the right, as also occurred in the Channel Islands during the German occupation. Likewise, the Falkland Islands drove on the right during the brief Argentinian occupation in 1982.[22]

Colonial powers have influenced the side on which countries drive, but not uniformly. Indonesia drives on the left because the pre-Napoleonic Netherlands drove on the left, but when the Netherlands changed to the right, it did not. Much of the British Empire drove on the left, but there were clear exceptions, Egypt being the main one. Gibraltar remains part of the old British Empire, but it changed to driving on the right in 1929, to be like neighbouring Spain. Portugal changed to driving on the right in 1928, as did much of its empire, including East Timor. However, when East Timor was invaded by Indonesia it changed back to driving on the left. Influence of colonial powers can be indirect. Japan is said to drive on the left because of the influence in 1859 of the British ambassador, Sir Rutherford Alcock (although the local explanation invokes the needs of the samurai warriors), and British influence in Shanghai meant that for many years China drove on the left. Occasionally, the side of driving is changed for principally symbolic purposes, as in Myanmar (Burma) where 'a new rule of the road [was] recently and suddenly imposed to supersede that inherited from colonial times'.[23]

Perhaps most striking is how relatively late in the twentieth century some countries changed from driving on the left. Panama only changed

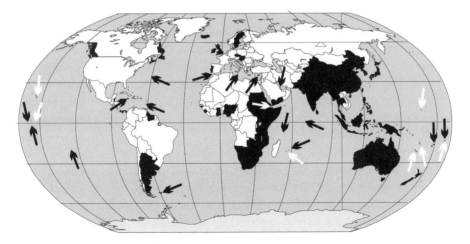

Figure 10.6 Countries driving on the right (white) and left (black) in the year 1919. White and black arrows point to islands that are too small to be seen at this global resolution.

in 1943, the stimulus being the construction of the Pan American Highway which otherwise would have been driven on the right for its entire length except for the short section through Panama. The Philippines, Argentina and Uruguay changed from the left to the right in 1945; China, Taiwan, and the Koreas in 1946. Subsequently, changes to the right occurred in Belize and Cameroon (1961), Ethiopia (1964), the Gambia (1965), Sweden and Bahrain (1967), Iceland (1968), Burma (1970), Sierra Leone (1971), Nigeria and Diego Garcia (1972), Sudan (1973), Ghana (1974), and South Yemen (1977).[24]

As Figure 10.6 shows, the situation in 1919 was very different from that at the end of the twentieth century.[25] Is there any clear pattern in the maps for 1919 and 2000? Some features are apparent. Many of the countries that continue to drive on the left are islands, often surrounded by neighbours who drive on the right. In the Eastern Caribbean, islands seem to alternate right and left, in part reflecting their British, French and Dutch colonial heritage. Even within Europe, the four remaining areas of left-driving are all islands. Britain and Ireland are well known; Malta less so. Most surprising is Cyprus, which drives on the left in both its Greek and Turkish parts, despite Greece and Turkey driving on the right. The result is particularly confused, since almost all cars are imported from Greece and Turkey and therefore are left-hand drive and hence poorly suited for driving on the left. The same problem occurs in the US Virgin Islands.

Apart from islands, the other clear pattern in the world map is the presence of a vast band of left-driving across southern and eastern Africa, across the Indian subcontinent, through the East Indies and into Australia. Many of these countries have borders with other countries driving on the left, and, when they do not, their borders are often in remote areas, such as the Himalayas or central Africa. These conditions do not apply in countries that have voluntarily changed their rule of the road, Sweden being the obvious example. By 1967, it was the last country in mainland Europe driving on the left and it had busy border crossings and a lot of international traffic.

The forces underlying the decision as to which side of the road to drive on can easily be seen in a simple model, a variant of what is known as a spin glass or an Ising glass. Although originally developed for describing magnets, it is actually a good model for social interactions where individuals are influenced by those around them. Anyone familiar with John Conway's game called 'Life', or with 'cellular automata', will recognise the basic approach.[26]

The model takes as its starting point that people in general, considered as individuals, communities, regions, or even countries, do not just make up their own minds and then press on regardless but, to a greater or lesser extent, change their behaviour in relation to those around them. Those who are geographically closest, their neighbours, have the greatest effect. What makes the situation really interesting is that surrounding countries are also changing their behaviour for the very same reasons. Everyone is looking over their shoulder at everyone else and deciding the best way to behave.

We can create a very simple world in which we can model this sort of situation. Figure 10.7 shows a map of this world, which we will call Chirenia. There are one hundred regions, each of identical size arranged in a ten-by-ten grid. The only unusual thing is that if one goes off the extreme right-hand side one immediately ends up on the extreme left-hand side, and if one goes off the bottom one immediately comes back on at the top. This has the useful mathematical property that every single region has exactly eight neighbours.[27]

Chirenia is a very rational and rule-governed world. When the motor car was invented in year 1 AF (After Ford), the people in each region realised it would be sensible if all the people in a region drove on the same side. However, Chirenia allows much regional autonomy and each eventually decides to choose whether to drive on the right or left by tossing a coin – heads for the right and tails for the left. The map of Chirenia shows those regions which have decided to drive on the right (shaded in white) and those that drive on the left (shaded in black). As chance would have it, slightly different numbers of regions decided to drive on the right and the left, fifty-nine regions (shaded in white)

Figure 10.7 Map of the regions of Chirenia in the year 1 AF, regions driving on the right shown in white and those driving on the left in black.

driving on the right, and forty-one regions (shaded in black) driving on the left.

Things soon got confusing for the Chirenese. Life was very civilised while they were only driving around within their own region, but as the number of cars increased and roads improved, so they started to cross the border into other regions. In consequence, they often had to change to driving on the other side of the road. After much negotiation, it was decided that the best thing to do was for each region to agree to change the side on which it drove if a clear majority of its neighbours drove on the opposite side of the road.

Think a bit about those neighbours. Every region of Chirenia has exactly eight neighbours, and there could be any number between nought and eight which drove on the opposite side of the road to themselves. The Chirenese decided that 'by a clear majority' meant six or more of the neighbours having opposite driving habits to themselves. So in the year 2 AF, the 'First Revision' occurred, as it became known. Every Chironese region looked at its neighbours, and seven regions decided that they should change the side on which they drove. The outcome, shown for year 2 in Figure 10.8, was that sixty-four regions now drove on the right, five more than before, and thirty-six drove on the left, five fewer than before (although two regions had changed to the less common situation of driving on the left).

Although it was hoped that, after the First Revision, everything would be settled, it was realised that still all was not well. Another revision was set up for year 3 AF, with the hope that a few minor adjustments would sort things out. The events of that year are still known in Chirenia as the 'Great Rule Revision'. Fifteen regions changed their rule of the road, so that now seventy-three drove on the right and twenty-seven on the left.

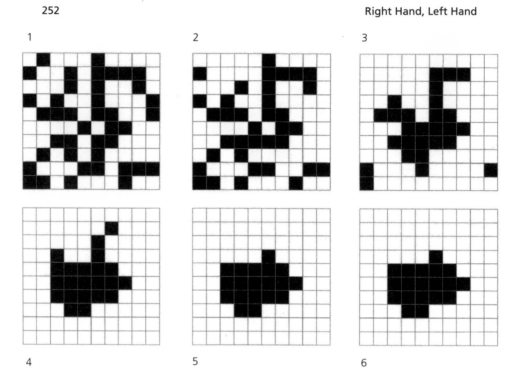

Figure 10.8 Driving on the right or left side of the road in the regions of Chirenia in the years 1 AF to 6 AF.

Although there were fears that the entire social infrastructure of Chirenia could be threatened by changes of such magnitude, a few far-sighted reformists, after looking at the map, decided that more changes could be allowed, and that they would be relatively minor.

The reformers were correct. The fears of a chaotic meltdown in year 4 did not materialise, only nine regions changing their side of driving. Chirenia was clearly dividing into two super-regions, the larger driving on the right, and the smaller driving on the left. Year 5 AF, or the 'Year of the Final Rule Revision', as it became known, was the last time the Chirenese changed their rule of the road, leaving eighty-one regions driving on the right, and nineteen on the left.

In year 6, a few ultra-conservative constitutionalists argued that further revisions should be considered and so, at great expense, the regions all again considered the possibility of change. None, however, found that a clear majority of its neighbours drove on the opposite side, so none changed their rule. The year-6 map therefore looks identical to the year-5 map, and unless the Chirenese constitution changes, so it will stay in perpetuity. The map for year 5 is stable. Chirenia is divided permanently between the Sinistralists and the Dextralists. Driving over these borders still produces a little confusion, but since most Chirenese prefer to stay close to home it is rarely a problem.

The driving habits of Chirenia are a surprisingly good simulation of the way in which countries of the world have decided to drive on the right or left. Although not all have made their decisions at the same time or for the same reasons, the main influence has been how their neighbours drove. The Chirenian model predicts that in the real world there are unlikely to be major shifts of driving habit in the future, with possible exceptions being Guyana and Suriname in South America.[28]

Just as with writing, so with driving there have been many attempts, usually North American, to explain why it is 'natural' to drive on the right. Many invoke the fact that most people are right-handed, but that has little going for it as a theory, since the same is the case in left-driving countries. Handedness clearly does, however, relate to the layout of the three pedals. Clutch, brake and accelerator (gas) are always arranged from left to right, with the left foot operating the clutch, and the right foot the accelerator and brake. It seems to be universal, irrespective of whether cars are right- or left-hand drive, and is unrelated to other cultural factors such as direction of writing. Whether it is the optimal way of doing things is another matter. The *Guardian* newspaper reported a while ago that motorists, presumably in automatic cars, who used the right foot for the accelerator and the *left* foot for the brake had reaction times a quarter of a second faster.[29]

The rule of the road seems to be subject to the same forces that have determined the direction of writing, with historical contingency being the main determinant. Functionalist arguments have tried, without much success, to claim it is better to drive on one particular side of the road, because it is inherently easier, better, safer, more efficient or more natural than driving on the other side of the road. There was, for instance, a claim in the 1960s that countries that drove on the left had a substantially lower accident rate than those that drove on the right, but I can find no support for this in more recent statistics. Each country has, to a large extent, made up its own mind about which side to drive, with occasional external interference, and the principal force for change has been the inconvenience of having neighbours who drive on the opposite side. It looks as though the world has now settled down into a steady state where the inconvenience of the occasional change at borders is less than the cost and difficulty of changing a complex system.[30]

Although I have concentrated on driving, I could have talked also about pedestrians, horses, ships, trains and aeroplanes. Each has its own complicated history, and there are dozens of theories explaining why actions are carried out on the left or right. The underlying principles, however, are the same as in driving. Interested readers should consult the detailed notes in my website. I will, however, finish this section on transport with a novel and extremely creative proposal, which, perhaps unsurprisingly, has met with no popular support. The Monster Raving Loony Party, an eccentric and now traditional part of any General Election in Britain, had a transport policy which it claimed, at a stroke, would have cured the problems on Britain's roads – cars would drive on the left and lorries on the right.[31]

Embracing, fighting to the death and playing tennis

The two examples we have looked at so far, the direction of writing and the side of driving, both involve social interactions, but mainly between large groups of individuals resulting in social conventions. Handedness is also important when two individuals interact. When people meet, the action of shaking hands is smooth and virtually automatic because everyone knows that the right hand is the one to extend. Without such a code, there would be uncertainty, as in the social kisses, which are now so prevalent but where the protocol is less well defined. In the arrivals lounge of an international airport, for example, a survey found that about two-thirds of people embraced rightwards (that is, clasping the other person to the right side of one's chest), and the rest leftwards, suggesting there is plenty of scope for confusion.[32]

Thomas Carlyle, with whom this chapter started, is credited with being one of the first to think about the problem of how handed individuals interact. Although he talked about the problems of being a left-handed mower, strictly speaking he should have spoken of the problems of being a *differently handed* mower – in a world of left-handers, the problem would be the right-handed mower. The question then once again arises as to why we are mostly *right*-handed. Carlyle started by noting the seeming universality of right-handedness, there being no people 'barbarous enough not to have this distinction of hands'. He then tries to answer a question that he himself describes as 'not worth asking except as a kind of riddle':

> Why that particular hand [the right] was chosen is a question not to be settled, not worth asking except as a kind of riddle; probably arose in fighting; most important to protect your heart and its adjacencies, and to carry the shield in that hand.

By an intriguing coincidence, his suggestion that we are right-handed because our heart is on the left of our body was also being put forward elsewhere in London, at almost precisely the same time, by Philip Henry Pye-Smith, a consultant physician at Guy's Hospital. He considered a group of human ancestors, half of whom were right-handed and the other half left-handed. They invented the shield, half carrying it on their left arm and half on their right. They went off to battle, and since their heart was better protected, more of those carrying the shield on their left arm survived; these, of course, being the right-handers. Pye-Smith, himself, was dismissive of his own theory, but it is actually a good one. Coming only a decade or so after Darwin's *Origin of Species*, it is a genuine evolutionary theory, and if its premises were correct it would have to be taken seriously. Its main problem is that the shield was invented long after right-handedness was the human norm, and thus the theory cannot be a serious contender for explaining right-handedness, but the underlying principle is sound – right- and left-handedness may have different advantages and disadvantages when individuals are competing or co-operating with one another.[33]

Modern humans, of course, rarely fight during ordinary everyday life, and certainly not to the extent of inflicting mortal wounds. There is, however, an institutionalised form of battle – sport. As George Orwell said, 'At the international level sport is frankly mimic warfare'; and even at local level, 'the most savage combative instincts are aroused', as one person tries to wound the chances of the other in direct competition. What happens when left- and right-handers compete in sport? Is either advantaged, and, if so, why? Although Carlyle and Pye-Smith speculated that left-handers might be disadvantaged when fighting, modern thought has concentrated on the idea that left-handers are advantaged, either because their brains are organised differently or because they are better adapted for sports and games. As always, a straightforward question becomes deceptively complicated when looked at in more detail.[34]

Left-handers can be advantaged or disadvantaged in sport if the sport itself is somehow asymmetric. The clearest example is baseball, which involves players running *anti-clockwise* from one base to another. In other words, the batter hits the ball and then runs towards first base, which is to the right. A right-hander starts with the bat held above the right shoulder, hits the ball by swinging the bat to the left, and ends up with the entire body turned anti-clockwise, with the bat over the left shoulder. After hitting the ball, the right-handed batter has swung round and is facing to the left, but then has to run to first base which is *to the right*. In the opposite direction, in other words. For a left-hander, everything is reversed; the bat starts over the left shoulder, there is a clockwise turn, and by the time the ball has been hit the player is facing to the right, and the left leg has already taken its first step towards first base. It is not surprising that left-handed batters are at an advantage, but one does not need to invoke any intrinsic advantage due to brain organisation: much can be explained by the asymmetric layout of the game. Baseball is not the only asymmetric sport. Sometimes the asymmetries are subtle, as in badminton, where the shuttlecocks have their feathers arranged clockwise, making them tend to go to the right, so that smashes are not equally easy from left and right of the court. Even racing is asymmetric, horses, greyhounds and people usually running counter-clockwise around race tracks, making injuries more common on one side of the body than the other. Occasionally, the rules of a sport explicitly disadvantage left-handers, as in polo where, mainly on the grounds of safety, the mallet has to be held in the right hand on the right side of the horse. Likewise, hockey sticks come only in a right-handed form and are held right-handed (and bullying off would be impossible unless both players held the stick the same way).[35]

Sometimes it is suggested that left-handers are disproportionately represented in sport because they have superior motor skills, particularly with ball control. The evidence here is limited, principally looking at the proportion of left-handers among top sports stars, but it seems to suggest otherwise. Amongst professional goalkeepers in football, for instance, left-handers occur in the same proportion as in the population as a whole, which suggests that left-handers are neither more nor less agile, have neither better nor worse reaction times, are neither better nor worse at attending to fast-moving objects or predicting their path; and so on. Left-handers are also no more common among professional darts players, again suggesting that they are not more precise at fine movements, nor more accurate, nor able to concentrate better; and so on. And again, left-handers are present at just the expected rate amongst professional ten-pin bowlers, again excluding many possible skill advantages for the left-hander.

Specific tests relating to the skills of left- and right-handed sportsmen are quite rare, but when they are tested they are typically found wanting, as in the claim that left-handers are better at fencing because of superior attention capabilities. Among left-handers in general, there is certainly little evidence that they are superior in motor skills. If anything, they are more likely statistically to be reported as clumsy or, in modern jargon, 'dyspractic'. Taken overall, there is almost no evidence that left-handers are better at controlling their muscles in the highly skilled actions that characterise most sports. So why are there some sports in which left-handers seem to do far better? The clue comes from looking at the sports themselves. They are sports such as tennis, boxing or snooker, in which there is direct competition between one player and another, rather than sports such as golf, ten-pin bowling or darts where, in effect, the battle is between the player and a target, be it hole, pins or board.[36]

The advantage of being left-handed in some sports comes from the simple fact that left-handers are rarer than right-handers. Think about tennis. The court is entirely symmetric and the rules are also symmetric, there being no hidden advantage to the right or the left. However, there is no doubt that at international level, there is an excess of left-handed tennis players. In tennis, the winner depends not only on who hits the ball hardest, fastest and most accurately across the net but also on who best predicts where the opponent will hit the ball (thus getting into the correct position to return it) and where the other player will find most difficulty in returning the ball. The ability to 'read' one's opponent is crucial, and depends mostly on having experience of many opponents. When two right-handers play tennis, each will try to read their opponent so that they place the ball on the backhand, which is more difficult to play than a forehand. What happens when a right-hander plays a left-hander? From the left-hander's point of view, nothing is different, a right-handed opponent being much like most of the other players he or she competes against. For the right-hander, however, things are different. Left-handed players are much rarer than right-handed, and so the right-hander has only about one-tenth the experience of playing left-handers as the left-hander has of playing right-handers. The left-hander, therefore, knows far more about their opponent's foibles than does the right-hander, which gives them a competitive edge. At the top level, that should mean that *more* than ten per cent of players are left-handed. How many more? Think about what would happen if fifty per cent of all top players were left-handed? The experience of top right-handers playing left-handers would then be exactly the same as the experience of top left-handers playing right-handers and there would be no advantage. In practice, one can expect an equilibrium somewhere between the population proportion of ten per cent left-handers and the no-advantage situation of fifty per cent left-handers; and that seems to be what happens, perhaps twenty per cent of top players being left-handed.

Similar advantages occur in other sports, such as baseball, where left-handed pitchers seem to have a similar advantage, and in cricket, where left-handed batsmen and bowlers are advantaged. Left-handers should also be advantaged in snooker and boxing (the proverbial 'southpaw') and are undoubtedly more successful in fencing. If left-handers have a tactical advantage because of their relative scarcity, then there may be other situations in which left-handers might be even more advantaged. In baseball the 'switch hitter', who can bat right- or left-handed, has an additional advantage because of the unpredictability of their shotmaking. In doubles tennis, there is a particular advantage to a pair in which one is right-handed and the other left-handed because, again, the opponents will never know quite which player is going to return the ball. In cricket, that advantage goes one stage further when a right- and a left-handed batsman are at the crease at the same time, because every time a single is run and the batsmen change ends, so the fielders have to move their positions, literally exhausting them as they keep moving around the pitch.[37]

If left-handers have such an advantage in sports like tennis and, in particular, in the near deadly sport of fencing, isn't it possible that we can turn round the argument of Carlyle and Pye-Smith? Although they suggested that it was *right*-handers who were advantaged in fighting, because their heart (on the left side) was protected, isn't it possible that fighting not only advantages *left*-handers but actually explains their very existence? Could the genes for left-handedness exist because of what biologists call frequency-dependent selection? Scarcity has its own value, in other words. Maybe. Theoretically, there is little doubt it is possible, as Anders Pape Moller and his colleagues have suggested quite forcefully. The question is whether, in practice, there has been sufficient fighting in human history to make it possible. Conceivably, there may have been, in the remote past, although there is no conclusive evidence.[38]

Wielding a scalpel and playing the violin

Sport and fighting are about competition, and left-handers seem occasionally to be advantaged, mainly for tactical reasons. What, though, about situations involving co-operation? Do left-handers find these problematic? Although there is little systematic research on the topic, one interesting area is the field of surgery, where teams of people work together carrying out complex manual tasks. Interest in the handedness of individual surgeons goes back to the time of Hippocrates, who recommended: 'Practise all the operations, performing them with each hand and with both together – for they are both alike – your object being to attain ability, grace, speed, painlessness, elegance and readiness.' Likewise Celsus, a Roman physician at the beginning of the first century AD, said a surgeon should be 'ready to use the left hand as well as the right'. The advice is good. Just as any right-hander would wish to be ambidextrous when they encounter that awkward DIY quandary involving the use of a screwdriver in a cramped right-hand corner, so surgeons probably wish the same when deep inside a body cavity.[39]

When several surgeons work together, each has to anticipate the movements of the other, ready to mop, swab, cut, apply diathermy or use forceps without the need for continual instructions. The problem was recognised by one of the most distinguished of nineteenth-century surgeons, Sir Benjamin Brodie, who commented, 'How much inconvenience would arise where it is necessary for different individuals to co-operate in manual operations, if some were to use one hand and some the other?' The only published study on handedness in surgeons was carried out by two medical students at St Mary's Hospital Medical School, who sent a questionnaire to the doctors in the hospital and found that whereas twelve per cent of the sixty-seven physicians were left-handed, *none* of the thirty-six surgeons were left-handed; a statistically unlikely result. It does therefore seem possible that surgeons are less likely to be left-handed.[40]

Surgeons are not the only ones who have to co-operate. Think about orchestral musicians playing the violin, viola, cello or double-bass. Each is played with a bow several feet in length, which, in fast passages, often moves at high speed. Given the risk of being hit by another player's bow, it is perhaps not surprising that string instruments in orchestras always seem to be played right-handed. Some musical instruments have to be played in a 'right-handed' way, the piano being an obvious example since the high notes are located at the right-hand end and the low notes at the left. Many instruments, however, are potentially 'ambidextrous', one such being the guitar. If the strings are put on a guitar in reverse order, then a left-hander can play them with no obvious disadvantage, Paul McCartney and Jimi Hendrix being well-known examples. Violins, cellos and double-basses can also be restrung in that way and played by a left-hander, as was done, for instance, by Charlie Chaplin in the film *Limelight*. Occasionally, left-handers have even been known to play 'backwards', the instrument strung as appropriate for a right-hander. The self-taught New Orleans jazz bassist Sherwood Mangiapane offers an example: 'He picked it up by ear and played it his own way,' said jazz historian Dick Allen. 'He played left-handed and it looked like he was playing it backwards. He didn't restring it. It was just the way he learned.' Even if solo violin players have occasionally managed to play left-handed, I have never seen a left-handed string player in a modern orchestra. Does that mean, then, that string players are mostly or even entirely right-handers? What about musicians playing other orchestral instruments, very many of which are designed to have the right hand doing something different from the left? Are they right-handed as well? In fact, they are not. In a survey of seventeen professional orchestras in the UK, nearly thirteen per cent of the musicians, irrespective of their chosen instruments, wrote with their left hand, somewhat *more* than would be found in the general population. Being left-handed does not seem to prevent one playing right-handed instruments at professional level in an orchestra. However, the need to co-operate in a social group means that left-handers play their violins in a right-handed way, just as they probably eat their dinner using a knife and fork in the same way as a right-hander.[41]

The body politic

If one types the words 'right' or 'left' into any search engine on the Internet, or looks for them in any of the major daily newspapers, or listens for them on the television news, then it is obvious that one of the commonest usages is for describing political views – 'left wing' versus 'right wing'. This modern usage has its roots in the decision made in 1789 by the officers of the new National Assembly in France to sit the radicals on the left-hand side of the chamber, the moderates in the centre, and the conservative nobles on the right of the presiding officer, although the terms only became consolidated in political use in France with the restoration of the monarchy after 1814. The first use of the term in English was by Thomas Carlyle (whose name seems writ through this chapter as if it were a stick of Brighton rock). In *The French Revolution* of 1837, Carlyle describes the chaos of the first National Assembly, in which one hundred or more of its 1200 members would try to speak at the same time.

> Nevertheless, as in all human Assemblages, like does begin arranging itself to like [in] rudiments of Parties. There is a Right Side (*Côté Droit*), and a Left Side (*Côté Gauche*); sitting on M. Le President's right hand, or on his left; the Côté Droit conservative; the Côté Gauche destructive.

The first use of 'left wing' in a recognisably modern political sense was as late as 1897 by the psychologist William James.[42]

Although somewhat simplistic, there is no doubt that the terms left and right, particularly when coupled with 'centre', 'hard', 'soft', 'far' or other adjectives, provide a useful measure of people's position on the political spectrum, although they fail to account for subtle variations in their political opinions. The label 'right' has been defined by the sociologist C. Wright Mills as 'celebrating society as it is: a going concern', whereas he sees 'left' as concerned 'with cultural and political criticism'.[43]

The use of right and left to describe political opinions raises all sorts of questions. Firstly, why did the terms catch on? Their use, as we have seen, is relatively recent, dating back to the end of the eighteenth century, and other terms have been used before that – for instance, at the beginning of the fourteenth century the Florentine Guelphs split into 'Whites' and 'Blacks'. However, as Norberto Bobbio has recognised, to be useful as a metaphor a description of politics has to be spatial, for political views are mostly about the *position* of one person relative to another. Black and white do not meet that criterion – and they also have the unfortunate effect that intermediates are all different shades of grey. Top and bottom would not work either, nor front and back.[44] In short, right and left are well qualified for the job.

Even if right–left is a useful spatial metaphor, is it anything but arbitrary? In other words, is it possible that, 'since the relative positions of the French deputies in relation to the president were only chance', left could just as well have described the conservatives? In a parallel universe, might the political terms for right and left have been the other way around? Probably not. Just as a key fact about handedness is that there is no society in the world which is left-handed, so it is crucial that left–right as a political distinction is used everywhere. Given the symbolic world described by Robert Hertz, the political use of left and right seems inevitable. Perhaps even those events in France that triggered the terminology were not quite due to chance – as a psychoanalyst has put it, 'But why did they sit on the king's right in the first place?' Symbolically, right represents the norm, the dominant, and the good, and hence it must also represent the status quo and the conservative. Likewise, it is hardly surprising that the radicals were to the left.

Support for the idea that the political usage of left and right is universal comes from an entirely different cultural tradition: the Tamil-, Telugu- and Kannada-speaking areas of South India, where from the eleventh until the nineteenth centuries there was a division into right- and left-hand castes. Although anthropologists dispute the precise meaning of the terms, there are striking similarities with modern usage. Legend says that the terms originated in a dispute between the Nagarattars and the Balijawars, probably at Kanchipuram, a town near Madras. The king arbitrated in front of the goddess in the temple, and the parties, who received betel and areca nuts from the king's left and right hands respectively, were ordered to live separately. The right-hand division was principally agricultural, and concerned with the ownership of land, whereas the left-hand division emphasised ritual purity. It could almost be a description of right and left in late twentieth-century politics, the right based on property while the left was racked by ideological disputes, preferring correctness to power.[45]

A very different view of left and right in politics takes as its starting point the fact that when people enter a room and sit down, their choice of where to sit is not made randomly. People who choose to sit on the

left-hand side tend spontaneously to make more movements of their eyes
to the right, and vice versa for those sitting on the right. Left- and right-
movers are said to have preferences for different academic subjects, and
to have distinct personality traits, left-movers exhibiting defence mechanisms involving the internalisation of conflicts, whereas right-movers
prefer to externalise conflict. It is possible, therefore, that the seating
plans of the French Assembly also reflected deep personality differences
between the leftists and rightists.[46]

When individuals describe their politics as some variant of right or left,
they inevitably do so *relative to other people*, right and left having no absolute political value. That may in part explain why people grow more conservative as they get older, society changing so that the radicalism of
youth becomes the status quo of later years. Because the politics of left
and right are relative positions, we can use voting patterns to ask how
people choose to describe themselves as left or right. The table below
shows the judgements of a professor of comparative politics concerning
the relative proportions of votes for right and left in sixteen European
countries over the second half of the twentieth century.

Table 10.1: Average share of the vote in sixteen long-established European democracies (%)					
	1950s	1960s	1970s	1980s	1990s
Social Democrats	34	32	32	31	30
Communists	8	7	8	5	4
New Left	–	1	2	3	2
Greens	–	–	–	2	5
Total left	**42**	**41**	**41**	**41**	**40**
Christian Democrats	21	20	19	18	15
Conservative	18	19	18	20	18
Liberal	9	10	10	10	10
Agrarian/Centre	7	7	7	5	6
Extreme right	1	1	2	2	6
Total right	**55**	**56**	**55**	**56**	**56**

It is a remarkable table in two different ways. Firstly, the overall proportions, particularly of the broad right and left groupings, show remarkable
stability across the years. Secondly, it is clear that there is a moderate but
consistent excess of right over left, there being almost three people on
the right for every two on the left. Why should that be? It might seem
intuitively natural that roughly half should be to the right and half to the
left. That is clearly not the case though. Why?[47]

The world is full of categories that are defined, in part at least, in relation to their antonyms: good–bad, up–down, happy–sad, rich–poor,
sweet–sour, true–false, and so on. Invariably, the first term in each of
these cases is seen by people as being 'evaluatively positive', to use the
psychological jargon. Asked to generate words spontaneously, people
generate more 'positive' words, and they do so more quickly. If one looks
at the total number of words used in written or spoken language, then
there are more positive than negative, and so on. In the psychology literature, this effect is called the 'Pollyanna effect', after the children's story
by Eleanor H. Porter in which Pollyanna is always able to look on the
bright side of any event. Pollyanna's first words in the book are: 'Oh, I'm
so glad, *glad*, GLAD to see you.' Later, she tells how her father had taught
her the '"just being glad" game', in which 'the game was to just find
something about everything to be glad about – no matter what 'twas'.[48]

Once it is looked for, the Pollyanna effect can be found in all sorts of
places. When filling in questionnaires, people are more likely to agree
than disagree. Similarly, they are more likely to answer exam multiple-
choice questions as true than false. We can now see why there should be
more right-ish voters than left-ish voters overall. The symbolism means
that right is the positive end of the scale, and hence it is more likely that
people will apply it to themselves than apply the less positive end of
the scale.[49]

The Pollyanna effect can readily be found in different types of data.
Can more precise predictions be made about it? They can, but before
doing that we need briefly to consider what is known as 'information
theory', developed by Claude Shannon in 1949 and a key theoretical
basis for much of computer science. Information theory starts out by
recognising that not all statements are equally informative. If I say, 'It is
raining outside,' you may not be particularly interested. Although it does
not rain every day in London, it does rain fairly often, so my statement
may therefore communicate nothing more than that it may be worth
taking your umbrella with you when you go out. If you live in a tropical
rainforest, where it rains almost every day, then my statement says even
less. It is a bit like being told that water is wet. It is true, but hardly adds
anything to your knowledge of the world. On the other hand, if you live
in the Kalahari Desert, where rain is extremely rare, then to be told that it
is raining will immediately capture your interest, for it is an exciting
event. Rare events, in other words, provide more information than
common ones. However, they are precisely that, rare. If each day in the
desert you asked whether it was going to rain, only to be told it was not,
then you would have to ask a lot of times in order to receive that rare but
useful information that it was. The average amount of information

depends therefore on how likely it is that something interesting will happen. What is surprising is that the maximum amount of information does not occur when there is a fifty-fifty chance of something occurring. Instead, it occurs when there is a rather smaller likelihood of an event occurring, at a value of about thirty-seven per cent. Interestingly, that number is a remarkably good approximation to the proportion of voters who are on the left rather than the right. The mathematics therefore suggests that we should not expect left and right to be equally prevalent in the population, and since, as we have seen in other contexts, 'right' is the norm whereas 'left' is the marked form, then it is the left that will generally be in the minority.[50]

This chapter has looked at a wide range of topics, exploring how social processes and interactions might relate to the usage of right and left, but there is one large area missing, and one which affects left-handers most directly – how society treats them. The next chapter will consider the attitudes of society to left-handers, and the problems of being a left-hander in a society of right-handers.[51]

☞ ☞ ☞ ☞ **11** ☜ ☜ ☜ ☜

KEGGIE-HANDER

The Sitwell family traces its origins back to 1301 in the area known as Hallamshire, now a part of Sheffield. They had a talent for making money – and for losing it. Originally rich on the basis of their ironworks, in the seventeenth century they were the world's principal manufacturers of iron nails. Eighteenth-century success at commerce was followed in the mid-nineteenth century by a bank crash, although only after a baronetcy had been conferred on Sitwell Hurt, who promptly became Sir Sitwell Sitwell, with the 'proud repetition of his name'. Rescue came with the discovery of coal under the family lands, and by 1892, when the fifth baronet, Osbert Sitwell, was born, the family was rich once more. Then, in the 1920s and 30s, the family showed a sudden and unexpected flowering of a rich aesthetic talent without familial precedent.

In 1900, John Singer Sargent was commissioned to paint a portrait that simultaneously showed the family's taste, wealth and good judgement. Osbert himself disliked it, and, indeed, did not like Sargent's work in general: 'he was plainly more interested in the appurtenances of the sitters and in the appointments of their rooms than in their faces'. Standing at the back of the portrait is Sir George Sitwell, the fourth baronet, with his arm around his thirteen-year-old daughter, Edith; in front is Lady Ida; and playing on the floor in the foreground are Sacheverell, aged two, and Osbert, aged seven. There is also the family dog, a black pug called Yum, which Osbert strokes with his left hand.[1]

It could have been coincidence or the demands of artistic composition that meant Osbert stroked the dog with his left hand, but it wasn't. Osbert was, in fact, left-handed, although there is little mention of this in his own writings. It isn't that Sitwell (Figure 11.1) wrote little about himself, for he published five volumes of an autobiography entitled *Left Hand, Right Hand!* They are strange volumes. Overblown, very dated, and of interest only as a period piece from a long dead age, it is difficult to

Figure 11.1 Sir Osbert Sitwell photographed in 1949 by Hamish Magee. He is probably looking at the manuscript of the final volume of his autobiography, *Noble Essences*. Turning the pages with the left hand probably reflects his left-handedness, although right-handers also sometimes do it that way.

believe that *The Times Literary Supplement* could have said, 'It would be hard to over-praise so strong and fine a work of literary art.' Now they seem hardly to be read, at least in my own university's library, where they are kept in the out-of-town store. Volume Two has been taken out only five times since 1964, and two of the volumes have not even been catalogued or bar-coded for the computer system in use now for many years. As far as I can tell, none is in print today, and one can see why. George Orwell put his finger on the problem for today's reader: 'It is to Sir Osbert Sitwell's credit that he never pretended to be other than he is: a member of the upper classes, with an assured and leisurely attitude which comes out in his manner of writing...'[2]

Despite fifteen hundred pages of autobiography, and even with a title like *Left Hand, Right Hand!*, Sitwell hardly mentions his left-handedness. By the end of the first volume, one even wonders if perhaps he was right-handed after all. The only unequivocal mention is when, at the age of seven, he was enrolled into a children's cricket team for a friendly match against the Yorkshire eleven, which was made a little fairer, since, 'They played, left-handed, against us, – and, as for that, I played left-handed against them, it being my natural bent...' Only in the second volume is there a clear mention. Sitwell had had a mysterious illness which kept him from school for several months. An up-and-coming young doctor, Bertrand Dawson, one day to be Lord Dawson of Penn and Physician to King George V, was called in:

He asked endless questions, no doubt summed up my character and then instead of advising treatment of a medicinal kind, merely recommended that, when I reached Italy, I should learn to fence – with my left hand...'

It was almost certainly good advice, even if without any formal medical justification. Sitwell's fencing skills, like his ability at all other sports, were, however, minimal:

> I manifested, I fear, no aptitude for fencing, but at any rate I was encouraged to use my left hand (one of the reasons of my crabbed handwriting being that, whereas it came naturally to me to employ my left hand, I was made to call my right hand into service instead)

Sitwell's handwriting was indeed appalling, particularly by the standards of his day: 'when I was twenty I wrote with the unformed hand of a boy of ten, but a hand that was illegible as well as childish'. Although, like Darwin, he put it down to heredity – 'the very bad handwriting with which nature and my ancestors endowed me (for it is said to be the most hereditary of all physical traits)' – it seems probable that the poor handwriting was in part the result of writing with the right hand rather than the left hand. At the age of twenty, he tried practising 'pot hooks', the descenders on handwritten letters like 'g', but to little avail. Ultimately, he made a virtue of necessity, and claimed that his poor handwriting forced him to think carefully about everything he wrote.[3]

Sitwell was hardly unusual in being a left-hander who was made to write with the right hand. A character in Dorothy L. Sayers' *Busman's Honeymoon*, published in 1937, is similar:

> 'Here you are, my lord,' said Sellon. He produced a box and struck a light. [Lord] Peter [Whimsey] eyed him curiously, and remarked:
> 'Hullo! You're left-handed.'
> 'For some things, my lord. Not for writing.'

Sellon and Sitwell would have been educated at much the same time. Their experience as left-handers was probably the norm for late Victorian times, and there is even a suggestion that it applied to that quintessential Victorian, Queen Victoria herself, who, although she wrote with her right hand, is said to have painted with her left and was therefore most likely a natural left-hander.

Less anecdotally, there is an account from 1880 of how at a school in Larbert, near Falkirk in Scotland, amongst the children aged four to seven years old, 'eight children had come to school left-handed'. Notice that phrase 'had come'; for they were not allowed to leave as left-handers, at least for writing. 'They were not allowed to use their left hands in writing or ciphering. Great trouble had, in fact, been taken to make them desist from using their left hands. Some of them had to be kept near the teacher at their writing, but in the playground they threw stones and played at bowls with their left hands.' The methods of making them 'desist' are not spelled out, but we are told that 'the teacher gives the child a great deal of trouble and perplexity', and that the 'left-handed child [is] forced to write with his right hand through fear of punishment'.[4]

The Victorians even invented a vicious leather device with a belt and buckles for strapping the left hand firmly behind the back. Other techniques were less brutal but had similar effects:

> I started school in 1925, when left-handed children were considered backward. There were only two of us in the class so we were put together … The treatment we had in school was to sit on your left hand and write with your right. It was impossible to keep your left hand still; so they tied the left arm behind your back … They shouted at us to use our right hand: then we were put in a corner and forbidden all games. We were told we were hopeless and everyone in the class repeated it.

Likewise:

> I started school during the first world war when most of the teachers were women and for a couple of years was allowed to write with my left hand and was happy to do so. When the war finished lots of men returned and took over the classes. Right away, the teacher came into the classroom and insisted that I should write with my right hand. He was very nasty about it and frightened me to death, poking his cane into my ribs, etc. … I shall never forget that teacher, although he has been dead for many years …

It was also clear that such pressure on left-handers did not come from the individual whims of the teachers, but were part of teacher training:

> At Bishop Otter Training College in Chichester [Doris Rayner] was threatened with expulsion for continually questioning the practice of children being forced to write right-handed. She and a fellow student … appealed to the college chaplain, Rev Jonathan Passe (who was left-handed) and he brought the issue before the governors. They accepted the students' protest and were among the first to abandon the requirements for students to write exams right-handed … On finishing College [she] was appointed to Wood Street Junior School. The headmaster

refused to accept her…because she was left-handed. She appealed to the council, who confirmed her appointment.[5]

Such practices were once highly prevalent – and one still hears occasional accounts of parents or teachers 'persuading' left-handed children to use their right hand. They are part and parcel of the experience of being a left-hander. Being different, being in a visible minority, and being in a minority with a host of symbolic associations, have inevitably affected left-handers. Given the opportunity and a sensitive audience, many left-handers will wax lyrical about sinistrality and the attitudes of the majority dextrals. Indeed, the modern world has provided rich opportunities to do precisely that with left-handers' clubs, special shops for left-handers, and, in recent years, Internet websites and discussion groups in which left-handers tell each other and the rest of the world about being left-handed and living as a left-hander in a right-handed world. Reading those websites does not, however, provide a particularly coherent picture of the experience of left-handedness in Western culture at the beginning of the twenty-first century. On the one hand, there are web-tales of the still continuing oppression and misery of a repressed but also irrepressible minority, yet, on the other hand, there is a sense that this is perhaps only a tiny proportion of the much larger population of left-handers, most of whom seem to suffer little if at all from their difference.[6]

Looking further afield, geographically and historically, soon reveals how discrimination against left-handers can take many forms from the systematic and the oppressive to the subtle but nonetheless effective. Among the most extreme are the Zulus of southern Africa who,

> if a child should seem to be naturally left-handed, pour boiling water into a hole in the earth and place the child's left hand in the hole, ramming the earth down around it; by this means the left hand becomes so scalded that the child is bound to use the right hand.

Less extreme are the Mohave of Arizona who,

> if the infant shows a tendency towards left-handedness while playing, it is encouraged to use its right hand because 'left-handed people are not liked' and 'people make fun of them'.

Hertz described how in the Dutch East Indies a Dr Jacobs 'often observed that native children had the left arm completely bound: it was to teach them *not to use it*'. In the past, left-handedness has not been allowed in Japanese or Chinese schools, and as recently as 1970 it was claimed that 'compulsory right-handed writing in schools [was the rule] in Spain, Italy, Yugoslavia, and in all the Iron Curtain countries with the honourable exception of Czechoslovakia'. Most notoriously of all, left-handedness was at one time simply illegal in Albania. In some African tribes, a candidate to be chief can be turned down entirely on the grounds of left-handedness, and likewise in the Jewish tradition, Maimonides was said to have enumerated a hundred blemishes which a rabbi must not show, of which left-handedness was one. Far milder is the situation in the Central Celebes (Sulawesi), where the Taradja accepted that left-handers were amongst them but merely said they were stupid. In Britain, it is easy to find old and new examples. In the seventeenth century, the physician John Bulwer, who first described sign language for the deaf, could say, 'this praevarication of acting with the Left hand in chiefe [was] an errour so grosse...' Even in the twentieth century, Oliver Lyttelton, the future Viscount Chandos, could describe a dinner in 1941 with the Governor of Malta, Sir William Dobbie, who, while discussing cricket, 'expressed his dislike of left-handed batsmen and bowlers, and said that he would beat a son of his who tried to bowl with the wrong arm.'[7]

Occasionally, being left-handed is seen in some ways as a virtue – although it still remains difference – there being, for instance, a Japanese proverb that a left-handed child may grow up to be a genius because of being different. Among the Gogo of Tanzania, the left-handed are ascribed a special ritual status, and Hertz described how, among the Dogon, left-handers are seen as being particularly powerful and skilful because they are the result of a fusion of two male twins (although that does also have the inauspicious effect of shortening their father's life). Left-handedness was certainly not a bar to success among the Arapaho Indians of eastern Colorado, whose greatest leader was known as Chief Left Hand, from his childhood habit of always using his left hand.[8]

To make sense of such a miscellaneous mass of often contradictory information, some theoretical perspective is needed, and this can be provided by the sociological concept of *stigma*. In 1963, the Canadian sociologist Erving Goffman published his masterpiece entitled *Stigma: Notes on the Management of Spoiled Identity*. It was, and remains, an influential book that in large part has been responsible for so many of the changes in the way that society responds to people with problems. That we no longer unthinkingly use harsh and hurtful terms such as cripple, spastic, lunatic, blind, deaf or dumb, is largely because of Goffman's influence. A string of more sensitive terms such as visually handicapped, hearing impaired, differently abled, or physically challenged cannot be written off as mere signs of 'political correctness' but instead reflect genuinely changed perceptions and ideologies resulting from Goffman's insights.[9]

The range of Goffman's *Stigma* was wide. The original Greek concept of a stigma was a visual sign, literally cut or burned into a person's body, which provided an immediate indication to the rest of society that an individual should be shunned, being criminally bad or morally unsafe – 'mad, bad or dangerous to know'. In modern medical usage, stigma are the characteristic physical signs of an illness. Goffman, however, broadened the concept, seeing underlying similarities between the social responses to very many conditions. Many stigmatising conditions involve visible disease, disability and deformity, such as disfiguring skin conditions, congenital malformations, blindness and amputations. What Goffman showed is that the same problems apply to people with hidden or masked physical conditions, such as a mastectomy or a colostomy; a history of past conditions such as epilepsy, mental illness or alcoholism; speech or other communication difficulties; or even purely social characteristics, such as homosexuality, having a criminal record, or being a prostitute. In short, it applies to anything that people might prefer other people not to know about, usually because it results in the individual no longer being treated as just an ordinary person. Social identity, the way others see us, can sometimes be so dominated and distorted by being different that the only identity remaining seems to be that of the condition itself. *Stigma* is about coping and living with stigmatising conditions, and about the responses of society and other people to those conditions.

To know whether left-handedness is stigmatising, we must look at the various characteristics that Goffman identified as a part of stigma and see how they apply to left-handedness. The range of conditions considered by Goffman is wide, and does not include only grossly mutilating and disabling conditions. A slight hearing defect, a bit of a limp, mild psoriasis, even bad breath, can all in the right circumstances be stigmatising. For a person in the public eye, even being unattractive can be stigmatising – and in Britain a senior politician described himself as 'too ugly to ever be Prime Minister'. So left-handedness cannot be excluded from being stigmatising on the grounds of not being serious enough.

Stigmatised conditions are deviations from a societal norm. There can be little doubt that right-handedness is the norm for most societies, both in statistical terms and in the provision and design of facilities. However, not all deviations from the norm are stigmatising – sometimes they may just be labelled odd or eccentric, and occasionally can even be viewed positively. To be stigmas, deviations must be viewed negatively by society, and as Hertz recognised, that has undoubtedly been true for left-handedness, particularly in many societies in the past. Individuals bearing a stigma can become aware of it from other people's reactions, a phrase such as, 'Oh, you are left-handed!' indicating deviation from the norm. That people never say, 'Oh, you are right-handed!' is itself indicative of the stigmatisation of left-handedness. Stigmas can vary in their impact upon people, but serious stigma – facial burns, a criminal record for murder, even being famous for appearing in a soap opera – can so dominate an individual's public perception that the real personality finds it difficult to emerge from behind the all-dominating stigma. Clearly, left-handedness rarely if ever produces a response that strong.[10]

Stigma is exacerbated by official labelling, in schools, hospitals and elsewhere, and the child labelled as needing special educational help because of left-handedness is more likely to end up stigmatised. Although it is tempting to see stigma as being the result of a condition itself, stigma also depends on the response of the person with the condition, the response of those around, and on societal norms. Many left-handers were brought up by sensitive, caring families and taught by teachers who neither exaggerated their sense of difference, nor ignored their occasional special needs, and for them there is little way in which the condition has been stigmatising. For others, their difference has led to oppression, as in those Victorian schools. Left-handedness becomes used by institutions and other people as an inappropriate explanation for all failings – every clumsy action reflecting left-handedness rather than mere inexperience or the inevitable accidents of everyday life – and it can then also become the basis of bullying and taunting. All such things can lead a person to the first stage in stigmatisation, which is embarrassment at having others find out about one's difference.

The next step in stigmatisation is shame, a feeling of being responsible for one's difference, through moral or other failing, or through insufficient willpower to correct it. The final stage in the process is stigma itself, in which the entire sense of identity, of self-worth, is perceived as spoiled or tarnished by being different. Though few left-handers reach the final stage, some undoubtedly are on the second step, and probably many at some time or other have been embarrassed by their left hand. Just as stigma is a social process, involving the interaction of the deviant with the rest of society, so the management of stigmas by those carrying them is also a social process. Stigmatised individuals often band together with others like themselves, both as a form of support and to campaign for a change in society's attitudes, although there is a risk that the rest of society sees such groups as confirming the nature of the difference. Clubs, societies and pressure groups for left-handers are found in many countries, and the most vociferous members see opportunities for political campaigning, as well as for more mundane activities such as running shops that provide products, such as left-handed can-openers, that ordinary shops perversely continue not to sell.

Taken overall, left-handedness can in some cases be regarded as a stigma. However, at the beginning of the twenty-first century it is mostly a stigma which is mild and of little consequence for most left-handers. As we have seen, though, it has not always been that way, and in less tolerant societies or sub-sections of society it probably still is not. The change in attitudes must, in part, be due to the dramatic increase in the numbers of left-handers during the twentieth century. As difference becomes less, so stigma also becomes less.

Stigmatised conditions invariably end up being used as derogatory terms, expressing disapproval in seemingly unrelated contexts. Left-handedness is no exception – one only has to think of phrases such as 'a left-handed wife' to mean a mistress, 'a left-hander or left-footer' to mean a Catholic, 'a left-handed bricklayer' to mean a Freemason, 'a left-forepart' to mean a wife, or the minimalist 'left-handed' to mean homosexual. A 'left-handed dream' is a bad dream, a 'left-handed opinion' is a weak one, 'to hear something over the left shoulder' is to misunderstand, 'to gain something over the left shoulder' is to gain it unfairly, and 'to be born on the left side of the bed' is to be illegitimate. Left-handed achievements are those which are not the real business of a person's life, as, for instance, when the playwright David Hare called his autobiographical prose jottings, *Writing Left-handed*. Likewise, the poet John Milton, in *The Reason of Church Government*, writing 'in the cool element of prose' rather than soaring in the high regions of poetry, said of himself, 'I have the use, as I may account it, but of my left hand.' Even science slips easily into such metaphors, so that substances with a negative refractive index which bend light in the opposite direction from usual are described as 'left-handed' materials. My personal favourite among all these uses is 'left-handed sugar-bowl', meaning a chamber pot. There is nothing new about such phrases, the Roman poet Horace, for instance, exclaiming *'Ego laevus!'* ['Silly me!' or, more accurately, 'Left-handed me!']. Just to emphasise the point, when the occasional phrase does refer to right-handedness, it refers to something that is socially *acceptable*, as in a 'right-handed fault', an error made with the best of intentions and not to be condemned, or the Latin term *dextram dare*, to give the right as a pledge of faith. Occasionally, the metaphorical use of left-handed can hide an altogether darker meaning, even a black joke. The principal war god of the Aztecs was known as *Huitzilopochtli*, which is usually translated as 'left-handed hummingbird'. It has been pointed out that 'hummingbird' is a rather cute name for a war god, and adding 'left-handed' makes the name completely bizarre. A metaphorical interpretation seems far more likely, the tongue and sound of the hummingbird echoing the appearance and sound of the rattlesnake. A left-handed hummingbird is therefore the deadly rattlesnake.[11]

Figure 11.2 Van Gogh's lithograph of *The Potato Eaters.*

A simple experiment shows how left-handers perceive their sense of difference. The lithograph of one of Vincent van Gogh's best-known pictures, *The Potato Eaters* (Figure 11.2), shows a group of Dutch peasants eating their frugal meal of a single large plate of potatoes. Does anything seem strange about the picture? People who know the oil painting of the same subject often say that the lithograph has a less subtle composition. The overall lightness may also seem wrong, as van Gogh himself commented to his brother Theo, in one of his several letters in April 1885:

> What you say about the lithograph, that the effect is woolly, I think too, and it is not my own fault in that the lithographer insisted that, as I had hardly left any white on the stone, it would not print well. At his suggestion I had the light spots corroded...

Occasionally, though, some people point out in amazement, 'You never get four or five left-handers all sitting in a room together!' They are correct, for van Gogh actually sketched five right-handed peasants, but their handedness was reversed by the inadvertent reversal during printing, which van Gogh had forgotten when he made the lithograph. He told Theo: 'If I make a picture of the sketch, I shall make at the same time a new lithograph of it, and in such a way that the figures, which I am sorry to say, are now turned the wrong way, come right again.' Statistical improbability apart, with five left-handers together in a room being very unlikely, the important message for our purposes is that, by and large, it is *left-handers* who notice the problem. Right-handers rarely comment.[12]

Right-handers seem to go through the world in blithe ignorance of the handedness of those around them. Irving Stone wrote *The Agony and the Ecstasy*, a biography of the artist Michelangelo who, according to some, was left-handed. However, 'During all the years I was researching and writing my book on Michelangelo I always assumed that he was right-handed,' said Stone, tellingly adding, 'Perhaps that is because I am.' Right-handers seem only to notice left-handers when they impinge upon them, bumping elbows at dinner, or using their mouse on the 'wrong' side of the computer. There can hardly be a left-hander who at some time hasn't heard the phrase, 'I didn't know that you were left-handed.' Left-handers, like any stigmatised minority, are good at picking up the tiny signs that indicate someone is 'one of us'. It is a bit like homosexuals, who are said to have a metaphorical radar for detecting other homosexuals (so-called 'gaydar'). Likewise, sinistrals instinctively identify their fellow left-handers, while the right-handed are hardly aware of them being in their midst. Paradoxically, though, there is a sense in which the invisibility of left-handers is a sign of their acceptance in the world. To be ignored is better than to be noticed and then discriminated against. In that Victorian schoolroom in Larbert, mentioned earlier, it would have been impossible to be a left-hander and not be noticed. In Goffman's terms, left-handers have managed to 'pass', to be accepted as they are, without their difference travelling ahead of them.[13]

Although left-handers notice the handedness of other people, that does not necessarily mean that their left-handedness is of central importance to their self-image or self-concept. A study looked at this by asking two large groups of college students and schoolchildren to 'Tell us about yourself'. Because the question was completely open-ended, and the subjects had no indication that handedness was being studied, left- or right-handedness only received a mention if it was of great importance to the individual's sense of their own being. Although, as expected, handedness was mentioned more by left-handers than right-handers, the vast majority of the individuals did not describe their handedness at all, only eight per cent of left-handed students and two per cent of left-handed schoolchildren mentioning it. That suggests most left-handers do not see their handedness as being of particular importance (although, of course, that does not meant there isn't a small group for whom it is of great importance).[14]

The blindness of right-handers to the left-handers among them does not mean that left-handers are treated entirely benignly at such times as they are noticed. As is so often the case, it is language which is so revealing. Osbert Sitwell noticed it in the Hallamshire dialect around Renishaw, near Sheffield, where he was brought up: 'if you are born left-handed, you are a "keggie-hander"'. It is one of very many such terms for left-handers used in the dialects of Britain.

The English Dialect Survey was initiated by Harold Orton of the University of Leeds and Eugen Dieth of the University of Zurich. Although it was originally mooted in the 1930s, the events of the Second World War intervened and the project was put on hold until July 1945, when Orton received a letter from Dieth, 'the answer to yours of May 14th 1940', which had bided its time on Dieth's desk 'until the day when things would be normal again'. Orton and Dieth realised, even in 1945, that the events of the war, increasing industrialisation, easier travel, population migration, the advent of radio and television and an increasing emphasis upon 'Standard English' would together mean the rapid extinction of many dialect phrases. Their survey, which Dieth called a 'linguistic Domesday Book', eventually took place between 1950 and 1961. Researchers visited 313 locations across England, many of them rural, interviewing local people and asking about some thirteen hundred everyday objects and events. The interviewees were mostly over the age of sixty, 'with good mouths, teeth and hearing', who had been born and bred in the local area. Half a million pieces of information were collected, along with many tape-recordings. The massive database was originally kept on cards, but it was computerised in the 1980s, with a comprehensive dictionary published only in the 1990s. It records the speech habits of the generation who were children in England before the Great War, a world redolent of *Lark Rise to Candleford* and Hardy's Wessex novels.[15]

The question on handedness was typical of many other questions in the survey: 'Of a man who does everything with this [show your left hand], you say he is [space for answer]. And with the other hand [space for answer]?' There are striking differences between the answers for the left and the right hand. Almost all the interviewees described a person using their right hand as a right-hander, the only exception being in Lincolnshire where the term 'cotmer-handed' was sometimes used. Left-handedness was another matter altogether, with eighty-seven distinct terms listed in the dictionary. Particularly striking is how localised are some of the terms. Keggie-hander, the term that Sitwell described in Hallamshire, or its variants, was found in Derbyshire, Leicestershire, Oxfordshire, Staffordshire, and Warwickshire. The richness and variety of English dialect terms for describing left-handedness can readily be seen in the maps in Figure 11.3.[16]

Many of these expressions can hardly be regarded as 'good' English, and some such as 'cunny-handed' and 'ballock-handed' are undoubtedly

Figure 11.3 Terms for left-handedness in different parts of England. There are six separate maps, for clarity, because the range of several words overlaps that of others.

what some dictionaries describe as 'vulgar'. Many, though, are still in daily colloquial usage; for instance, my mother still uses the term 'cack-handed', which I learned from her as a child in what was then Middlesex. Although the origin of many terms is obscure, the origin of this is scatological, 'cack' being a country term for excrement. Its use gives a frisson of the dunghill when suddenly used by the late Lord Annan, in his otherwise urbane, witty, and sophisticated book, *The Dons*: 'Students will graduate cack-handed unless they are taught how to relate their own specialism to every other…' Other terms for left-handed refer to the genitalia, 'cunny' (or more politely 'cony') for the female pudenda and 'ballock' from the Anglo-Saxon 'bealluc' for testicle. Some phrases are foreign imports, such as 'kay-handed', which is used in Northumberland and is thought to come from the Scandinavian. One can only wonder why, in Wiltshire, left-handers are known as 'Marlborough-handed', although some reference to the nearby public school must be possible. The term 'south-pawed', whose etymology is conventionally described as coming from early twentieth-century American baseball lore, is also a dialect term from Cumberland, and is almost certainly far older than baseball. There is a certain geographical logic here, Cumberland in the north of England using 'south-pawed', whereas further south, in Lincolnshire, 'north-handed' is used. Perhaps the most difficult terms to understand come from Norfolk, where logic goes awry: despite 'left-hand afore' quite sensibly meaning left-handed, 'right-hand afore' also apparently has the same meaning. Finally, it is perhaps worth mentioning that the richest variety of terms is found in Yorkshire with at least nine different expressions.[17]

Just as in England, so in Scotland there is a wide range of terms for left-handedness, as was shown in a similar survey conducted at the same time: the Linguistic Survey of Scotland. The situation was so complicated that the editors published two maps, one displaying the common terms for 'left' and the other common terms for 'handed', both of which vary in different regions of Scotland. To get a sense of the complexity of terms, combine one of the words from the first column in the table below with one from the second – very many of the possible combinations actually occur.

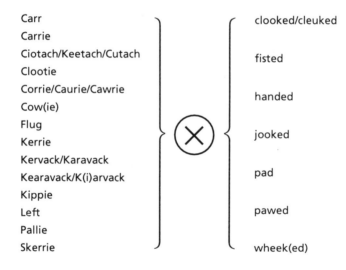

Carr	clooked/cleuked
Carrie	
Ciotach/Keetach/Cutach	fisted
Clootie	
Corrie/Caurie/Cawrie	handed
Cow(ie)	
Flug	
Kerrie	jooked
Kervack/Karavack	
Kearavack/K(i)arvack	pad
Kippie	
Left	pawed
Pallie	
Skerrie	wheek(ed)

There are numerous other terms, such as maeg-handed (Shetland), pardie-pawed (Orkney), corrie-juked (Sutherland), garvack (Ross and Cromarty), maukin (Banff), flukie (Aberdeen), pallie-euchered (Angus), dirrie (Fife), honey-pawed (Dunbarton), cooter (Bute), Fyuggie (Ayr), skibbie (Lanark), skellie-handed (East Lothian), and flog-fisted (Kirkcudbright).[18]

The most frequent complaint of left-handers concerns the problems of living in a right-handed world. Scissors are a particular bugbear. For instance, Jennie Cook of Chingford remembered how:

> When I was about 8, one of the teachers asked if someone could help cut out some circles. I volunteered. When I handed her one of my circles she looked annoyed and told me crossly that I wasn't cutting it properly ... It wasn't until recently that it occurred to me that when you use right-handed scissors, that unless you look directly over the top of the blades, that you can't see the line you are cutting.

Scissors are such an everyday item that most right-handers do not even realise that they have a handedness. Open a pair of scissors and lay them on the table so that they make a cross with the handles towards you. The top blade runs from the top left to bottom right. Turn them over and they still do the same thing. Scissors are chiral objects and no amount of rotation will make a right-handed pair of scissors into a left-handed pair. Those made for right-handers are designed so that when they are held in the right hand the eyes see the two blades coming together, thus enabling cutting to be precisely controlled. More subtle is that the design also helps push the blades together as they are used.[19]

To appreciate the frustrations of a left-hander, consider the kitchen scissors I recently bought at Sainsbury's. 'Left- or right-handed use' they proudly trumpeted, in a rare consideration of the needs of left-handers. The note, 'Designed by Robert Welch MBE, Royal Designer for Industry', helped to remove any lingering doubts. However, what the left-handed Prince William would make of products from this 'Royal Designer' is difficult to know. Admittedly, the handles of the scissors are symmetric, so they can be held both by right- and left-handers, but that doesn't mean they can easily be used by left-handers, because the blades cross in a way appropriate for right-handers, and, as Jennie Cook would have told Robert Welch, that makes it almost impossible for a left-handed user to see the object they are cutting. So the scissors are not for left-handers after all. The blurb on the wrapper is a hollow gesture, signifying nothing. The handles may indeed be suitable for right- and left-handers, but this is about as relevant as suggesting that a fashion garment is suitable for men or women on the basis that both men and women have two arms and two legs, ignoring the other very obvious physical differences between the sexes.

When visiting my mother, I told her this story of the not-so-ambidextrous scissors, and she showed me her own 'left-handed scissors', made by Wilkinson Sword. The handles were moulded to take a left thumb and fingers comfortably, and a right-hander would find it nigh-on impossible to put their right hand into them, but once again the blades were still right-handed. The manufacturer had modified the cheap part of the process, producing some special left-handed moulded plastic handles, but had not undertaken the more expensive modification, retooling to produce properly left-handed blades. Having said that, there undoubtedly is now an increasing number of properly left-handed scissors available in shops.[20]

If, in the words of Thomas Carlyle, 'Man is a tool-using animal... Without tools he is nothing, with tools he is all,' then without left-handed tools a left-hander risks being nothing. Certainly there is little doubt that left-handers mean nothing to many of the designers of domestic and industrial tools and appliances. Power saws spray sawdust into the face of the left-hander; microwave ovens have controls on the right and doors that open towards the left; automatic drinks-dispensers again have controls on the right; and so one could go on. The electric kettle shows what can be achieved. The cordless kettle is both safer and more convenient, particularly if there is a transparent water-level indicator on the side. In the past, however, most kettles had the indicator on the left side of the handle, meaning that it could only be viewed when holding it in the right hand. The solution is so simple that it almost defies explanation as to why it took so long to put an indicator on each side, as is now done with many kettles.

Whether such indifference to one in ten customers is wilful neglect or simple hand-blindness is not clear. I had high hopes of finding out, when, after a long search, I finally discovered a book entitled *The Left-handed Designer*. Unfortunately, Seymour Chwast, its author, is a graphic designer, and had nothing at all to say about the design needs of left-handers. I turned instead to a book by the engineer Henry Petroski, *The Evolution of Useful Things*, which expertly describes the complex design histories behind such mundane objects as the paper clip, the zip fastener, and the knife and fork. His final chapter muses on the reasons for success and failure in design, and he takes the unmet needs of the left-hander as a paradigm case. Progress in engineering and design comes from failure and anticipation of failure. If designers forget to consider whether left-handers will be able to use a product, and subsequently fail to observe that the product is disproportionately unusable among left-handers, then it will continue to be made with right-handers in mind. However, part of the problem also lies with left-handers themselves, who are not only a little too good at adapting to a right-handed world but also, as Petroski says, 'Nor do they seem as a rule to express any pressing need for special left-handed devices.' That, though, does not mean they will not use and buy such products if they are available, as Petroski found when a left-handed friend enthused about a left-handed Swiss army knife, which not only had a left-handed corkscrew but also blades that could be opened with the left hand.[21]

Left-handers have not only been ignored by designers, engineers and manufacturers, but are often also forgotten in schools. A survey of twenty-seven schools in a London borough found that, although most did provide some forms of specialist left-handed equipment, in particular left-handed scissors, they provided little else. Hardly any schools had books on left-handedness, either for pupils or teachers, and few teachers had been on specialist training courses, particularly for the teaching of writing. Only a third of schools even knew how many left-handers there were in the school. The problem is not confined to schools. In a parliamentary debate in the House of Commons in 1998, the Under-Secretary of State for Education confessed, 'I had not given proper thought to [left-handedness] before,' and admitted that the government not only kept no statistics on left-handedness but, even more scandalously, had no statistics at all on overall numbers of children with special educational needs. She did, however, concede that teachers are expected to know which children in their class are left-handed, and that they should, in particular, seat them appropriately so that elbows are not bumped while writing. That may seem a small step, but it is undoubtedly progress.[22]

Left-handers are a people without history, being mostly ignored and invisible within museums and archives. While curating an exhibition on left-handedness, Nigel Sadler wrote to all UK museums asking whether their collections contained artefacts specifically for or used by left-handers. Only two were able to provide any such information from their catalogues. One was the Museum of London, which had a left-handed medieval carpenter's side-axe, and the other was the Royal Armouries, a museum of theirs in Leeds having a display case devoted to left-handed weapons. Left-handed artefacts undoubtedly do exist, specifically created for the special needs of left-handers. For instance, antiques experts say that there were nineteenth-century left-handed moustache cups, although Sadler was unable to find one, not least owing to the inadequacy of most museum records.[23]

A characteristic of the last quarter of the twentieth century was the development of pressure groups supporting virtually every special interest group. It is surprising, then, that no systematic lobby emerged protesting against what *The Economist* called 'the peculiar degradation of that last great mass of the still submerged: the left-handed'. People have tried to set up such groups, but attempts have so far been almost entirely without impact in effecting change. Despite the division of the political world into 'left' and 'right', the truly left have no political voice. Indeed, they do not even have a decent name for the problems they face. Gerald Kaufman suggested 'dextrism [for that] assumption of superiority which litters the English language with discriminatory phrases', and *The Economist* tried to introduce 'handism' and 'sinistrism', in protest at the domination of language by dextral usages. They dryly commented that 'the only recorded sinistral success in an essentially one-sided cold war was the introduction … in the medical profession, of the term rectum for a little regarded part of the human anatomy'. You may gather from such a joke that *The Economist's* campaign was not over-serious, its two articles on handedness each being published as light-hearted Christmas editorials. Any discussion of the needs of left-handers seems almost immediately to be beset by rather poor puns about left-handers' rights or the Universal Declaration of Human Lefts. Rarely has the association of left-handers' rights with political correctness been better parodied than by Theodore Dalrymple in his novel, *So Little Done*. The passage deserves a lengthy quotation:

> The only safe thing to be in the Department was a victim: and even the left-handers banded together and demanded a meeting to make everyone aware of the difficulties left-handers faced in a right-handed world. They said that the latest research had proved that left-handers lived ten years fewer than right-handers, and that therefore they were entitled to early retirement, especially as much of the excess mortality was accounted for by their increased susceptibility to accidents brought about by equipment designed solely for the convenience of the right-handed, such as scissors. They demanded that henceforth at least the same proportion of equipment in the Department be adapted for left-handed use as the proportion of left-handers in the staff or the population as a whole, a reasonable demand, they said, after several centuries of attempted suppression of left-handedness by parents and teachers, who had tried to change children's handedness as if it were merely a matter of moral failure rather than of neurology and hence an integral part of a child's personality. The more extreme among the left-handed lobby demanded that an even higher proportion of equipment be adapted to left-handed use, claiming that a significant number of so-called right-handers were really left-handed, having been forced to change their preference in childhood, and that, with a modicum of official encouragement, they could be returned to their true identity, and hence to personal wholeness. They also said that restitution was only just and reasonable after so many centuries of oppression by the right-handed.
>
> Left-handed scissors made their appearance for the first time in the Department, and a monitoring group was set up to check that they were

available easily to those of the staff who might need them. But even this did not satisfy the lobby, which had scented blood: it pointed out that all the handles of the toilet cisterns were for use by right-handers, and demanded the installation of left-handed cisterns. And then it moved on to what it called *handedness-biased language*, the use of which it wanted to eliminate from the Department: terms such as *sinister* and *gauche* which carried derogatory connotations concerning left-handers and left-handedness, and were thus deeply wounding. Even the past participle of the verb *to leave* became suspect, since leaving is often sad and unhappy, and it was officially recommended that it should be avoided whenever possible. *He left his flat* should henceforth be written *He vacated his flat* or even *He leaved his flat*.

In humour such as this, we perhaps glimpse the real reason why left-handers will never fully achieve their rights. Overly serious left-handers are deemed not to have realised that sinistrality is one of the gods' best jokes against humankind, and in failing to see this they commit the ultimate sin of not having a sense of humour, of not being a good sport.[24]

There are, nonetheless, simple, straightforward and genuinely serious questions about the rights of individuals in society who happen to be different. The philosopher John Rawls, in his influential book *A Theory of Justice*, devised a good way of thinking about whether or not something is just. Imagine a world in which you are to be brought back at random in one of several different forms; for example, maybe as a man or maybe as a woman. If you are currently a man and you would be unhappy to find yourself as a woman in that future life, then this suggests that women are presently being unjustly treated. Similarly, if you would be unhappy to find yourself as a member of an ethnic minority or in a wheelchair, the principle again applies. The same question can be applied to handedness. If you are a right-hander and would feel aggrieved, upset or disadvantaged by being brought back as a left-hander, then left-handers have a genuine grievance. To put it another way, imagine a truly radical party of the left, with a left-handed Prime Minister and a left-handed Cabinet, which enacted legislation saying that all scissors had to be left-handed, that writing would be from right to left, that machinery and tools of all sorts should be built only in their left-handed forms. One imagines that right-handers would protest at such changes, but in that case left-handers also have a legitimate protest against the present situation.[25]

A sinister possibility, if one will excuse the phrase, is raised by new genetic technologies, particularly if the assumption is correct that handedness is determined by a single gene. The millennial editorial published in the *Guardian* on 1 January 2000 predicted that the human genome project would soon lead to 'working out the genes or combinations of genes which are responsible for musicality, schizophrenia, *left-handedness*, height, athletic prowess and aggression' (emphasis added). How such information might be abused can be seen in the dystopic science-fiction film *Gattaca* (1997) – the title comes from the base pairs of DNA – in which genes determine destiny. The left-handed hero Vincent is prevented from achieving his dream of going into space because of his inferior genes, one of which is for left-handedness. The plot involves an elaborate subterfuge, which not only means substituting blood, urine and DNA from someone with 'good' genes, but also involves learning to be right-handed. Disaster at the last moment is only narrowly avoided when a sympathetic security guard ignores the crucial slip as Vincent provides an important urine specimen: 'For future reference, right-handed men don't hold it with their left. It's just one of those things.' Lest such a scenario seems to be in the distant future, sperm and ovum banks in the United States are already providing catalogues which describe donors, including 'religion of the donor … sexual orientation, *left or right handedness*, language skills, musical skills, favourite color and favorite sports' (emphasis added). There is undoubtedly scope here for what one might call 'genism', perhaps anticipating a situation in which, as Vincent says in *Gattaca*, 'I guess no-one orders southpaws any more.' Perhaps, though, there is hope in the theory I proposed earlier – that the left-handed gene exists because it confers creativity and special talents on those who carry it – in which case left-handers could also be at a premium.[26]

Many of the problems that left-handers experience in society are due to erroneous thinking about left and right, errors which are surprisingly widespread. The next chapter will be devoted to some of these. Both it, and the chapter after, are of necessity collections of miscellanea, often interesting but inevitably lacking a coherent theme. The reader more interested in following the big story about symmetry and asymmetry in the universe may prefer to jump straight to chapter 14.

VULGAR ERRORS

In 1646, during the English Civil War, Sir Thomas Browne (Figure 12.1) published one of the more curious classics of English literature, the *Pseudodoxia Epidemica*. Only three years earlier, in 1643, Browne, a practising doctor in Norwich in East Anglia, had published the *Religio Medici*, his reflections on religion, medicine and life, which was an immediate success. *Pseudodoxia Epidemica* was similarly successful, reprinting six times and being translated into French, German and Dutch. Sub-titled *Enquiries into Very Many Received Tenets, and Commonly Presumed Truths*, the book as a whole is more usually known as *Vulgar Errors*. It is not so much a book to read as an encyclopaedic catalogue of many errors that were 'epidemic' at the time. Nowadays, we would probably think of them as 'memes' – cultural mutations or urban myths, spread by word of mouth, and maintained by their appealingness to the human mind.[1]

The literary style of *Vulgar Errors* is far from easy for a modern reader, the late W. G. Sebald describing how,

> Browne wrote out of the fullness of his erudition, deploying a vast repertoire of quotations and the names of authorities who had gone before, creating complex metaphors and analogies, and constructing labyrinthine sentences that sometimes extend over one or two pages, sentences that resemble processions or a funeral cortège in their sheer ceremonial lavishness.

The book, though, is marvellous in its 'enumeration of many unconnected particulars'. A predecessor was the *Erreurs populaires* (or *Popular Errors*) published in 1578 by Laurent Joubert, a doctor at Montpellier, where Browne himself would study in 1631. Joubert's book also has a marvellous range, much concerned with medicine, sex and child-rearing, such as 'Whether it is true that one can tell from the knots in the cord of the

Figure 12.1 Engraving of Sir Thomas Browne (1605–1682).

afterbirth how many children the woman who just delivered will have', 'that baths people make at home are too hot' and 'whether it is right to say that pear with cheese is bound to please'. It's hardly surprising also to find, 'Whether it is good to stop children from using their left hands'.[2]

Browne's book sits on the cusp of an intellectual revolution, between the humanistic scholarship of the Renaissance, with its emphasis on a detailed reading of classical literature and the citing of sources, and the scientific revolution that was just beginning. Although Browne never became a fellow of the newly founded Royal Society – it is said his prose was too flowery – he was well acquainted with some of the major scientists of the time, including Robert Boyle, who discovered the gas law which is now named after him, and William Harvey, the discoverer of the circulation of the blood. In his house in Norwich, in what he called an 'elaboratory', Browne carried out some of the first experiments in chemical embryology, probably inspired by the embryologist Fabricius, who worked in Padua, where Browne once studied.[3]

Browne assessed the vulgar errors on 'the three determinators of truth – authority, sense and reason'. That is very much also the approach of modern science: *authority* – citation of the opinions, thoughts and experiments of earlier scientists; *sense* – observation and the collection of experimental evidence through the senses; and *reason* – logic and argument. Having said that, Browne did not apply them equally, and his conclusions often supported his own beliefs, which were those of the Protestant established church. Dr Samuel Johnson, in his *Life of Sir Thomas Browne* of 1756, is particularly critical, pointing out that:

Notwithstanding his zeal to detect old errors, he seems not very easy to admit new positions; for he never mentions the motion of the earth but with contempt and ridicule, though the opinion, which admits it, was then growing popular, and was surely plausible, even before it was confirmed by later observations.[4]

Several of the errors discussed by Browne were explicitly concerned with handedness, in its various forms. This chapter will therefore begin with one of his examples, before looking at some more modern examples, which no doubt would have fascinated him.

'That a badger hath the legs of one side shorter than of the other'

Browne tells us that this opinion although 'not very ancient, is yet very general', particularly emphasising that it is 'assented unto' by those who behold and hunt badgers daily. Although not mentioned by Browne, the belief seems to come from the idea that having legs of different lengths would allow badgers to escape by running along the furrows of ploughed fields, the longer leg in the furrow and the shorter leg on the ridge. It is certainly a charming image. It should not come as a surprise that 'the brevity [is] by most imputed to the left', for left–right symbolism inevitably means that if one side is shorter it must inevitably be the left.[5]

Browne had little sympathy for the idea, and marshalled all three of his methods to attack it. Firstly he tells us that an authority, Aldrovandus, 'plainly affirmeth there can be no such inequality observed'. Secondly, he half-heartedly attempted to look for the difference himself – 'for my own part, upon indifferent enquiry, I cannot discover this difference'. Particularly interesting is his third method, the use of reason and argument. While accepting that in some cases animals do have legs of different length, he emphasises that this is invariably a difference between front and back, rather than left and right. In passing, he admits that lobsters often have the 'chely or great claw of one side longer than the other', but he then asserts that this is not properly a leg, but better regarded as a form of arm. He goes on to argue that a badger having legs shorter on one side would be a 'monstrosity' that is 'ill contrived', and that this condition would sorely interfere with an efficient movement, and hence 'the imperfection [of the idea] becomes discoverable'. Intriguingly, Browne does not mention what to modern thinkers would be the obvious objection to the idea: that symmetry is necessary for going in a straight line, and that an animal with different leg lengths would tend perpetually to go in a circle (and, indeed, would do so even in ploughed furrows unless the legs precisely matched the ridges and furrows). Such an argument would be typical of much of modern evolutionary thought, with a theoretical analysis of the selection pressures imposed by the world as well as the likely advantages and disadvantages of the outcomes, which is as important as actual data.[6]

The next example is also drawn from Browne, although with some modern twists as well. After that, all of the errors will be modern.

'That the heart is on the left side'

If this really were an error, then much of the present book would be misguided or plain wrong, so a closer examination is worthwhile. Browne himself is dogmatic that the heart is actually in the centre:

> That the heart of man is seated on the left side is an asseveration, which, strictly taken, is refutable by inspection, whereby it appears the base and centre thereof is in the midst of the chest.

The French playwright Molière would have loved the idea. In his comedy *Le Médecin malgré lui*, he used the possibility that doctors might not even agree on the position of the heart to mock the medical profession:[7]

> *Géronte*: It seems to me you are locating them wrongly: the heart is on the left and the liver is on the right.
> *Sganarelle*: Yes, in the old days that was so, but we have changed all that, and we now practise medicine by a completely new method.

In 1993, the Edinburgh Science Festival commissioned an interview survey as part of its pre-conference publicity. Its press release said, 'Two thirds of the public appear not to know the position of the heart...the majority subscribed to the common misconception that the heart is on the left side of the body.' The percentages answering Left, Centre and Right are set out in Table 12.1.

Table 12.1: Edinburgh Science Festival survey on the position of the heart

		Position of heart (%)		
		Right	Centre	Left
	Overall	20	36	43
Social class	AB	15	50	35
	C1	14	38	48
	C2	16	41	43
	DE	28	27	45

The results are intriguing. 'Centre', the correct answer according to the writer of the press release and to Sir Thomas Browne, was particularly common in people of higher social class, and presumably also of greater education. So 'centre' seems to be the more sophisticated answer. Also remarkable, and ignored by the press release, is the high proportion of people who said the heart was on the *right* side. Although this might reflect gross anatomical ignorance, it may also in part be due to right–left confusion.[8]

What is the basis of the idea that the heart is not on the left but is central? It is probably several fold, as in many social phenomena, but part of the problem lies in the complexity of our anatomy. In *Popular Fallacies*, a twentieth-century version of *Vulgar Errors*, A. S. E. Ackermann took up this question and accepted that both the most muscular part of the heart, the left ventricle, and the main artery, the aorta, are on the left, so that the heartbeat is felt most clearly on the left side. He then muddied the water, though, by pointing out that a vertical slice downwards through the breastbone would have more of the heart on the *right* than the left. The answer, in the end, depends on what you mean by 'left', 'right' or 'centre'. Sir Thomas Browne gave the example of a sundial, in which a sloping piece of metal, the gnomon, is attached at the centre and reaches to the edge, and yet is still described as 'in the middle'. On that basis, since the heart reaches from the centre of the chest to the left-hand side, it might just be said to be 'in the centre'. That, though, smacks of semantic gamesmanship. If something is truly in the centre, then it looks the same in a mirror – there is a symmetry in other words. But Figure 1.1 in the first chapter shows that this is clearly not the case for the organs of the chest. The heart is not in the centre, but on the left.[9]

Even if subtle semantic distinctions explain Browne's rather eccentric claim that the heart is central, how do we explain all the people in the survey who thought that the heart was not on the left? Much of the problem is that people's understanding of anatomy is often very poor, as can be seen in the diagrams of Figure 12.2, in which people were given an outline of the human body and asked to show the location of various organs, including the heart.

Many people are very uncertain what is going on inside their bodies, having little idea, for example, that the stomach is on the left or the gall bladder and liver on the right. Striking also is that so many of the organs appear much smaller than they truly are. The heart, for example, is a fairly large organ, weighing about 300 grams – it needs to be that large to pump sufficient blood around the body, day in and day out. There is no way that such a large chunk of meat could lurk in one tiny corner of the chest. It has, in other words, inevitably to be fairly near the mid-line.

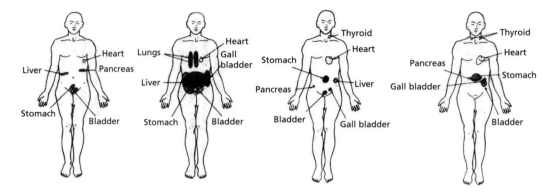

Figure 12.2 The position of various organs of the body as understood by members of the public.

That may be one reason why the more sophisticated survey respondents suggested that the heart is in the centre.[10]

What about the pain from heart disease? Everyone has seen those dreadful B-films in which a dire plot can only be rescued when, in a stressful emotional situation, a character suddenly puts his or her hands on the left side of the chest and falls to the floor, dead from a heart attack. Surely the pain from a heart attack is left-sided? No, for here also are myths. The pain from angina – that gripping, almost strangling, chest pain felt on walking upstairs or uphill – and the pain of myocardial ischaemia (a heart attack or coronary), with its deep, crushing quality behind the sternum, or breastbone – 'like having an elephant sitting on your chest', as patients sometimes describe it – are typically felt in the centre of the chest. The pain is central because the sensory nerves of the heart are the left *and right* vagus nerves, reflecting the heart's origin as a tube in the centre of the body. Heart damage therefore stimulates both the right and left vagus nerves, and the pain is felt in the body's mid-line. When chest pain is indeed left-sided, particularly if far to the side and just beneath the left breast, then the diagnosis is usually 'atypical chest pain', a very different kettle of fish which rarely has anything to do with angina or coronary artery disease, and may well have a psychological origin.[11]

Taking everything together, the truth about the location of the heart is probably best summed up, albeit completely inadvertently, in the title of a recent book: *The Heart is a Little to the Left*.[12]

That identical twins can show mirror-imaging

This is mostly myth but does have some residual truth. There is no doubt that identical twin pairs often have one left-hander and one right-hander, but, as we have seen in chapter 7, that can easily be explained by the workings of the *C* gene. Pairs of twins can also be found in which one has a normal left-sided heart and the other has *situs inversus*, with the heart on the right, and the other internal organs also reversed. This, though, can readily be explained by genes such as the *iv* gene or the genes for Kartagener's syndrome, which were described in chapter 5. Identical twins have also been described in which other parts of the body are mirror-images, such as the face or the teeth, or the hair whorl on the back of the head, which goes clockwise in about three-quarters of people and makes the hair part on the left.

Because identical twins come from a single fertilised egg, it is tempting to suggest that, as that egg splits into two, so the right and left halves become mirror-images of each other. There are many things wrong with such a theory, not least that many identical twins, particularly those sharing membranes, result from splitting occurring very much later, often eight or ten days into development, when the embryo already comprises millions of cells. The major problem for any theory of mirror-imaging is that the mere existence of pairs in which, say, one is right-handed and the other left-handed cannot prove the existence of mirror-imaging. Ordinary single children are left-handed in about ten per cent of cases, and, therefore, genetically identical twins can also on occasion show different handedness (see chapter 7). If mirror-imaging does occur, then its signature ought to be that left-handedness should be more common in identical than in non-identical twins. However, despite many studies, there is no evidence to suggest that identical twins are particularly likely to be left-handed. That sounds the death knell for mirror-imaging as an explanation of handedness in twins.[13]

That the left testicle hangs lower because it is heavier

I cannot resist this one, simply because I discussed it in perhaps the most notorious paper I have ever published. The testicles are indeed asymmetric, and in most men the right testicle is higher than the left. As the eighteenth-century art historian Johann Winckelmann commented, concerning Classical Greek and Roman sculptures: 'Even the private parts have their appropriate beauty. The left testicle is always the larger, as it is in nature.' I'd been aware of this quotation for a while, and when travelling around Italy as a student I took the opportunity to check it out, wandering around museums, notebook in hand, looking at several hundred classical stone scrota – and, it must be said, getting some very strange looks in the process. Winckelmann was correct. In the majority of statues, the right testicle was higher and the left testicle was larger (at least on a brief visual inspection). The only problem is that in flesh-and-blood men not only is the right testicle higher, it is also slightly *larger*. So why had those astute and assiduous Greek observers of beautiful young men got it so wrong? The answer lies in a triumph of theory over observation. The function of the testicles was not at all clear to the Greeks. Aristotle, for instance, suggested that the testicles acted as weights which kept open the ducts where the seed was discharged, and also tensioned the entire body, among other things causing the voice to deepen during puberty. Since the weight of the testicles was seen as a crucial part of their physiological function, it is hardly surprising that the lower testicle, the one which had pulled down further, also had to be the heavier – simple mechanics but, unfortunately, incorrect biology. My brief paper on the topic, published in the prestigious science journal *Nature*, itself a commentary on an elegant paper showing that the asymmetry of the testes is even found early in fetal life, achieved a minor notoriety.[14]

That left-handers die younger

'Lefties die young' was the title of an editorial in the *Guardian* newspaper on 5 February 2001. It is a popular myth, and is almost certainly wrong, but that has not stopped it having wide and continuing currency. The myth began in 1988, when psychologists Diane Halpern and Stan Coren published a brief paper in *Nature* suggesting that left-handers died earlier than right-handers. They had looked at an encyclopaedia of baseball players, which recorded the handedness of each player and their age at death, and found that the left-handers died at a younger age than the right-handers. A further paper from the same authors in 1991, in the *New England Journal of Medicine*, described a group of recently deceased left-handers who had died at a younger age than a comparable group of recently deceased right-handers. The findings were well publicised, despite being entirely the result of a statistical artefact. Much of the problem is shown by the huge size of the supposed difference in lifespan of right- and left-handers – more than seven years. If correct, it would be the most substantial influence on human lifespan known to modern public health medicine, equivalent to smoking 120 cigarettes a day and dwarfing everything else in preventative medicine. Somewhere there had to be an error.[15]

The first thing to realise is that there is nothing wrong with the basic data – when they die, left-handers are indeed younger than right-handers. Paradoxically, however, that does not mean that left-handers have a lower life expectancy than right-handers. The problem is a well-known one in epidemiology and is peculiar to studies known as 'death cohorts', which take a group of people who died at the same time and look backwards at their lives. In contrast 'birth cohort' studies, which are free of problems, take a group of people born at the same time and look forwards through their lives. Everyone in a birth cohort study has their twentieth, fortieth and sixtieth birthday on the same date, whereas in a death cohort there are large differences in the date when people have their birthdays. Whereas a twenty-five-year-old would have had their twentieth birthday five years ago, the twentieth birthday of a ninety-year-old would have been seventy years ago. Many things were very different seventy years ago, including society's attitudes to left-handedness. In Western societies, the proportion of left-handers was much lower at the beginning of the twentieth century than at the end, as was seen in Figure 9.1. On average, then, left-handers have been born later in the century than right-handers. It is hardly surprising, therefore, that left-handers who have recently died were born later in the century, and will thus appear to have died at a younger age. The problem of the death cohort is best seen if one looks not at the age at death in people who have just died, but instead at the current age of living people. Just as dead left-handers died younger than dead right-handers, so living left-handers are, on average, younger than living right-handers, the difference in age being precisely the same in each case. Handedness is, in effect, a marker for being born later in the twentieth century. To make the point more clearly, people who read Harry Potter books are younger than those who do not. Ask the relatives of a group of recently deceased people whether their loved one had read Harry Potter and inevitably one will find a younger age at death in the Harry Potter enthusiasts, but that is only because Harry Potter readers are younger overall. There is no need for a government health warning on the cover of Harry Potter books.

That left-handers suffer more from immune disorders

'Some scientific research has…linked left-handedness with problems of the human immune system,' said the *Observer* newspaper of 4 February 2001, while also discussing the suggestion that left-handers might live less long than right-handers. To be fair the article also said, 'the field is highly controversial and inconclusive'. The idea that left-handers are more vulnerable to allergies and immune disorders goes back to a study published in 1982 by neurologists Norman Geschwind and Peter Behan. Geschwind thought he had noticed a higher rate of left-handedness in his patients with immune disorders. He and Behan tested the idea by distributing questionnaires to left-handers attending a London shop that specialised in items for left-handers. They compared these left-handers with a group of right-handers and found a higher rate of immune disorders in the left-handers. Geschwind and Behan's article instantly attracted much media attention, and another myth was born.[16]

Geschwind subsequently nurtured his myth with a series of three long and dense papers, later published as a book, which put forward a complex theory relating left-handedness to a massive range of other conditions. Central to the theory was the idea that people showed different levels of the male hormone testosterone during fetal development, and that testosterone affected both brain development and the development of the immune system. So complicated was the theory overall that Phil Bryden and I could only summarise it in a complex diagram reminiscent of Crewe railway junction. The problem with an exceedingly complex theory, particularly one which invokes something that is essentially unmeasurable – such as the levels of a hormone early in fetal development – is that everything seems to be explained by it and no one observation can easily prove it wrong. Indeed, a particular problem for medicine and science is that erroneous ideas much more easily enter the literature than leave it. Huge numbers of negative findings are often insufficient to dislodge a single positive result if enough people believe in that positive result. So it was with the Geschwind and Behan study.[17]

Nevertheless, the Geschwind theory did have an Achilles heel, because it said that left-handedness and immune disorders, both relatively easy to measure, are associated. Many researchers carried out studies searching for the link. By 1994, Phil Bryden and I had unearthed eighty-nine different studies, carried out by twenty-five teams of researchers, and involving over 21,000 patients with immune disorders and 34,000 controls. Overall the results were pretty clear. Left-handers showed no systematic tendency to suffer from disorders of the immune system. Why, then, had Geschwind and Behan gone so wrong? The most likely answer is that by carrying out their study in a shop for left-handers, and assessing disease only on the basis of self-reporting by left-handers themselves, they had inadvertently introduced a bias.[18]

That Neanderthals were left-handed

Homo sapiens is not the only species of early modern human to have walked the Earth. For many hundreds of thousands of years in Europe, there was at least one other species, *Homo neanderthalis*, Neanderthal man, who became extinct less than thirty thousand years ago. Despite the large number of fossils that have been found, as well as evidence of stone tools and other artefacts, much controversy surrounds the differences between *Homo sapiens* and the shorter, more heavily built Neanderthals, with their heavy eyebrow ridges and coarser facial features. Many myths have also grown, one of which has been put forward by Stan Gooch in his eccentric book *The Neanderthal Question*. Gooch suggests that Neanderthals, as well as being shorter, brown-eyed and short-sighted, with dark curly hair – all of which distinguished them from the more numerous, taller, blue-eyed, longsighted, fair-haired Cro-Magnons – were also entirely left-handed, in contrast with the right-handedness of the Cro-Magnons. Gooch also suggests that modern humans resulted from interbreeding of Neanderthals and Cro-Magnons. Modern humans who are short-sighted or left-handed are therefore revealing their Neanderthal origins. It is a theory completely devoid of supporting evidence, and what little evidence we do have suggests that Neanderthals had the same brain asymmetries as modern humans, and therefore would have been right-handed. Indeed, Stephen Jay Gould has commented that 'we are fairly confident that righthandedness predominated in Neanderthals'. What is interesting about Gooch's theory, as Robert Hertz would have recognised, is the way two contrasting human species have been assumed to differ in a host of other ways, including inevitably right and left, resulting in a powerful myth-generating engine.[19]

'Neanderthal' is sometimes used as a term of abuse today. Gooch goes one step further, linking Neanderthals not only with left-handedness but also with left-wingedness: 'The Labour Party, not just metaphorically, but quite literally, is the Party of Neanderthal. The Conservative Party, not just metaphorically, but quite literally, is the Party of Cro-Magnon. Labour MPs in Britain, for example, are on average shorter than Conservative MPs.' Given that the best evidence that Gooch could then muster was the prominent eyebrows of some senior trade-unionist Labour Party members of the time, it is clearly time to make one's excuses and leave the discussion...

That the two hemispheres of the brain are responsible for 'Dr Jekyll and Mr Hyde'

Dr Henry Jekyll, in his final confessionary letter at the end of Robert Louis Stevenson's *The Strange Case of Dr Jekyll and Mr Hyde*, describes how he 'drew steadily nearer to that truth by whose partial discovery I have been doomed to such a dreadful shipwreck: that man is not truly one, but truly two'. If 'man' is truly two, an ape and a saint, a Jekyll and a Hyde, locked together inside a single body, then the temptation is great to suggest that one must be in the left hemisphere and the other in the right. Indeed, the sacred–diabolic dichotomy has forced some to say the

sacred is located with language in the left hemisphere, for 'in the beginning was the Word' (John 1:1). Theologically, it is then a small step to ask
if the split-brain patient has one soul or two, and to conclude that, if
each person has only one soul, then that soul resides not in Descartes's
central pineal, but in just one of the two hemispheres, the left.[20]

Psychotherapy of various persuasions has sometimes also been seduced
by the idea of two brains, seeing the warring hemispheres as suitable for
hypnosis or other more mechanistic interventions aimed at influencing
one half while the other half conveniently keeps quiet. Whether the therapies work is difficult to ascertain, particularly given that non-specific
placebo effects can be so powerful in psychotherapy. More problematic
still is to assess whether any of the changes produced are due to an altering of the neurological relationship between the right and left hemispheres. Probably not, is the answer. 'Right' and 'left' are mainly being
used as metaphors in such therapies for describing different ways of
thinking about the world; one relatively logical and verbal, and the other
more intuitive.[21]

Jekyll and Hyde, to the layman, are often associated, quite erroneously,
with schizophrenia. The comparison should properly be with multiple-
personality disorder. In schizophrenia, the split is not between two separate personalities but rather between different aspects of mind, such as
between thought and feeling, or feeling and action. Multiple-personality
disorder might look as if it could be explained by one personality being
in one hemisphere and the other in the other. That is tempting until one
realises that many cases of multiple-personality disorder have more than
two distinct personalities, and then the limitations of the right–left
dichotomy become apparent, for where are the third, fourth and further
personalities to go? The possibility that there might be additional personalities was something that Dr Jekyll himself reflected on: 'I say two,
because the state of my own knowledge does not pass beyond that point.'

That hooked writing indicates which side of the brain has language

'Just now Harry was jotting down a note, and you couldn't help but be
aware that he was left-handed, because he was the type of awkward left-
hander who hunches way over and curls his shoulder, arm, wrist and
hand into a pretzel shape as he writes' is how Tom Wolfe, in *A Man in
Full*, described the hooked or inverted writing position (Figure 12.3).

About thirty per cent of left-handers use the hooked position when
writing, the most usual explanation being that they do it either to avoid
smudging the ink or to make it easier to see what they are writing. There

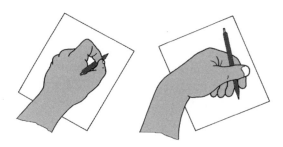

Figure 12.3 The normal (left) and the hooked, or inverted, writing position (right) in a left-hander.

is also a tendency for parents who write in the hooked position to have children who also use the hooked position, suggesting that imitation could be important. More difficult to explain in any of the theories is why perhaps two to three per cent of *right*-handers also use the hooked writing position.

Because thirty per cent of left-handers are hooked writers and thirty per cent of left-handers have language in the right hemisphere, an influential theory from the 1970s was that the hooked position was a marker of language in the right hemisphere. Given the difficulty in the days before brain scanning of knowing in which hemisphere language was located, the idea rapidly gained currency, and was soon found in textbooks of psychology and psychiatry, as well as in popular accounts of handedness. More systematic studies, however, have simply been unable to find any evidence that hooked left-handed writers are more likely to have language located in the right side of the brain than non-hooked left-handers. The reason for the hooking therefore remains a mystery, although it does seem to have some advantages in terms of mechanical efficiency.[22]

That left-handers are more intelligent on average

A typical claim is that 'left-handers are twice as likely to qualify for membership in Mensa, the high-IQ society'. Although one study did find a higher rate of left-handedness in members of Mensa, the result was probably unreliable, not being found in another similar study. The question is best studied by looking at measures of intelligence in large-scale studies of representative populations, of which there have been two. I was involved in one, using data from the UK National Child Development Study, which had looked at over 11,000 children at the age of eleven. The result was clear enough in statistical terms. There *was* a difference in intelligence, but it was *right-handers* who were slightly more intelligent. Taking the two studies together, the effect was so tiny as to be of interest only to statisticians. Compared with an average IQ of 100, left-handers had an average IQ a mere half-point or so lower, at 99.5 – of no practical consequence. It was thought that this result might reflect the statistical excess of left-handers in the population who are severely retarded, but that could not be the explanation since the difference remained even when those with IQs less than eighty were removed from the calculation.[23]

That left-handers are more creative on average

This myth takes many forms, but it resurfaced recently in a study of company directors of traditional and dot.com companies, with the dot.com entrepreneurs supposedly being 'twice as likely to be left-handed or ambidextrous'. Although there are recurrent claims of increased creativity in left-handers, there is very little to support the idea in the scientific literature, and most of the popular literature merely cites anecdotes about Leonardo, Holbein and Paul McCartney, ignoring the fact that for every McCartney there seem to be another nine talented rock musicians who are right-handed. Although there is one much-cited study in the scientific literature finding an excess of left-handers among architects, it is the old, old story in laterality research that the finding doesn't seem capable of replication. This, then, seems far too little on which to hang a theory of any consequence. Having said that, it is possible, as suggested in chapter 9, that if left-handers are more variable in their cerebral organisation, then *some* of them could be more creative, although there would also be a large number of creative *right*-handers from the same source.[24]

That the right half of the brain is underused in Western societies

This idea seems to have originated in the 1960s, with the deeply flawed concept of 'hemisphericity' – the idea that many people use only one side of the brain for solving a problem, and that, with proper training, we can voluntarily will the other side to solve problems and thereby become more creative, imaginative, relaxed, attentive and just about anything else we care to name. It spawned such books as *Drawing on the Right Side of the Brain* and *Right Brain Sex*, both of which are probably adequate self-help manuals for people who wish to draw better or be sexually more fulfilled, so long as one accepts that the right–left part of the story is mere metaphor and has nothing at all to do with the two large lumps of wet stuff in the left and right halves of our skulls.

The idea that the right half of the brain is underused in Western society (compared, of course, with the mystical East) has about as much going for it as that other recurrent urban myth that most adults use only ten per cent of their brain to carry out everyday tasks. This misconception appears frequently in everyday discussions, even among otherwise well-informed people, yet it has not a hint of factual basis. Brain-scanning studies invariably show that all of the brain is active most of the time. Nor is there any suggestion that, overall, the left half is used more than the right half. The most charitable explanation for such claims is that people find them vaguely reassuring, perhaps being persuaded that they have the potential after all to be creative geniuses, if only they could 'put their mind to it'. The problem is that if their brain really were ten times more powerful and they did 'put their mind to it', they would soon realise that the idea is half-baked in the first place.[25]

That Picasso was left-handed

Picasso was the most famous, prolific and creative of all modern painters, reinventing himself stylistically throughout his life. Left-handers therefore proudly proclaim him as a left-hander on a myriad of enthusiastic websites written for their fellow left-handers; and there are websites and books produced by academic researchers that say the same thing. There is only one problem in all of this – Picasso was *not* left-handed. He was indisputably right-handed. Picasso had a love affair with the camera throughout his life, at various times at least five photographers living in his house and photographing him extensively. In the post-war period, we have photographs of him in almost every possible situation: signing his name, drawing in a sketchbook, scratching on to 35 mm slides, making an aquatint, cutting out rolled clay, shooting a revolver, turning a skipping rope for his children, and, of course, painting – painting on canvases, painting on ceramic dishes, painting sculptures, painting with a 'light pen' on to photographic film, and painting on glass for the famous film by Clouzot, in which the camera is placed behind the glass. In every single case, without exception, he holds the pen, knife, brush, pencil or stylus in his *right* hand. There is also a picture in 1937 of him painting his master work, *Guernica*, brush in his right hand. Going back further, there is a self-portrait in oils from 1906 with the palette held in the left hand (as would be expected for a right-hander), and a photograph of his studio in 1901 where his brush is in the right hand. There simply seems no doubt that Picasso was a right-hander.[26]

So why does everyone say Picasso was left-handed? Certainly no-one cites any documentary evidence that Picasso claimed to have been a natural left-hander who was forced to change his writing hand, nor are there any claims that he used both hands equally when painting. The explanation is distinctly unedifying, and tells us much about the ways in which science is often carried out, books are sometimes written, and countless web pages are compiled. Sir Thomas Browne would have recognised it all too well. Much knowledge in books, science journals and web pages comes not from direct experience but from 'authority'. In other words, someone else said it, which makes it true, so someone else says it in turn. Just as people could say badgers had legs that were longer on one side than the other without checking by looking at badgers, so people say Picasso was left-handed without checking any of the thousands of photographs of him. All academics know the problem – it is called secondary citation. Not going back to primary sources, to the raw data, to the original accounts, but instead saying merely what others have already said. I am aware of having nearly made that mistake several times in writing this book, but have fortunately checked the originals before embarrassing myself too much. Even so, there are probably mistakes made in ignorance, some caused by relying on authority. I just hope I haven't previously claimed in some forgotten article that Picasso was a left-hander. No doubt someone will love to tell me should they find it.[27]

Where, then, did the Picasso myth originate? The earliest mention I can find of it is in 1966 in *The Left-hander's Handbook*, by James T. de Kay, where it is mentioned twice. In contrast, there is no mention of Picasso being left-handed in Michael Barsley's *The Left-handed Book*, also published in 1966. Barsley does mention other famous left-handers and it is unlikely that he would have omitted someone as famous as Picasso from his generally thorough account. Nor is Picasso mentioned in Barsley's 1970 book *Left-handed Man in a Right-handed World*, although de Kay is acknowledged in the preface. Likewise, Vilma Fritsch's scholarly *Left and Right in Science and Life*, first published in German in 1964, does not mention Picasso's left-handedness, despite her describing Picasso painting a picture on glass where right and left are reversed. Picasso's left-handedness looks like a 1960s myth, probably originating in de Kay's book and perhaps inspired by the then prevalent idea that the right brain is more creative. That de Kay is a prime suspect is seen in another error. In 1979, he published *The Natural Superiority of the Left-hander*, which not only claims that Bob Dylan is left-handed – a claim now also found on many websites – but also includes a drawing of Dylan playing the guitar left-handed. The only problem is that in every photograph I have ever seen Dylan plays the guitar right-handed. Two such mistakes begin to look like carelessness or worse.[28]

That Einstein and Benjamin Franklin were left-handed

The conclusions of this section are pretty predictable after the previous one. The world's most famous scientist, and one of America's founding fathers – one can see that left-handers would want to claim them for their own, as indeed is done by many websites and popular books.

The case of Einstein is straightforward. Just as Picasso is probably the most photographed artist ever, so Einstein is probably the most photographed scientist. There are many pictures of him writing with his right hand, as well as striking a match with his right hand. No documentary evidence that I am aware of suggests Einstein was either a natural left-hander or a 'switched' left-hander. He is therefore right-handed until proven otherwise. It is, as with all historical figures, possible that he had been forced to change his writing hand, but to claim he was left-handed on that possibility alone is to create what philosophers of science call an 'irrefutable hypothesis'. All the evidence supports the idea that he was a right-hander, and that is the only scientifically justifiable conclusion. When or where this myth arose I am not sure, but the earliest example I have found is in Rae Lindsay's *The Left-handed Book* of 1980. As we are about to see, that is a less than reliable source.[29]

Benjamin Franklin is slightly more complicated, and once again de Kay, in his 1966 book, makes the earliest claim that I can find. There is no doubt that Benjamin Franklin, in 1779, published a witty and much-quoted little piece with the title of 'A petition of the left hand: to those who have the superintendency of education'. Written as a letter, and signed by 'The Left Hand', it complains about the left hand's poor treatment in comparison with her right-handed sister – 'She had masters to teach her writing, drawing, music and other accomplishments, but if by chance I touched a pencil, a pen, or a needle, I was bitterly rebuked.' Although often taken as a statement that Franklin was himself left-handed, there is no other evidence supporting such a conclusion. The logic of presuming that because Franklin wrote about the lack of training of the left hand, he must also have been a left-hander is as sensible as presuming that John Stuart Mill was a woman because he argued for votes for women, or that Abraham Lincoln was a slave because he was opposed to slavery. Much more likely is either that Franklin believed in ambidexterity or that the petition was merely a practical joke, similar to his 1768 *Petition of the Letter Z*, which argued that the last letter of the alphabet was neglected despite the fact that it is 'of as high extraction and has as good an estate as any other letter of the Alphabet'. The argument, however, seemed to be settled in *The Left-handed Book*, where there is a photograph of an oil painting of Benjamin Franklin writing with a quill pen held in his left hand. The painting, however, is the famous one by Mason Chamberlin in the Philadelphia Museum of Art. The dénouement of the story is pretty obvious. The original painting shows Franklin as *right*-handed. Lindsay, or perhaps a picture editor, thought the story would be more convincing if Franklin were shown as left-handed, and the picture was reversed.[30]

The cases of Einstein and Benjamin Franklin are similar to those of Picasso and Bob Dylan. Much quoted as left-handed, there is in each case no documentary evidence supporting the claim, and strong visual evidence to the contrary. Nevertheless, the myths will probably live on for years to come. As George Eliot said in *Adam Bede*, 'Falsehood is so easy, truth so difficult.' These memetic mutations, these cultural lies, would make an interesting study for historians of popular culture interested in the evolution of erroneous ideas. It's good, though, to know that occasionally the truth does emerge, as in the case of the novelist James Michener, nominated by *Southpaws International* as one of its 'Southpaws of the year'. Michener wrote back to say that the only thing he did with his left hand was use it occasionally for scratching his right elbow.[31]

That Kerrs and Carrs tend to be left-handed

The Scottish surnames Carr and Kerr come from the Gaelic *caerr*, meaning awkward, and the expressions *ker-handed* and *carry-handed* are common terms for left-handedness. Popular belief in Scotland is that Kerrs and Carrs are left-handed, and that they built their castles with 'left-handed' spiral staircases, to advantage them in fighting. An anonymous poem describes the Kerrs:

> But the Kerrs were aye the deadliest foes
> That e'er to Englishmen were known
> For they were all bred left-handed men
> And 'fence against them there was none.

The legend of the left-handed Kerrs and Carrs seemed to be supported when, in 1974, the Research Unit of the Royal College of General Practitioners published a report claiming that nearly thirty per cent of modern Kerrs and Carrs were indeed left-handed. If true, it would be extremely interesting, because surnames are inherited in the same way as the male Y chromosome, through the paternal line. A student and I tested the claim by randomly selecting people called Kerr or Carr from the London telephone directory (together with a control group with other British surnames) and asking them about their own handedness and the handedness of members of their family. There was no evidence at all that Kerrs and Carrs are more often left-handed. So where did the Royal College of General Practitioners go wrong? What happened is an object lesson in how not to do research. The study was progressing slowly until a newspaper report described the study, upon which many Kerrs and Carrs immediately wrote to the College. Unfortunately, it was only left-handed Kerrs or Carrs who bothered to get in touch. The sample was biased and, in consequence, the conclusion completely wrong.[32]

That there are tribes which are mostly left-handed

This myth keeps recurring, appearing for instance on the Internet discussion list 'alt.lefthanders', where it was claimed that 'As many as 70 per cent of the Punjab tribe of Fiji used the left hand by preference. There is also a consistently higher occurrence of left-handedness in aboriginal tribes of North and South America, South Africa and Australia.' If this were true, it would be very important for our understanding of right-handedness and brain lateralisation. The idea seems to have come from *The Left-Handers 1997 Desk Calendar*, which perhaps contains more errors per square inch than any other source I know. It actually conflates several sources. Sir Daniel Wilson in his book *The Right Hand: Left-handedness*, published in 1891, does suggest, on the basis of a report by the correspondent of *The Times* who visited Fiji in 1876, that 'left-handed men are more common [amongst the Viti or Fijians] than among white people', although he has little doubt that right-handers are still in the majority. Wilson also cited a report from the *Medical Record* of 1876 that seventy per cent of the inhabitants of the Punjab, in India, are left-handed, although he doubted its accuracy, dismissing it as the 'chance observation of a traveller'. Since a large proportion of the Fijian population originated in the Indian subcontinent, one can see how the two stories perhaps became merged.

Other similar stories keep surfacing with the same regularity as tales of the Loch Ness monster. A recent one comes from Diane Paul who, in *Living Left-handed*, claimed that 'In 1990 Russian scientists discovered that 75 per cent of Taimyr natives living in Russia's Arctic north were left-handed.' As usual, no reference is given. Some claims, such as the one that Vasco da Gama on his voyage to Calicut said that the people of Melinda were all left-handed, originated in a simple linguistic confusion. Going back yet further, there is a story that Alexander the Great found a country where the inhabitants were all left-handed, believing that the left hand should be more honoured because it is closer to the heart (Alexander would have taken note of this, since he himself was reputedly left-handed).

There are two common denominators of all such stories. Firstly, they are never supported by any solid or reliable data, always being mere travellers' tales. In the scientific literature, I know of no society where even twenty per cent of the population have reliably been shown to be left-handed. Secondly, and just as important, people *want* to believe in such stories. Our credulity reflects a recurrent desire – one that Hertz understood – to impose symbolic symmetry on a world that seems asymmetric; and a mirror-society populated by left-handers would do precisely that. This does not mean, however, that such societies exist. They are almost certainly figments of our imagination or fulfilments of subconscious wishes.[33]

That clocks go clockwise because sundials go clockwise

'Why do clocks go clockwise?' is a popular question on websites, particularly sites that encourage children and young people to be interested in science. Invariably, the answer given is that since in the northern hemisphere, where clocks were invented, the sun rises to the east, travels round to the south, and sets to the west, then the shadow on a typical domestic sundial goes clockwise. As a reader of *New Scientist* then put it, 'it was only logical for the sweep of the hands to move in the same direction'. However, as is so often the case, 'only logical' conceals a mass of confusion and complexity.[34]

Early clocks were massive, heavy, expensive affairs, and a neglected triumph of Renaissance technology was, over five centuries, to reduce these vast, inaccurate early clocks to precise, small, reliable timepieces that could be carried or even worn. The size and expense of early clocks meant there might be only a single turret clock in a town, where it would be placed high on a tower so that everyone could see it, along with a bell that meant it could be heard even when out of sight. Before that time, sundials had fulfilled the same role, being placed high on walls so that they could be easily seen by many people and from a distance, in contrast to small, horizontal sundials that could only be read close by. An elaborate and sophisticated vertical sundial from Queens' College, Cambridge can be seen in Figure 12.4. A moment's glance shows the problem of the sundial explanation of why clocks go clockwise. Sundials on vertical walls have to have their hours arranged *anti*-clockwise. Since vertical sundials were the obvious models for the first public tower-clocks, the 'logical' explanation of why clocks go clockwise falls apart completely.[35]

The standard explanation for clocks going clockwise has another more serious problem, because there is always an implicit assumption that clocks have always gone the *same* way. In 1443, the Florentine painter Paolo Uccello painted in fresco a clock dial high on the inside west wall of the Cathedral in Florence (Figure 12.5). This clock is strange in several

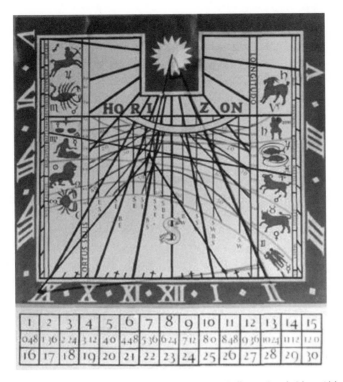

Figure 12.4 The vertical sundial in the Old Court, Queens' College, Cambridge. Although often attributed to Sir Isaac Newton, the original sundial was erected in 1642, the year of his birth, and the revised dial put up in 1733, after he had died.

ways. Firstly, it has twenty-four hours, whereas the eventual, admittedly rather peculiar, standard for clocks was to have twelve hours on the dial and for the hands to go round twice during the day. The twenty-four hours also followed the system known as *hora italica*, with XXIV meaning the hour of sunset rather than midnight, a system that survived into the eighteenth century. The layout of the numbers is also unusual, with 'I' at the bottom rather than at the top. But the most unusual feature of this clock is that it goes anti-clockwise.[36]

Uccello's anti-clockwise clock is far from unique. Just down the road from the Cathedral in Florence is the church of Ognissanti, where, in about 1481, Botticelli painted a marvellous fresco of Saint Augustine in his study. On the shelf behind the saint is a small, domestic clock, complete with visible cogs and wheels, and a face exactly like that painted by Uccello, with the Roman numerals going anti-clockwise, and a single hand that points vertically downwards exactly between the hours I and XXIIII. There were also other types of clock around in the Italian *quattrocento* and *cinquecento*. Some went round once in twelve hours but in an

Figure 12.5 The clock painted in fresco by Paolo Uccello on the inside of the Cathedral in Florence.

anti-clockwise direction; others went round four times in twenty-four hours, marked with the digits I to VI; others, again, went round once in twelve hours but were marked I to VI and then I to VI again; still others went round once in twenty-four hours but were marked I to XII and then I to XII again. No doubt there are yet more variations.[37]

Uccello's clock has been used in economics as an example of what is called 'positive feedback' or 'lock-in'. There are often competing designs for new technologies, recent examples being operating systems for computers and videotape formats. Sometimes, one design is so self-evidently better that its competitor rapidly becomes obsolete, as when thermionic valves in radios were replaced by transistors. On other occasions, however, the relative advantages are minimal or even non-existent. Random fluctuations can then result in one technology being adopted despite it not having any obvious advantage, simply because it is slightly more common, meaning that somewhat more people have it, more people hear about it, and more people can teach other people how to use it. Once one technology gains a greater share of the market, there are resultant economies of scale and it becomes cheaper to produce, further advantaging it over its rivals. As a result, one technology can almost completely displace the other. This phenomenon was first described in 1890 by the Victorian economist Alfred Marshall, in his *Principles of Economics*:

'whatever firm gets a good start' will eventually win, despite the fact that it can be utterly random which firm, technology or design gets the good start. Displacing an existing system can then be almost impossible, despite a mass of obvious absurdities – once again think of the QWERTY keyboards on millions of computers worldwide. The problem is not even restricted to humans. Ant colonies find the same thing – if a longer, less efficient route to food happens to be used first, then the colony continues using it, despite the fact that in other situations ants can optimise extremely difficult problems.[38]

In the Renaissance period, clocks were a newly emerging technology the design of which was yet to be standardised. Uccello's clock of 1443 shows precisely that. The first mechanical clock had probably come into existence by about 1271 AD, and a century and a half later there was still no consistency in the direction of the clock, the position of the numbers on the dial, the writing of the numbers or even the number of hours to be displayed. Within another century, this had all changed as the clock and watch industry developed internationally. By 1550 clocks were becoming mass-produced domestic objects, and a clockwise dial with twelve hours and two rotations in a day was the norm.[39]

So why do clocks go clockwise? The most probable answer is that clockwise happened to get there first, probably owing to chance alone, and the system became locked in to that direction. No other explanation is needed. Having said that, there is a little evidence that the human brain is more attuned to clockwise rotation than to anti-clockwise direction, one study finding that, although memory for direction of rotation is generally very bad, there is consistent evidence for a bias towards reporting rotation as clockwise, whatever the actual direction. Clocks might then go clockwise because our brain, in some sense, goes clockwise. What is clear is that clocks do not go clockwise because sundials go clockwise.[40]

That mirrors reverse left and right but not up and down

A character in a short story by Jorge Luis Borges declares that 'mirrors and copulation are abominable, because they increase the number of men'. Look at yourself in a mirror, and where once there was only one person, now there are two, for someone very like you stares back from the glass. However, if you move your right hand, the person in the mirror moves their left hand; if you wink your right eye, they wink their left eye. That extra person's clothes are also switched, with a breast pocket on the right rather than the left, and buttons done up in reverse manner. Even a flower in the buttonhole or medals on the chest are on the right rather than the left. All the objects in the room are also reversed – the writing on books and newspapers is in mirror-writing, pictures are reversed, and so on. Seemingly more mysterious, perhaps even monstrous, is that nothing else in the mirror seems to have changed. In particular, top and bottom are still where they should be, and front and back don't look any different either. So hasn't the mirror reversed left and right? No is the answer, although it is convenient to talk as if it has. It is such a hoary old chestnut of a problem that I have left it until last.[41]

The Greek philosopher Plato discussed the problem in the *Timaeus*, as also did the Roman poet and philosopher Lucretius in his long poem *De Rerum Natura*, and the problem has continued to intrigue, provoke and infuriate since then. As is so often the case, the physicist Richard Feynman seems to have had a particularly clear insight into the problem, which we will return to later.[42]

Conventional mirrors of the sort used for shaving or washing do not, in fact, reverse left and right, but instead reverse *front* and *back*. That may seem strange, but it at least explains why we sometimes talk of mirrors producing an image that is 'back to front'. A conventional ray diagram shows what happens to the light coming to our eye from an object seen in a mirror (Figure 12.6). The right-hand end of an image is still on the right and the left-hand end is still on the left. However, the front and back of the object are very clearly reversed.

Describing a mirror as reversing front and back is good enough, given the way we usually use mirrors, but it is more accurate to say that a mirror reverses the dimension at right angles to its surface. When we look into a vertical, wall-mounted mirror, we do indeed see back to front, but as Umberto Eco points out, if we should 'more generally be used to mirrors fixed horizontally to the ceiling, as libertines are used to them', then we would know that horizontal mirrors reverse top and bottom, and not front and back, for those are the directions at right angles to the mirror. Likewise, if you stand sideways on to a mirror, and look at the mirror out of the corner of your eye, right and left are now properly reversed. Intriguingly, though, in each of these situations the person in the mirror is always left–right reversed – waving their left hand back if we wave our right, and so on.[43]

So why does the mirror *appear* to reverse right and left. The answer to that lies not in optics but in psychology. When we describe objects in the mirror, we use the terms, top–bottom, front–back, and right–left. Once any two of these are specified, then the third is also defined. Because we are two-legged organisms, which stand upright (so that our head end, at the top, is defined as the part further from the centre of the planet), and we have eyes on one side of our head (so that front is the direction in which our eyes are looking and back is where we can't see), then those

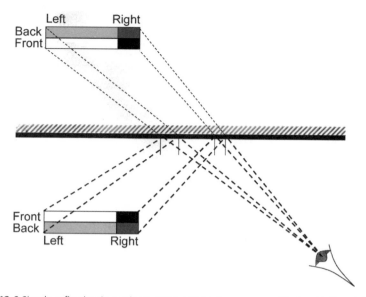

Figure 12.6 Simple reflection in a mirror. Front and back are reversed but not left and right.

are the two primary dimensions. Right–left can only be defined after top–bottom and front–back have been identified. The mirror always reverses one dimension, but that can only be described once top–bottom and front–back have been identified, and therefore whatever dimension a mirror actually reverses physically it will always be *described* as right and left being reversed.

Another way of thinking about mirror-reversal is to try an experiment. On a piece of paper write **Right** on the right-hand side and **Left** on the left-hand side. Hold the paper up to a mirror and, as expected, everything is reversed – the writing is back to front, and the word **Right** is now to the left of the word **Left**. Left and right seem to have been reversed (Figure 12.7).

Now do the experiment again, but this time write **Right** and **Left** on a transparent piece of glass, and hold the piece of glass straight up to the mirror. Figure 12.8 shows that when you do this *the writing in the mirror appears exactly the same as that written on the glass itself*. The original and the mirror-image are visible at the same time and both are identical.[44]

Do the experiment once more, but this time make sure that you make all the same movements in picking up the piece of glass that you made in picking up the piece of paper. You will notice that you have to cross your arms over, and in the process the piece of glass has been rotated 180 degrees around the vertical axis. Now, the writing in the mirror is reversed, as was the writing on the paper.

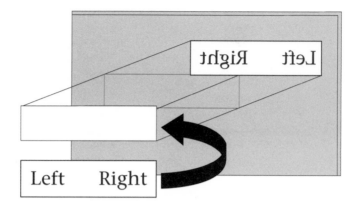

Figure 12.7 When a piece of paper is looked at in a mirror, it is itself first reversed in front of the viewer, and hence the writing appears reversed.

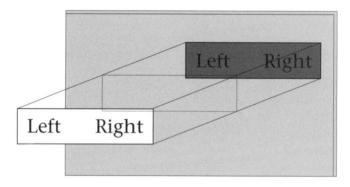

Figure 12.8 Writing on a piece of transparent glass or perspex is not reversed in a mirror if the glass or perspex is itself not turned around in front of the mirror.

It should now be clear that mirrors reverse writing because we turn the writing around before looking at it in the mirror. When we don't do that, as with the writing written on glass, then it is clear that nothing has been reversed left and right. That, quite simply, is why the writing on the paper is turned around in the mirror – because it has been turned around; and turning it around was necessary because, since the mirror reverses front and back, you would otherwise have only seen the back of the paper which was blank, rather than the front on which was the writing. As ever, the physicist Richard Feynman put this clearly:[45]

It is the same with a book. If the letters are reversed left and right [in a mirror], it is because we turned the book about a vertical axis to face the mirror. We could just as easily turn the book from bottom to top instead, in which case the letters will appear upside down.

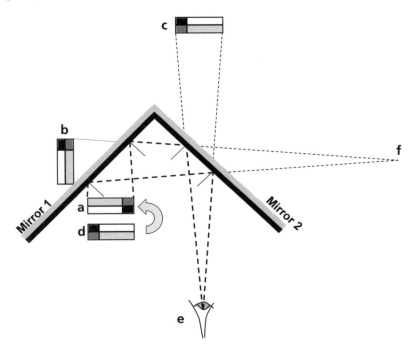

Figure 12.9 An object reflected in two mirrors does not appear to be reversed.

The mirror experiment shows that right and left are the poor relations of top–bottom and front–back when describing the world. If there is a change in just one dimension, we always prefer to describe in terms of left–right, because that is regarded as the least important way. As Feynman implied, if we turn a book upside down we will be well aware of having done it and there will be no mystery, but because we reverse it left to right by rotation, we barely even notice what has happened and are then surprised at the result.

The final twist in this story is that there *are* mirrors which do indeed reverse left and right. It is easy to make one by taking two ordinary mirrors and putting them exactly at right angles to one another. Look at yourself, and you will find that when you wink your left eye the person in the mirror also winks their left eye. Figure 12.9 shows what is happening: the light from the object *a*, is first reflected in Mirror 1 (and if Mirror 2 were not present its image, *b*, would be seen from *f* as 'mirror-reversed', although, as we have already realised, this means that front and back are actually reversed). After bouncing off Mirror 1, the light now passes to Mirror 2, where it is again reflected, and enters the eye, *e*, which sees a virtual image, *c*. Of course, just as before, in order to hold the object *a* up to the mirror, we had to turn it around horizontally from its original position, *d*, and *d* looks identical to the image *c*. The image in the mirror, *c*, is therefore showing *d* (and *a*) as it truly is, with its left on its left and its right on its right. Or so it seems. Actually, because Mirror 1 and Mirror 2 are at right angles, the second reflection has not switched around front and back of the object but instead has switched its left and right. The image has now been inverted *twice* but it looks as if there has been no change. An even number of reflections makes an image look the same as itself, because the mirror-image of a mirror-image is indistinguishable from the original. Our eye and brain have no way of telling if something has been reversed and reversed again, or is simply unchanged.[46]

A double mirror also has the intriguing property that if it is rotated through ninety degrees, then it turns the person *upside down*. To see why, rotate the diagram of two mirrors through a right angle. A double mirror is very different from a single mirror, which, because it is entirely flat, does not produce any difference in the image simply by being rotated in its own plane. The properties of double mirrors were known to Lucretius, and also to Plato, who after discussing how right and left are normally reversed by a mirror, went on to say:[47]

On the other hand, right appears as right and left as left when the visual stream is reversed at the point of coalescence, as when the surface of a mirror is concave and transfers the right side of the visual stream to the left and the left to the right. The same mirror turned lengthwise again, makes the face appear upside down, turning the ray top to bottom.

As a final teaser, what happens with *three* mirrors. Imagine you are in a room in which both of the walls have been tiled with mirrors, as has the ceiling? When you look into the top corner of the room, will your right and left hand be the correct way around or reversed?[48]

A chapter like this not only exposes a range of conceptual errors in thinking about right and left, but it also dips the toe into all sorts of topics, some of which are trivial but nevertheless fascinating. One could fill a whole chapter with them, or even a whole book. I will restrict myself to a brief chapter.

☞   ☜

☞ ☞ ☞ ☞ **13** ☜ ☜ ☜ ☜

THE HANDEDNESS OF MUPPETS

On my bookshelves there is a whole shelf devoted to popular books on left-handedness; and on the World Wide Web there are hundreds of sites about left-handedness, with the number growing all the time. The books and the sites, usually written by left-handers for left-handers, tell and retell the core modern myths that define the experience of being a left-hander. Many are opinionated, many are frankly wrong. Often they have an almost perversely light touch, peppered with bad puns and cartoons, as if left-handers are themselves embarrassed at even discussing the fact of their difference. Lorin Elias, of the University of Saskatoon in Canada, recently reviewed the books and sites, and pointed out that if in the past left-handers have comprised 'a relatively silent minority', with only the occasional book devoted to their experiences, now the Web 'has resulted in a virtual explosion of available information (most of which is advertising)'. Despite much of the material being scientifically unsophisticated, or just plain wrong, the Web does 'provide its mainly left-handed audience with helpful hints and all the left-handed trivia they could ever ask for'. So, if trivia is asked for, trivia is what this chapter will be about. But not mere trivia about left and right for trivia's sake. There are no lists here of second-division soap stars who happen to be left-handed. Instead there are non-trifling trivia which I hope you will find amusing, worthy, informative, or surprising.[1]

The Muppets are left-handed

The Muppets are cute puppets who appear in a successful series of TV programmes; and they are mostly left-handed. They were devised by Jim Henson, who was said to be left-handed. It's tempting, therefore, to say that he was trying to create that world about which so many left-handers fantasise – one in which everyone is left-handed. Not so. Henson may have been left-handed, but almost all the other puppeteers who were helping him were right-handed. Puppets are difficult to control, with the most difficult part being the head, which is therefore given to the puppeteer's right hand, which is the more skilful. The next most difficult part to control is the hand, which is therefore allocated to the puppeteer's left hand; and so muppets are left-handed.[2]

Why the BBC test card is back to front

In 1967, the BBC needed a 'test card' for its new colour broadcasts, to be put on the screen when programmes were not being transmitted so engineers could adjust the quality of the picture. The picture chosen is still used today, and despite being shown only in the middle of the night, is still well known. A BBC engineer, George Hersee, used a now-famous photograph of his eight-year-old daughter Carole, who sits on the left of the screen, wears a red hair-band in her long hair, and is playing noughts and crosses on a blackboard with a piece of chalk held in her right hand. Carole, however, was left-handed, and the picture was left–right reversed at the request of an unknown person high up in the BBC who felt it was inappropriate to have her shown as left-handed.[3]

The Roman emperor Commodus was left-handed

The Roman emperor Commodus was one of the worst of all Roman emperors. Despite being the son of the tolerant Stoic philosopher and emperor Marcus Aurelius, 'every sentiment of virtue and humanity was extinct in the mind of Commodus', as Edward Gibbon put it in *The Decline and Fall of the Roman Empire*. Commodus was also, like the emperor Tiberius, left-handed, a fact which, perhaps not surprisingly, none of the 'lists of famous left-handers' seems to feel worth emphasising. Although he despised all academic learning, he was a superb javelin thrower and archer, 'and soon equalled the most skilful of his instructors in the steadiness of the eye and dexterity of the hand'. His appetite for blood found its most perfect outlet when the emperor himself entered the gladiatorial arena, fighting no less than seven hundred and thirty-five times as a *secutor*. As Dion Cassio tells us, 'he held the shield in his right hand and the wooden sword in his left, and indeed took great pride in the fact that he was left-handed'. His success may well, in part, have reflected the strategic advantage seen by left-handers in certain competitive sports; but being an absolute, total and ruthless dictator probably helped a little as well.[4]

The sad tale of left-handedness in the cinema

Despite being left-handed, the Roman emperor Commodus was portrayed in the multi-Oscar-winning film *Gladiator* as right-handed.
Admittedly, it is not much of a goof given that hardly any other facts in
the film were historically correct. Left-handers are rarely portrayed as
such in films; Joan of Arc, for instance, was portrayed as right-handed by
Ingrid Bergman in the 1949 film of the same name, and was conspicuously shown as such on the poster, holding up her sword. Alexander the
Great, reputedly a left-hander and the subject of a number of films, is also
portrayed as right-handed. Only when forced to do so by the obvious
subject matter of the film does cinema get left-handedness right. An
obvious example is *The Babe Ruth Story* (1948) about the baseball player
Babe Ruth, whose fame was all the greater for his being so good and also
left-handed. Occasionally films try really hard and still get it wrong. *The
Left-handed Gun* (1958) was a rare and apparently worthy exception, in
which Paul Newman played Billy the Kid, the gun-slinger, as a left-
hander. I say 'apparently worthy', because the film was in error. The
legend of Billy the Kid being left-handed is based on the only photograph
that is undisputedly of him, taken in Fort Sumner in 1880. The original
was long lost, but the copies looked pretty convincing as he stood there,
Colt revolver on his left hip and a Winchester carbine in his right hand.
Pretty convincing, but unfortunately wrong. When the original finally
turned up in 1986, over a century after Billy's death, it was found to have
been produced using an old technique known as tintype or ferrograph.
Tintypes reverse the image during processing, as can be seen on close
inspection of the original picture, the buttons and belt buckle being back
to front. Figure 13.1 shows Billy as he actually was, a right-hander after
all. Another celebrity to knock off the lists of famous left-handers.[5]

In the film *Titanic* they only built a set of half the boat

To save costs, an unlikely concept for one of the most expensive films
ever made, the makers of *Titanic* built only one half of the doomed ship,
and they then gave an impression of the entire ship by left–right reversing many of the shots during post-production. That, however, meant
that many of the props and costumes had to be produced in two versions,
one normal and the other mirror-reversed, so that labels, such as the
names on sailors' hats, 'WHITE STAR LINES', had to be inverted to say
'ƧƎИI⅃ ЯAƚƧ ƎTIHW'. Likewise, jackets and coats had to button the other way.
One major decision was *which* side should be built. In the end, it was
decided to build the starboard side, because the prevailing winds on the

Figure 13.1 The tintype (ferrograph) photograph of Billy the Kid, taken in 1880. The right-hand image shows the original print in which Billy looks left-handed. In the image on the left, the original has been left–right reversed to show Billy as he actually looked, holding his rifle with his left hand, and with the six-shooter ready to be used by his right hand. That the original fer-rograph is reversed left and right can be confirmed by looking at the buttons on the waistcoat.

film set then meant that the smoke from the funnel of the static ship would appear to be blowing towards the rear of the boat. That, however, produced a particular problem for the major scene in which the passengers are boarding the ship at Southampton, since embarkation traditionally occurs from the port side (hence the name). The entire scene was therefore subsequently reversed during editing, producing confusion for everyone during the filming of what came to be known as 'flopped' shots. As the producer, James Cameron, said, 'There were some real instances of pretzel logic, though…"It is right? Then it's wrong. It's wrong? Good!"' Although all of the artefacts were left–right reversed, the actors themselves, particularly the many extras, who were mostly right-handed and right-footed, could not be reversed, and therefore appear as left-handed and left-footed in the flopped scenes. An interesting challenge is to go through the film in slow motion, watch the scenes on the port side of the ship, and see if one can tell that this is a mirror world. In particular, look at the extras waving goodbye, and at the ship's hawsers.[6]

How not to eat in Arabia Deserta

Near the beginning of the film *Lawrence of Arabia*, T. E. Lawrence is being escorted across the desert to meet Sherif Ali by Tafas el Raashid. Tafas offers Lawrence some Bedouin food, which Lawrence gratefully accepts and eats with the left hand, the hand which no Arab would use for eating, but which they would instead use principally for cleaning themselves below the waist. It is unlikely that Lawrence, who knew the Middle East extremely well, would have made such a cultural gaffe. The subsequent murder of Tafas by Sherif Ali also has no basis in historical fact, Lawrence, in the *Seven Pillars of Wisdom*, describing how Tafas returned unharmed to his village. But 'accuracy' and 'Hollywood' are words that every day drift further and further part.

On rare occasions, artists tell us which hand was used to produce a work of art

In 1934, Salvador Dali produced what he called an 'espasmo-graphisme', which was an aquatint engraving entitled *Onan*. In the lower right-hand corner there is a message which explains, in French, how the picture 'was obtained with the left hand whilst with the right hand I masturbated myself'. It is not one of Dali's greatest works, and one can only presume the normally dominant and highly skilled right hand was otherwise committed to the task requiring greater sensitivity and finesse.[7]

Mirror-writing

In his novel *Jack Maggs*, Peter Carey describes Maggs writing a letter:

> He wrote, *Dear Henry Phipps*, in violet-coloured ink. He did not write these words from left to right, but thus: ꙅqqiⁿ9 ʏⁿи9H ⁿɒ9ᗡ . He wrote fluidly, as if long accustomed to that distrustful art.

Most right-handers, and Maggs seems to be right-handed, mirror-write with their left hand. That is probably because the part of the brain controlling the left hand is the mirror-image of the brain controlling the right hand, with the two brain areas connected by the large fibre bundle called the corpus callosum. Any action causing movement in one hand therefore sets up a mirror-image copy in the other. That is why right-handers are often surprised at how easy it is to mirror-write their name in large letters on a blackboard using their left hand, if their right hand is also writing in the usual direction at the same time.[8]

An early case of spontaneous mirror-writing was reported in about 1700 by Rosinus Lentilius, who described a soldier from Nordlinga who, after his right arm had been hacked off in a battle, started to mirror-write with his left hand. Mirror-writing seems to be commoner in left-handers, Sir Paul McCartney, the left-handed Beatle, describing how, 'When I was a kid I seemed to do everything back to front. I used to write backwards and every time the masters at my school looked at my book, they used to throw little fits.' In a large study carried out in America before the First World War, spontaneous mirror-writing occurred in about one in every 2500 schoolchildren. All of the cases were left-handed, and since at that time about four per cent of the children were left-handed, then one per cent or so of left-handers were spontaneously mirror-writing. As was the habit at the time, these left-handers were probably being taught to write with their *right* hand, so when they wrote with the more skilled left hand they produced a mirror-image. Mirror-writing is also quite common in right-handers after a stroke paralyses the right side of the body.[9]

Occasionally, cases are reported of 'double-mirror writing', in which the person writes not only in the mirror direction, from right to left, but also upside down. It was described by Clérambault as *écriture en double miroir*, and it looks like this.

Like a mirror-image reflected in another mirror, the two transformations have made the writing become the right way round again, and it can now be read by turning the page around.[10]

There have been two famous mirror-writers. Lewis Carroll, the author of *Alice in Wonderland* and *Through the Looking-Glass*, wrote letters in mirror-writing, although they were only occasional set pieces to amuse children who corresponded with him. Quite the most famous mirror-writer is Leonardo da Vinci, whose extensive notebooks, written for his own use, are almost entirely in mirror-writing.[11]

The world's most famous left-hander, Leonardo da Vinci, may have been born right-handed

Leonardo as a left-hander is almost a legend, the genius of the notebooks seemingly confirmed by their mirror-writing, imparting a sense of 'difference', of coming from a different mental world from other mortals. That he used his left hand is indisputable. His friend, the mathematician Luca

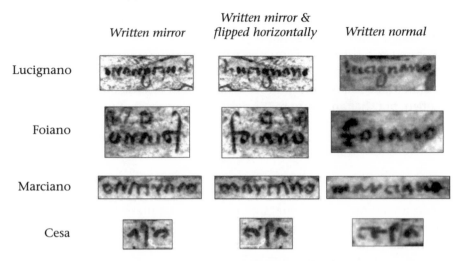

Figure 13.2 The names of four Tuscan towns from Leonardo's maps of Val di Chiana. The names in the left-hand column are from the draft version of the map, and show Leonardo's usual mirror-script written from right to left. The names in the right-hand column are from the final version of the map, where the writing, unusually for Leonardo, is written conventionally, from left to right. These are most easily compared with the mirror-script versions by looking at the middle column, which shows the mirror-script flipped horizontally, so it reads from left to right.

Pacioli, described him writing with his left hand, and his drawings show the characteristic backward-leaning hatch marks of a left hander, \\\\\, rather than the forward-leaning hatchings typical of a right-hander, /////. Although Leonardo mostly wrote in the mirror direction, he occasionally, when other people needed to read the writing, wrote in the normal direction. In 1502, Leonardo drew two maps of Val di Chiana in Tuscany, one with the city names written in mirror-script and the other, probably the finished product, written conventionally. Comparing them, particularly if the mirror versions are 'flipped over' as in Figure 13.2, shows that Leonardo's mirror-script is a near perfect mirror-image of his normal handwriting.[12]

Most people who write in mirror-script with the left hand have been trained to right in the regular direction with the right hand. Might Leonardo, therefore, have been a right-hander who, because of injury, had to change to writing with his left hand? Even Leonardo's earliest-known notes, written at the age of twenty-one – an age when secrecy would probably not have been a particular concern – are written in mirror-script, suggesting that Leonardo had little choice in the matter. It is possible that his right hand was damaged at some time; at one point Leonardo 'thanked God for having escaped from murderers with only one hand dislocated'. There are also suggestions that some early drawings were drawn with the right hand. An early drawing of a hand, which it is suggested is most likely Leonardo's own right hand drawn in a mirror, does also show a clear deformity of the fingers. Overall, it seems quite possible that Leonardo was brought up and learned to write right-handed, but that an injury to his right hand meant that for the rest of his life he drew with his left hand, and wrote mirror-script with his left hand.[13]

Asymmetry is often misspelled

The word 'asymmetry' is what linguists call a geminate, because one of its letters, the 'm', is doubled, the term coming from the Latin for a twin. Such words are often spelled wrongly, and a search on any scientific database will find many examples of '*assymmetry', '*assymetry', and '*asymetry'. The most commonly misspelled word of this sort is probably 'accommodation', rendered as '*accomodation' and '*acommodation', although '*capuccino' rather than 'cappuccino' seems to be spreading rapidly. As an undergraduate in the 1970s, I misspelled 'asymmetry' as '*assymmetry' while writing to Michael Corballis, and his gentle admonishment stopped me ever doing it again: 'you should notice asymmetry itself has an asymmetry which you have neglected'.[14]

A tornado in the bath

Tornadoes almost always spin anti-clockwise in the northern hemisphere and clockwise in the southern hemisphere, although there are rare exceptions. The reason is the Coriolis force, named after the nineteenth-century French physicist Gustave Gaspard Coriolis. As the Earth spins, so the air and the water on its surface also spin. Because a point on the equator of the Earth moves faster through space than does a point close to the poles, any large mass of fluid in the northern hemisphere will be moved faster near the equator, and that will make it turn anti-clockwise. In contrast, in the southern hemisphere it turns clockwise. Hence tornadoes, and ocean currents, turn in opposite directions in the two hemispheres. None of that is controversial, and can easily be seen on pictures from meteorological satellites. What has caused much dispute and endless debate is whether the Coriolis force matters in the small volumes of water found in an ordinary bath tub. One side of my bath is further north than the other; the two sides are therefore moving through space at slightly different speeds, which could impose a twisting force on the bathwater. So, in theory, the water is more likely to go down the plughole anti-clockwise in the northern hemisphere and clockwise in the southern hemisphere, with presumably an even chance of going either way for a bath exactly on the equator. Whether it actually does this is still an open question waiting to be solved by intrepid and patient experimentalists, although the general feeling seems to be that any tiny effect of the Coriolis force would be swamped by other factors.[15]

Wash your knickers on the left side

A reader of *New Scientist*, quite contrary to conventional scientific stereotypes of boring, geekish men in white coats, bought a pair of silk knickers, enclosed with which she found the seemingly uninterpretable instructions, 'To maintain the fine texture we recommend washing the silk on the left side.' Another reader clarified the meaning. The knickers were made in Switzerland, where the German word *links*, as well as meaning 'left' also means 'inside out'.[16]

Left- and right-handed pine trees

Pine trees in the northern hemisphere often have a conspicuous anti-clockwise spiral visible in the lines of the grain running down their trunk. North of the equator, trees tend to have more leaves on the southern side, because more light comes from that side. Winds, which mostly come from the west, push harder on the leafy southern side of the tree, which twists the tree, and so it turns anti-clockwise. It's a nice theory, and it looks even nicer when one finds that pine trees in New Zealand tend to have clockwise spirals. The real test, though, is to take a normally anti-clockwise northern hemisphere pine and grow it in the antipodes. Unfortunately, they carry on growing anti-clockwise, which suggests that genetic factors probably have a large role to play as well.[17]

The right- and left-mouthed cichlids

The cichlids are an extremely diverse family of fish, hundreds of different species of which are found in the Great Lakes of East Africa. A particularly peculiar one is *Perissodus microlepis*, which lives in Lake Tanganyika and eats scales from the flanks of other fish using the highly specialised adaptation of a mouth that points to one side so that the fish's teeth can grab the scales. Whether a fish has a left- or a right-facing mouth is determined genetically, with right dominant over left. The proportion of right and left types varies from year to year around an exact fifty-fifty mixture, owing, it seems, to frequency-dependent selection, where there is an advantage to being rarer. Imagine most fish are right-mouthed and eat scales only from the left side of other fish. Their prey would soon realise this and concentrate on avoiding attacks from the left. Left-mouthed fish would then be at an advantage, since prey would not notice them, and their numbers would correspondingly increase, eventually resulting in an excess of left-mouthed fish. The prey would then focus attention on right-sided attacks, so that now the rarer right-mouthed fish would have the advantage and their numbers would increase. In the long run, therefore, half the fish will be right-mouthed and the other half left-mouthed, as is the case.[18]

The oystercatcher and the right-handed mussel

Oystercatchers are birds with a long straight beak, which is used for hammering at the shells of molluscs such as mussels in order to smash them open. It is hard work and an oystercatcher may have to hammer six hundred times on a mussel before it gets its dinner. Oystercatchers are therefore sensitive to anything that makes them more efficient at opening mussel shells. If you have a taste for *moules marinières* you may have realised that the two shells of a mussel are not quite symmetric in shape, although you are less likely to have noticed that in many mussels the right side of the shell is slightly thinner, by about 0.036 mms on average. It isn't much, but oystercatchers care about these things because they need to hammer three per cent less often on that side before the shell cracks. They have developed, therefore, a subtle strategy whereby firstly they try to detect which side is thinner and then preferentially hammer on that side. If they can't tell, they hammer on the right side anyway, on the basis that this tends to be the thinner side.[19]

The British Royal Family has several left-handers

A recent photograph showed Prince William, the elder son of Prince Charles and the late Diana, Princess of Wales, on his 'gap year' in Chile, using his left hand to wash a floor. Several years earlier, he was photographed using his left hand to sign the register at Eton, so his left-handedness is not in dispute. Prince Charles is also commonly mentioned on lists of famous left-handers, although on a rather flimsy basis. William's great-grandmother, Queen Elizabeth the Queen Mother, is left-handed. William's great-grandfather, King George VI, wrote with his right hand and also had a severe stammer, which has been put down to him being a natural left-hander forced to write with his right hand. George VI was the only member of the Royal Family to appear at Wimbledon as a competitor, in the men's doubles in 1924, when he played left-handed; he and his partner were eliminated in the first round. George VI's own great-grandmother, Queen Victoria, was also said to be left-handed, and although she wrote with her right hand she painted with her left. One of her grandsons, Prince Albert Victor, the Duke of Clarence and son of the future Edward VII, was apparently also left-handed (see the next section).[20]

Jack the left-handed Ripper?

The five women who were victims of Jack the Ripper all had their throats cut from left to right, which suggested to police and coroners that the murderer was left-handed. Over the years, the list of suspects has been long – one book names thirty-seven – but most are highly speculative. Frustratingly, there seem to be few for whom the handedness has been ascertained. An exception is the Duke of Clarence, Queen Victoria's left-handed grandson, who died early, owing in one theory to neurosyphilis contracted from a prostitute, on whom it was speculated he subsequently took his revenge.[21]

Six left-handers have been president of the United States

Of the forty-two men who have been American president, six have been left-handed, the first being James Garfield, the twentieth president, who took office in 1881 and was assassinated the same year. The next was Harry S. Truman, thirty-third president from 1945 to 1952, who wrote with his right hand but always used his left hand for the president's throw that inaugurated the baseball season. Gerald Ford, thirty-eighth president from 1974 to 1977, was a left-hander who gained a reputation for clumsiness, seemingly due to a tendency to turn anti-clockwise while the right-handed aides around him turned clockwise. Whether or not that lost Ford the election, subsequent White House staff took no chances. A member of staff for the fortieth president, Ronald Reagan (1981 to 1989), a left-hander who wrote with his right hand, described how, 'The trick we use is to let the president stand out about a half step in front of us, or if that's not possible, then to give him a full step of distance on either side. In that way, no matter which way he turns, he doesn't tangle with anyone else.' The forty-first president, George Bush Snr, in office from 1989 to 1993, was a left-hander who, in 1992, fought a presidential election in which all three candidates were left-handed, including the third-party candidate, Ross Perot, thus ensuring that the next president would be left-handed. The victor was the forty-second president, Bill Clinton, in office from 1993 until 2001. Despite being the fourth left-hander in the White House in twenty years, Clinton's handedness still surprised his staff, such as the strategic advisor Dick Morris, who described how the president,[22]

> ...tinkered with each word over the phone and finally faxed back to me my draft with his extensive handwritten notes. The president is left-handed. I had never before seen a lefty's check marks, which start with a downward stroke and then go up and out to the left instead of to the right. I couldn't figure out what they were. In my next call that night, I asked him, and he chuckled, saying, 'That's the correct way to make a check.'

Popular works on left-handedness make much of the fact that five of the eleven post-war presidents have been left-handed (forty-five per cent), and, in particular, that two-thirds of the past six have been left-handed. As first sight it does seem remarkable, but some reflection soon suggests the clustering is as likely as not due to chance. Of the nineteen men who have been president since 1900, only five have been left-handed; that is twenty-six per cent. If left-handers really were more likely to gain high office, then there should also be an excess of left-handed vice-presidents, but among the fourteen vice-presidents since 1900 – and excluding the five who went on to be president – only Nelson Rockefeller was left-handed, a proportion of only seven per cent. Before 1900, there was only one left-hander amongst the twenty-three presidents and their twenty vice-presidents (2.3 per cent). Overall, there have been seven left-handers out of seventy-six presidents and vice-presidents, which is an unremarkable 9.2 per cent.

Two right-handed American presidents learned to write with their left hands

During his stay in Paris as American envoy, Thomas Jefferson, later to be the third president of the United States, formed a close relationship with Mrs Maria Cosway, an Anglo-Italian artist. Although events are far from clear, it seems probable that, on 16 September, he and Maria visited Le Désert de Retz, a fashionable 'wilderness' not far from Paris, where Jefferson fell and either fractured or dislocated his right wrist. Normally a prolific letter-writer, for a week or two afterwards he had to make use of an amanuensis. On 5 October, he wrote a brief note to Maria with his left hand, using the right hand only to sign the letter. Until the middle of November, he continued writing with his left hand, although brevity was the order of the day. In a letter of 22 October, he alluded to the problems of left-handed writing and, tantalisingly, made an elliptical comment on the perhaps compromising way in which the wrist was injured: 'How the right hand became disabled would be a long story for the left to tell. It was by one of those follies from which good cannot come, but ill may.' Speaking with his heart as well as his head, he also said, 'My right hand presents its *devoirs* too, and sees with great indignation the left supplanting it in a correspondence so much valued. You will know the first moment it can resume its rights.'[23]

Woodrow Wilson, who in 1912 became twenty-eighth president of the United States and in 1919 won the Nobel Peace Prize, suffered a series of strokes throughout his life owing to high blood pressure, the most serious being in 1919, while he was still in office. The first was in 1896, when he was forty, and still a professor at Princeton. On 27 May, three days before sailing to Glasgow for the university's celebrations of the jubilee of Lord Kelvin's chair, Wilson developed weakness and clumsiness of the right arm. Quite remarkably, he almost immediately, and seemingly with no need to practise, began writing with the left hand in a fluent and legible script, only a little different from his normal right-hand script, and he continued using the left hand for writing and typing until April 1897. A rare exception to the use of the left hand was in a letter in January 1897 to his wife Ellen, the last paragraph of which was written with the right hand since, as Wilson explained in a phrase reminiscent of Thomas Jefferson's assertion of the rights of the right, 'to do anything else would be too much like *kissing* you by machinery'.[24]

Mobile phones may cause right-sided brain tumours

Brain tumours usually occur equally often on the left and the right. There are controversial claims that brain tumours can be caused by the electromagnetic radiation from mobile phones heating the part of the brain next to where the phone is held. The evidence is weak, but an intriguing finding is that brain tumours in mobile-phone users are more likely to be on the right side, the side that the majority of phone users hold their phone. Whether or not the result is replicated, lateralisation will play a useful part in supporting or refuting the theory that cellphones cause tumours.[25]

The side of a fool's heart

The Authorised Version of the Bible contains a highly enigmatic passage in the book of Ecclesiastes: 'A wise man's heart is at his right hand; but a fool's heart at his left.' Clearly, this has no literal truth, but it is difficult to discern any metaphorical meaning. Something must have been lost in translation from the Hebrew, but what? The New English Bible offers no additional insight, instead adding extra confusion: 'The mind of the wise man faces right, but the mind of the fool faces left.' Again, this cannot be meant literally since a mind cannot face right or left. A third translation seems even less useful, mixing the hand and the heart: 'A wise man's understanding is at his right hand, but a fool's heart is at his left.'

I wondered about this passage for a long while and then serendipity intervened. While looking up something else in the *Encyclopaedia Judaica*, I at last found an explanation that gave the phrase a half-decent meaning. The translation in the encyclopaedia was: 'The wise man's mind (tends) to his right (i.e. to what brings him good luck), and the fool's to his left.' In other words, the wise man does what in the past has brought him luck, and since the best predictor of the future is the past, this will probably continue to bring luck in the future. It has nothing to do with the heart at all, nor with right and left.[26]

Pictures look very different when viewed as mirror-images

The art historian Heinrich Wölfflin described how paintings and drawings look very different when looked at in a mirror, an effect he demonstrated using Rembrandt's engraving *The Three Trees*. Figure 13.3 shows the picture as it was printed and the mirror-image. Look at the two versions and try to decide what is so different about them – for they undoubtedly have a very different feel.

In the top version, 'the group of trees at the right gives an impression of energy', whereas in the lower version, 'the trees are devaluated and emphasis now seems to rest on the flat, extended plain'. Art historians have often suggested that pictures are 'read' from left to right because Western viewers usually read text in that direction. That may be true in part, but it is also probable that the left half of space attracts greater attention because of the right parietal lobe. Either way, our eyes are first attracted to the left side of the picture and then move across towards the viewer's right side, giving the different emphases seen in the Rembrandt engravings. In Figure 13.3, it is actually the top version which is the 'correct' one, although, of course, Rembrandt must have drawn on to the etching plate what we see in the lower, 'reversed' version of the picture. Somewhat surprisingly, although ordinary viewers say pictures such as those in Figure 13.3, or indeed any painting and its mirror-image, look different, they cannot reliably decide which is the original and which the mirror-image, unless they have seen the picture before.[27]

Figure 13.3 Rembrandt's engraving of *The Three Trees*. One of the versions is as the picture was printed, and the other has been mirror-reversed. Look at the text to find which is which.

The two halves of a face do not look the same

In 1930, William E. Benton patented his 'Reality mirror', a cunning piece of silvered metal that allowed one to look at a photograph of a person and see how they'd look with two right halves to their face or two left halves. One of the photographs that Benton provided was of the author Edgar Allan Poe (Figure 13.4), and Benton describes how the right side reveals 'the conscious or obvious side of Poe's face and nature was honest, direct, reserved, dependable and business-like', whereas the left side shows 'the subconscious or hidden side of Poe's face and nature was

Figure 13.4 A photograph of Edgar Allan Poe, in its original form (centre) or as a composite of two right cheeks (left) or two left cheeks (right).

tragic, creative, plotting, cruel, sensual'. 'There is a little of Dr Jekyll and Mr Hyde in each of us', as Benton says.

There is no doubt that the composite of the two right cheeks looks very different from that of the two left cheeks, although this, in part, is because Poe's parting is on the left side, resulting in very different hair styles in the two versions. Also, the light is coming from Poe's left side, meaning that the shadowing of the two eyes is very different. However, even if one takes such factors into account there is still no doubt that the right cheeks differ from the left cheeks. Most people recognise a face made from two right halves, but not a version constructed of two left halves. The reasons are complex, resulting not only from asymmetries in the face itself, and in the muscles used for expression, but also from the fact that our right hemisphere is specialised for looking at faces, looking more at the left side of space and hence seeing more of the right cheek.[28]

The sun shines from the left

Shadows in paintings provide a sense of depth. Although the sun is usually the source of light in pictures, it often cannot be seen, its position being inferred from the pattern of the shadows. Most often, the light in a painting comes from above and somewhat to the left. Since the sun is typically the source of the light, it is hardly surprising that the light comes from above, but why the left? The usual explanation is that, because artists are mostly right-handed, they work with the light to their left so that their hand doesn't throw a shadow on the page. That may, however, not be the entire explanation. The effect of the direction of light can be seen in Figure 13.5 which shows various effects of shadowing. Each circle, like a half-moon, is strongly illuminated on one side and in shadow in the other. This can give a sense of depth, the circle looking either concave, like a little depression, or convex, like a ping-pong ball.

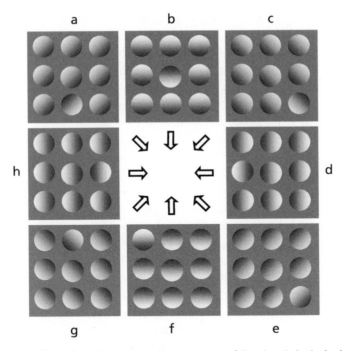

Figure 13.5 The effects of shadowing. In each square, one of the nine circles is shaded differently from the other eight. In (b) it looks as if the light is shining directly from above, with eight of the circles 'sticking out' as rounded bumps, whereas the ninth, in the centre, is going inwards to form a little dip. In (f) the light seems to be coming from below so that now only one circle sticks out. Turn the book upside down and the effects will be reversed. In (d) and (h) the light comes directly from right and left, and it is difficult to see which circle is the odd one out. In (a), (c), (e), and (g) the light comes from one of the diagonals. If you are having trouble seeing the blobs in depth, they may seem clearer if you look at them slightly to one side, rather than directly.

One of the circles, in each case, is shaded in the reverse way to the others and it should 'jump out' at you. The odd one out is easier to see when the light is directly from above or below rather than from the sides. However, careful experiments find that when the light is slightly from the left, the odd one out is seen more easily than when the light is from the right. Our brains seem to presume that the sun is shining from the left.[29]

The left-sided Madonna

Women prefer to hold a baby on their left rather than right side. The obvious explanation – that most women are right-handed and want to leave their right hand free – has to be wrong because left-handed women also hold babies on their left side. Lee Salk, the New York psychologist who first systematically described this, suggested that the child was closer to the mother's heart on the left side, which may have had a soothing effect. Mothers also talk differently to babies depending on the side the child is held, having a lower-pitched, quieter and more soothing voice when the baby is held on the left, and a higher-pitched, louder and more arousing voice if the baby is on the right. The ultimate test for such theories will be to look at the child-holding behaviour of women with *situs inversus*, whose heart is on the right.[30]

Salk also noticed that Renaissance paintings of the Madonna and Child usually showed the child held on the left side, and he suggested that this was a universal portrayal of a mother and child. It is a nice theory, and I remember describing it to a close friend as we walked through an Italian art gallery. The early Renaissance paintings in the first rooms, dozens of which were of a Madonna and Child, showed the child on the left in almost every case. However, further on in the gallery we reached the *cinquecento*, and the theory suddenly seemed to work much less well. Back home, a systematic search through the library confirmed my impression – in 1250 almost every such painting had the child on the left, but by the period 1450 to 1550 a majority of paintings showed the child on the Madonna's *right*. The proposed explanation is complicated, far from trivial, and involves the changing perception of the Madonna in Catholic theology, which is known as the Cult of the Virgin Mary. Salk went wrong because he used a single book of paintings called *The Christ Child in Devotional Images*. When I got this book from the library, I found the full title was *The Christ Child in Devotional Images in Italy During the Fourteenth Century*, the period when most Madonnas were still depicted as holding the child on the left.[31]

The right foot is more ticklish than the left foot

Studying tickle scientifically is not easy. However, there is now a standard 'tickle apparatus', and it has been used among other things to provide scientific proof that it is more difficult to tickle oneself than to be tickled. The 'tickle apparatus' – the phrase sounds terribly grand and creates images of a sophisticated robot, metal fingers ready to reach for one's most sensitive regions – is actually very straightforward. Three times, once every second, a pointed nylon rod, ten centimetres long, and with a tip less than a millimetre across, is stroked across the sole of the foot. Subjects then say how tickled they were, on a scale from one to five. The surprising thing is that, in most subjects, the right foot proves to be more ticklish than the left. Why this should be is not at all clear, but the difference doesn't seem to relate to handedness or footedness.[32]

Corns are worse on the right foot

Mr H. G. du P. Gillett, a chiropodist in Walthamstow in North London, surveyed thousands of corns on the toes of his patients. Most corns were on the fifth toe, and they were worse on the right foot than the left. The fourth toe was the next most common site and there also the corns were worse on the right side. In contrast, the much rarer 'webbing corns' that occur in the space between the two smallest toes were equally common on right and left. The vulnerability of the right foot is presumably due somehow to most people being right-footed.[33]

Sixteen left feet and true

The Anglo-Saxon measurement of length of land was the 'perch' (or 'rod' or 'pole'), which, in a somewhat more precise form, existed into the twentieth century, when it was defined as $5^{1}/_{2}$ yards. Its original definition was the total length of the left feet of the first sixteen men to leave church on Sunday morning. Those left feet must have had an average length of nearly $12^{1}/_{2}$ inches, which seems on the large side.[34]

The sinister unicorn

The narwhal has a single ivory tusk, two or three metres in length, which protrudes straight ahead of the animal. The tusk is actually a tooth, a massively extended upper incisor, and, as Herman Melville said in *Moby Dick*, 'it is only found on the sinister side, which has an ill effect, giving its owner something analogous to the aspect of a clumsy left-handed man'. More peculiarly still, it always shows a quite tight left-handed spiral. No specimen ever seems to have been found with a right-handed spiral. Rare specimens have been described in which there are two tusks, one on each side, but, even in these, both tusks have a left-handed spiral. The most famous explanation for this 'very singular and exceptional thing' is that of the biologist D'Arcy Thompson, who, in his book *Growth and Form*, says that there must be some consistent twisting force to produce such a perfect spiral. The tooth that comprises the tusk grows throughout the animal's life, and the theory is that a slight but continual torque produced by a small left-handed curvature of the narwhal's tail could produce such a force, the animal in effect continually rotating around the inertial mass of its ever-growing tusk. Although it is a nice theory, it is almost certainly wrong, since fetal narwhals, which do not have a horn, still have an asymmetric skull, the blow-hole being placed somewhat to the left of centre. This asymmetry in the skull may well help accentuate the animal's hearing, as occurs in some owls, where having ear canals of different shapes and at different heights allows vertical location to be perceived.[35]

Specimens of narwhal 'horn' may have helped create the medieval legend of the mythical unicorn, with its single, straight, white horn. In medieval tapestries, such as the famous series of *The Lady and the Unicorn* at Cluny in Paris, the unicorn's horn clearly shows a spiral which is very similar in appearance to that of the narwhal. However, the master who created these tapestries clearly did not copy a particular specimen of narwhal's horn, as the direction of the twist in the tapestries is inconsistent, four of the six being left-handed and the other two right-handed.[36]

The Ambidextral Culture Society

Throughout history, some people have dreamed of ambidexterity; a world where either hand is capable of any task. Plato eulogised Scythian warriors who could pull a bow with either arm, and said 'it ought to be considered the correct thing that the man who possesses two sets of limbs, fit both for offensive and defensive action, should, so far as possible, suffer neither of these to go unpractised or taught'. Lurking at the heart of the ideal of ambidexterity is the belief that human behaviour is infinitely malleable and hence infinitely perfectible. Like Plato, J. B. Watson, the founder of the school of psychology known as behaviourism, suggested that a child is born to have two equally proficient hands, except that,

> Society soon thereafter steps in and says, 'Thou shalt use thy right hand.' Pressure promptly begins. 'Shake hands with your right hand, Willy.' We hold the infant so that it will wave 'bye-bye' with the right hand. *We force it to eat with the right hand. This in itself is a potent enough conditioning factor to account for handedness.* [Emphasis in original]

If social pressure, alone, makes us right-handed, then training and practice ought also to make us ambidextrous, and, in so doing, improve the health, welfare and economic efficiency of humankind. The ambidextral movement reached its apogee at the end of the nineteenth century, the Victorian novelist Charles Reade making the case forcefully and describing how 'every child is even and either handed till some grown fool interferes and mutilates it'. The movement reached its height of absurdity in the Ambidextral Culture Society, founded by John Jackson. Robert Baden-Powell, the founder of the Boy Scouts, described the military advantages of ambidexterity in the introduction to Jackson's book on the subject, published in 1905. He also described his own ability to write equally well with each hand and signed his introduction with right- and left-handed signatures. Even Baden-Powell, however, could not carry out the final step in Jackson's path to full ambidexterity, that of writing entirely different things with each hand at the same time, although he did admit it would have helped enormously with 'the heavy pressure of my office work'. Such was the high moral tone of the ambidextral movement, and so great were the benefits promised, that inevitably there was soon a backlash against it, a host of social and physical pathologies being blamed on 'the ambidexterity crank [who] is deserving of a more severe punishment than any other of our many criminally insane'.[37]

A poem about the left hand

I know of only one poem about the left hand. It was written by Jack Anderson, a right-handed American poet, who, without espousing ambidexterity, nevertheless emphasised the neglect of the left hand.[38]

Toward the liberation of the left hand

I want to have my right hand,
the hand I write with, the hand
I use to eat and point with,
the hand with which I shake hands,
that hand, I want it chained.
                              Then
I am left with my left hand,
the dumb sinister one I
keep about me but barely
allow.
        It's gauche, it's faulty,
it paws and grabs, not always
getting things in hand: it can
break as well as hold.
                        Leery
of that hand, I never know
what it has in it, but fright
excites me, I want to scare
myself silly, to give me
the creeps, to moan when I see
how my left hand, faltering,
handles itself.
                  Though it can

hardly write two words, I'll let
it forge my signature, I'll
let it, if it wants, call it
-self me.
 Me, I want to live
on a dare, daring myself
to scrape through with one hand tied
and the other running wild.

Novels very rarely have left-handed heroes

Michael Barsley in *Left-handed Man in a Right-handed World* could only come up with one book in which the hero was a left-hander, Ian Hay's *'Pip': A Romance of Youth*. First published in 1907 at the end of the golden Edwardian age before the slaughter of the Great War, and republished by Penguin just as another World War was starting, it is a novel where 'cook' teaches one about the world, where fathers are absent medical geniuses who understand nothing about practical life, where boarding schools are the norm and people do things 'for the good of the House' or make themselves 'a bally ass', where the cricket pitch and the golf course provide excitement and moral discipline, where swains court damsels, and where coherent character development means nothing. Its sole claim to interest is that its hero is left-handed; or rather, as the author puts it, his left hand 'was constantly usurping the duties and privileges of its fellow, such as cleaning his teeth, shaking hands, and blowing his nose: – literal acts of *gaucherie*'.[39]

If one wants a serious piece of literature with a left-handed hero then that means a novel equally as unreadable as *'Pip'*, albeit one still admired in academic circles though probably read little outside them, Goethe's *Wilhelm Meister*. At a crucial moment in the story, as Wilhelm is sitting with a young woman, we read, 'I was…directed to hold out the little finger of my right hand: she placed her own against it; then with her left hand, she quite softly pulled the ring from her finger, and let it run along mine' – where it became irretrievably stuck. Wilhelm had to remove the ring from his right hand, and that was made much easier, because, 'By good luck, I was left-handed, as indeed, throughout my whole life, I had never done aught in the right-handed way.'[40]

If left-handed heroes are thin on the ground, villains may be somewhat more common, mainly because they allow a neat twist in the plot of a detective novel, which is considerably helped by most right-handed readers never considering that a character might be left-handed. The tradition may, in part, go back to Jack the Ripper, but it is also found in the greatest example of the detective genre, Arthur Conan Doyle's Sherlock Holmes. In *The Boscombe Valley Mystery*, Watson, as usual, is given all the important information but fails to use it. The inquest was told, 'the posterior third of the left parietal bone and the left half of the occipital bone had been shattered by a heavy blow from a blunt weapon', and as Holmes says, 'How can that be unless it were by a left-handed man?' The handedness of detectives seems less often to be stated, although Guido Brunetti, the Venetian detective created by Donna Leon, is right-handed. Here, once more, the old symbolisms show themselves: the villains are left-handed and those who catch them are right-handed.[41]

There are hardly any good jokes about right and left

Left–right jokes may be rare because people so confuse right and left that they risk getting the punchline wrong. I'm fussy about jokes, having the old-fashioned idea that they should be funny. I don't, therefore, count puns on the meanings of right, such as those awful T-shirts proclaiming, 'As the right hemisphere of the brain controls the left side of the body then only left-handers are in their right mind.' Nor do I count twee substitutions of 'left' wherever the word 'right' is found in a word, as in a book by Michael Barsley, which refers to the 'copyleft'. My sole candidate for a right–left joke is therefore a graffito still occasionally seen on lavatory walls: 'I'd give my right arm to be ambidextrous.' It's still pretty awful but it has just the right provenance for ending a chapter of trivia.[42]

Although this chapter has been essentially light-hearted, the rest of the book has explored deeper concepts concerned with the importance of handedness and ambidexterity, asymmetry and symmetry. After so much discussion of asymmetry, it is time to take a look at the thing without which asymmetry would be meaningless – symmetry.

MAN IS ALL SYMMETRIE

The more one thinks on it, the stranger it seems to devote a whole book to the difference between right and left. Could you imagine a book similarly devoted to the difference between up and down or front and back? Something about the difference between right and left particularly concerns us. The explanation lies not in the nature of asymmetry, but in the more primitive and mathematically profound idea of *symmetry* that pervades the human view of the world. George Herbert, the seventeenth-century divine and poet put it well:

> Man is all symmetrie
> Full of proportions, one limbe to another,
> And all to the world besides...

Our symmetric arms, legs and body tell a story to us of 'all the world'; of how the universe is, how it was created, and how it ought to be. We might call it the symmetry story – perhaps better called the symmetry myth or the symmetry legend – a story lurking deep within the human psyche that gives left–right differences their power and impact, adding symbolic value wherever such differences are found and sometimes misleading even hard-headed scientists. Maybe it is even the 'symmetry instinct'.[1]

'Symmetry' is a Greek word and concept, its meaning captured in the dictionary definition: 'correct proportion of the parts of a thing; balance, harmony'. As George Herbert intimated, symmetry and proportion mean much the same thing, with proportion often carrying the additional implication of being well proportioned or properly proportioned. Socrates, in his search for 'the Good', invoked a potent intellectual trinity of Beauty, Symmetry and Truth, and since then, there has been a presumption that where one of these is found, the other two will be lurking

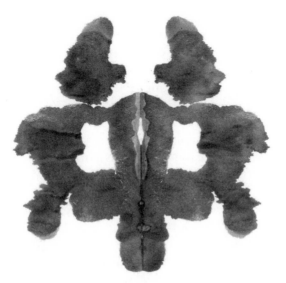

Figure 14.1 'Ink-blot' figure similar to that developed by Rorschach as a projective test of personality.

close by. Consider the paragon of goodness described in Jeanette Winterson's *Oranges Are Not the Only Fruit*, where all of these meanings are invoked:

> Once upon a time, in the forest, lived a woman who was so beautiful that the mere sight of her healed the sick and gave a good omen to the crops. She was very wise too, being well acquainted with the laws of physics and the nature of the universe. She was perfect because she was a perfect balance of qualities and strengths. She was symmetrical in every respect.

Symmetry as truth and beauty is a fiction that has misled many in love, war and science. Symmetries may reveal truths, and truths may be beautiful, but there is little that is contingent in those interrelationships.[2]

Modern usage, particularly in biology, uses symmetry almost entirely to mean 'bilateral symmetry', or 'mirror symmetry', where one half of an object is the mirror-reflection of the other half. Examples are everywhere – one need look no further than one's own face or hands. Consider also the famous ink-blots, first published in 1921 by the Swiss psychiatrist Hermann Rorschach (Figure 14.1), and used as a projective test of personality, albeit one that has remained controversial. It is hardly surprising that many people, particularly those with a scientific or biological background, see biological images in the Rorschach figures, since bilateral symmetry itself defines so many animals.[3]

Bilateral symmetry is but one of a much wider group of symmetries that underpin whole areas of mathematics. There is one very simple idea behind the general concept of a symmetry – it is any way of changing an object so that, at the end of the process, the object is apparently unchanged. Mathematically, a symmetry is a *transformation*. With bilateral symmetry, the transformation is provided by mirror-reversal. Look at a Rorschach-type ink-blot in a mirror and it appears the same. Change without change is the signature of symmetry. The only transformation that leaves the ink-blot looking the same is a mirror-image, but other geometric objects can be manipulated in other ways and yet still look the same. A square, for instance. Rotate it ninety degrees around its centre and it still looks the same, because it has a symmetry of rotation. In other words, it has four symmetries, because it can be turned through 90, 180, 270 or 360 degrees and remain unchanged. A square, however, has other symmetries, because it can be mirror-reflected down either of its diagonals or horizontally or vertically across its middle, and also be unchanged. Reflection and rotation apply to a single object. When dealing with sets of objects, there are two other transformations that leave the appearance unchanged: translation and glide reflection. Figure 14.2 shows translational symmetry. Pick up the picture, slide it along one leaf length and it will then overlie the image exactly, giving the appearance of no change. Glide reflection is seen in the sort of footprints that people leave in the sand (Figure 14.3). Take a footprint, mirror-reflect it, slide it along and put it down; and so on, step by step. Reflection, rotation, translation and glide reflection are the building blocks of symmetries, and at a deeper level each of them can be seen as a type of reflection, since rotation, translation and glide reflection can each be built out of several separate reflections.[4]

Patterns that cover surfaces have a particularly wide range of symmetries, as can be seen in Figure 14.4, which shows some of the many designs found in Roman mosaic floors, or in Figure 14.5, from Zaire, which shows a complex central pattern with a glide reflection, as well as several distinct border patterns. Throughout history, and across the world, people have covered everyday objects in patterns – one only has to look at wallpaper, carpets, curtains, dresses or ties to see examples. Lines, circles, blobs, and innumerable other shapes have been repeated, rotated, reflected and painted in different colours, using all of the various types of symmetry that mathematicians have described. The Hungarian mathematician, George Pólya, showed that there are exactly seventeen *types* of pattern that completely cover a surface (see Figure 14.6). All seventeen were known to the artists who decorated ancient pots and temples, created Roman mosaic floors, decorated the Alhambra, worked with

Figure 14.2 An example of the mathematical process of *translation*, seen in a carved wooden border from Norway.

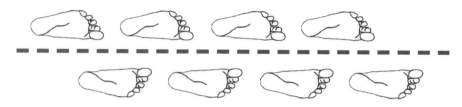

Figure 14.3 Human footprints showing the mathematical process of *glide reflection*.

William Morris on his wallpapers, and so on. Humans are pattern-making animals, and despite the best efforts of minimalist designers and architects, people have an almost compulsive desire to cover anything and everything with patterned decoration, this coming, as the art historian Sir Ernst Gombrich has noted, from our inherent sense of order. We should hardly be surprised, then, if many scientific theories also gain their sense of order from the deep symmetries contained within them – it is as if they also are covered in patterns. Equally, we should not be surprised if a love for symmetry sometimes misleads, most often because theory has dominated over experiment, as in one of the great experiments of the early age of electricity, to which we now turn.[5]

One of the great surprises in early nineteenth-century science was provided not in London, Paris or Berlin, or any of the other great centres of scientific thought, but in Copenhagen. Hans Christian Ørsted was a scientific outsider and did the sort of experiment that perhaps only outsiders dream of doing. The discovery of electricity had, almost literally, galvanised scientific Europe, and there was a belief everywhere that somehow electricity and magnetism were interconnected. Ørsted's forte was building batteries that could pass large currents through a thick piece of wire, and he used these to look for an effect upon a nearby magnet. Hanging a magnet so that it was aligned with the Earth's magnetic field, he placed a large piece of wire at right angles underneath, and then threw the switch to see if the current passing down the wire would make the magnet move (see Figure 14.7). He performed variations on this experiment for eight years, without success – the magnet didn't move. However

Figure 14.4 Examples of different patterns found in Roman mosaics.

Figure 14.5 Raffia cloth from Kuba in Zaire.

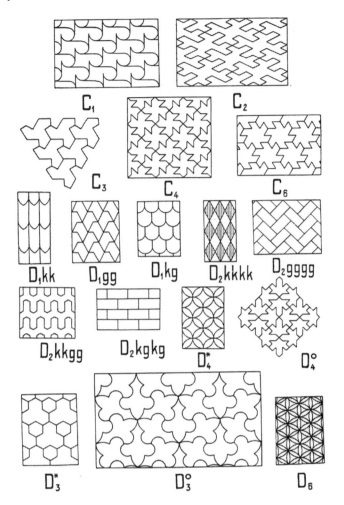

Figure 14.6 The seventeen mathematically different types of pattern that can be used in wallpaper, described by Georges Pólya.

simple, it was the wrong experiment and it was a belief in symmetry that stopped Ørsted doing the right experiment. The experiment he kept repeating is seen in Figure 14.7a, with the wire running at right angles to the magnet. Eventually, in the spring of 1820, Ørsted tried another experiment, seen in Figure 14.7b, placing the magnet parallel to the wire. To his amazement, the magnet swung to the right, clockwise. Within months, Ørsted was known throughout scientific Europe: in London, Sir Humphrey Davy described the experiment to Michael Faraday, who immediately repeated it, and in Paris the results were rapidly extended by Ampère.[6]

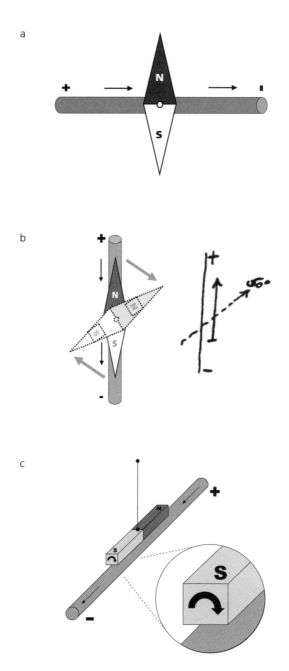

Figure 14.7 a, The experiment which Ørsted repeated over and over again, in which the magnet did not move. The magnet, with its north and south poles marked, is suspended over the wire, which is shown carrying a conventional current (→) from positive to negative. **b,** The successful experiment in which the magnet is parallel to the wire; the diagram on the right is from Ørsted's experimental notebook of 15 July 1820. **c,** A three-dimensional view of the successful experiment showing the hidden asymmetry due to the electrons within the magnet.

It does indeed seem to be a remarkable result. Over sixty years later, in 1883, Ernst Mach wrote of 'the intellectual shock...experienced when [he] heard for the first time that a magnetic needle lying in the magnetic meridian is deflected in a definite direction away from the meridian by a wire conducting a current being carried along in a parallel direction'. Mach was clear that the problem lay in 'the principle of symmetry employed by Archimedes [which] may lead us astray'. Look at Figure 14.7b. With no current flowing, the apparatus looks entirely symmetric (and, indeed, geometrically it *is* symmetric, as can be seen by putting a mirror alongside it). Why, then, should the magnet move *to the right*? How can symmetry result in asymmetry? Ørsted had no idea, though in a paragraph he added to an article he was preparing for publication, but eventually scratched out, he came close to the correct answer: 'The electromagnetism reveals to us a world of secret motions...It is not improbable that even magnetism involves certain rotations.'[7]

There are indeed 'certain rotations' that result in the apparatus looking symmetric even though it actually is not. Chen Ning Yang, one of the Nobel Prize-winning physicists who finally showed that the physical world is not left–right symmetric, explained Mach's problem: 'A magnet is a magnet because it contains electrons making loop motions.' The electrons in the magnet all spin in the same direction to create the magnetic field. Figure 14.7c represents the spinning electrons symbolically on the end of the magnet, revealing that the apparatus is not symmetric, despite appearing so. Suddenly, it is less surprising that the magnet turns to the right rather than the left, because the apparatus itself is asymmetric.[8]

That Ørsted took eight years to make his discovery is probably because he made the same mistake as everyone else, presuming that to find an effect the apparatus had to be asymmetric (Figure 14.7a) rather than symmetric (Figure 14.7b). The correct experiment was not performed because it was assumed that it couldn't possibly work. To make such a mistake once was unfortunate. That the same mistake could be made a century and a half later seems incredible, yet, as we saw in chapter 6, a similar assumption concerning symmetry was made in the world of nuclear physics. It took Wu's cobalt-60 experiment to refute this, showing that parity was not conserved and that, as Lee and Yang predicted, the weak force was asymmetric and electrons left-handed. Something deep inside our minds apparently wants to insist on symmetry.[9]

Symmetry has undoubtedly been a powerful tool for twentieth-century physics. Physics had always dreamed that the structure of the universe would be revealed through the structures of mathematics. Kepler, the early seventeenth-century astronomer, had an almost mystical belief in mathematics, seen in his idea that the distances between the orbits of the six planets, Mercury, Venus, Earth, Mars, Jupiter and Saturn, would reflect the five perfect Platonic solids, the tetrahedron, cube, octahedron, dodecahedron and icosahedron. It was a beautiful theory that was doomed once the planetary orbits were measured accurately. The first great success of a theory of symmetry in predicting the nature of the physical universe was in 1905, with Einstein's theory of relativity. The key concept of Einstein's theory is *invariance*: namely, that the laws of physics will apply in the same way across a wide range of situations; for instance, in all directions in space. Invariance is an idea closely related to symmetry, in that a situation looks identical after a series of transformations. Using the powerful tools of group theory, physicists not only explained a wide range of phenomena but, almost miraculously, also predicted new phenomena. The prediction and subsequent discovery of a whole menagerie of exotic sub-atomic particles, starting with the positron, was the crowning glory of the approach, the root of which was described by Richard Feynman as 'a fact that most physicists still find somewhat staggering, a most profound and beautiful thing... that, in quantum mechanics, for each of the rules of symmetry there is a corresponding conservation law'. Translation in space means that momentum has to be conserved; rotation means that angular momentum has to be conserved; and translation in time means that energy has to be conserved.[10]

Much of the argument from symmetry is an argument about logic and the structure of geometry. If a theory says that I can stand in front of you and, at the same time, stand behind you, then such a theory is wrong, not as the result of any experimental testing but because geometry prohibits an object simultaneously being in two different places. Just as theories requiring that two and two make five are wrong, so theories that misunderstand geometry must also be wrong. In the same way, symmetry has clarified a mass of misunderstandings and restricted the types of experiments that might usefully be carried out. Where symmetry is potentially misleading is where, instead of being used as a *constraint* on theories, it is instead seen as a *requirement* of the way physics ought to work. On occasion, science gently, almost unwittingly, slides across that boundary. Richard Feynman described a conversation he had in 1956 about the presumption that parity was conserved:

> I was sharing a room with a guy named Martin Block, an experimenter. 'Why are you guys so insistent on this parity rule?... What would be the consequences if the parity rule were wrong?' I thought a minute and said, 'It would mean that nature's laws are different for the right hand and the left hand, that there's a way to define the right hand by physical

phenomena. I don't know that's so terrible, although there must be some bad consequences of that, but I don't know.

Feynman had an interesting answer to why 'those guys' were so insistent on symmetry: 'Symmetry is fascinating to the human mind... We have, in our minds, a tendency to accept symmetry as some kind of perfection.'[11]

Feynman's phrase 'in our minds' is the clue to the potency of the symmetry principle. As the so-called anthropic principle reminds us, the only scientific laws that physicists and cosmologists can create, use and understand are those that the human brain can process. Like any calculating machine, the human brain does some tasks superbly, some poorly, and some not at all, and for some problems it even systematically goes wrong. The dependence of styles of thought and perception on the way the brain is organised is seen in Mach's principle, described in chapter 4, that to *perceive* asymmetry, an organism's brain must *be* asymmetric. Physicists may love symmetry, but, as psychologists Michael Corballis and Ivan Beale have pointed out, 'A perfectly symmetrical physicist could neither have discovered nor even suggested the non-conservation of parity.'[12]

The human love of symmetry, and the truth and beauty associated with it, may be the intellectual equivalent of a visual, or cognitive, illusion. Physicists, like us all, find it difficult to avoid the idealist conception of 'the Good', which can be traced back to Socrates. Symmetry is seen as the royal road to truth – what Abdus Salam described as 'an inner harmony, a deep pervading symmetry'. The problem is that symmetry can be too powerful and too seductive, as the philosopher Baas van Fraassen has described.

> Symmetry takes the theoretician's hand and runs away with him, at great speed and very far, propelled solely by what seemed like his most elementary, even trivial, assumptions... Symmetry arguments have that lovely air of the *a priori*, flattering what William James called the sentiment of rationality.

Although the assumption of symmetry often seems trivial, it is nonetheless still an assumption and requires testing against hard experimental data. Those data are often not collected because of the deceptive beauty of symmetric theories.[13]

The human mind might like symmetry but Nature does not always provide it. As a result, even if symmetries are apparently violated, the mind uses various theoretical gambits to restore the otherwise lost symmetrical purity. George Johnson described the need for these devices in his book, *Fire in the Mind*:

We perform these heroic feats of imagination to preserve what we believe in our heart of hearts to be true: that the universe is ultimately as symmetrical as the music we make, the diamonds we carve, something harmonious to the human mind...We, the pattern finders, the pattern makers, instinctively long for symmetries.

Although Lee and Yang showed that parity was not conserved in physics, it did not take long before it was suggested that even if parity (P) and charge (C) were not conserved, there was a deeper symmetry, CP invariance, which was found by reversing both left and right *and* positive and negative. Richard Feynman illustrated the concept with the mirror-image of an ordinary clock, its dial back to front, its screws having left-handed threads, its springs wound the other way than usual, and so on. Such a mirror clock would not keep exactly the same time as a normal right-handed clock, because parity is not conserved. However, if the left-handed clock were made of anti-matter, then it would keep precisely the same time as a normal right-handed clock made of ordinary matter. Symmetry is restored. This seemed an elegant idea, except that experimentalists soon showed it failed in the face of hard data. Nothing daunted, an even more complex beast, CPT invariance was suggested by the theorists, in which charge and parity can be seen as symmetric if time (T) is also taken into account; left, positive and forwards in time being equivalent to right, negative and backwards in time. The left-handed, anti-matter clock would have to run backwards to keep correct time.

So far, CPT invariance does seem to work, although the key experimental tests are still awaited. The theoretical rewards for CPT invariance are potentially great, providing, among other things, an explanation of why the universe is full of matter rather than anti-matter. At the same time, it also satisfies an apparent psychological need among theoreticians: 'Physicists seeking the symmetry of the universe will not be happy until everything is accounted for and the sums balanced,' as physicist Frank Close has put it. In other words, there is still an assumption that the universe *has* to be symmetric, or, more accurately, supersymmetric, supersymmetry being the latest addition to the theoretical circus, linking together the two very disparate families of sub-atomic particles, fermions and bosons, and, in the process, predicting the existence of a myriad of new particles with such exotic names as squarks, selectrons, gravitinos and winos. The novelist Vladimir Nabokov would have recognised the approach, 'for what is your cosmos but an instrument...which, by an arrangement of mirrors, appears in a variety of symmetrical forms?'[14]

Physicists are not alone in their search for symmetry, although there is no doubt that they are by far the most subtle and sophisticated in their theorising. Wherever humans find asymmetry, they try to render it more symmetric. The Toraja tribe of Celebes (Sulawesi), for instance, use their right hand for almost all tasks, but also believe that everything is reversed for the dead, who, they believe, not only pronounce words backwards instead of forwards, but also always use their left hand. Anthropologists call this 'symbolic reversal', and it can be represented as an equation, *Right:Left :: Alive:Dead :: Forwards:Backwards*, which is read as 'Right is to Left *as* Alive is to Dead *as* Forwards is to Backwards'. The reversal, in some sense, restores symmetry to an asymmetric system, in the same way that, for physicists, charge–parity or charge–parity–time are once more symmetric. Indeed, substitute *Positive:Negative* for *Alive:Dead* in the Toraja's scheme and the equation is very similar to that of CPT invariance. The same process can also be seen in the Beaker I Indo-Europeans, whose burial positions were discussed in chapter 2. Men were buried on their left side, with their head to the north and facing east, whereas women were buried on their right side, with their head to the south and facing west: *Male: Female :: North:South :: East:West :: Left:Right*. Symbolic reversal is also well illustrated in a short story by Edgar Allan Poe, about poor young Tony Dammit, whose mother believed a regular flogging was the best way to make a child good. However,

> …poor woman! She had the misfortune to be left-handed, and a child flogged left-handedly had better be left unflogged. The world revolves from right to left. It will not do to whip a baby from left to right. If each blow in the proper direction drives an evil propensity out, it follows that every thump in an opposite one knocks its quota of wickedness in. I was often present at Toby's chastisements, and, even by the way in which he kicked, I could perceive that he was getting worse and worse every day.

The plot is easily summarised: *Right:Left :: Good:Bad :: Evil out:Evil in :: Flog:Don't flog*.[15]

Symbolic reversal is found even in biochemistry. In the 1930s, for instance, Kögl suggested that the 'unnatural' D-amino acids were found in proteins in cancers; the symmetry this time expressed in the equation *Left:Right :: Natural:Unnatural :: Health:Cancer*. Likewise, in the 1980s there was a suggestion that the unnatural, left-handed form of sugar, L-glucose, could be used as a sweetener that would not be metabolised, and that would therefore allow the obese to become slim: *Right:Left :: Metabolised:Not metabolised :: Fat:Thin*. If something one way round is bad, then perhaps its mirror-image must be good.[16]

The most obvious and important example of symbolic reversal in biology comes from the nineteenth century, after Broca had shown how, for most people, language is located in the left hemisphere of the brain. In France, that led to a real theoretical problem, because the symmetry of the brain had been assumed by the seventeenth-century philosopher Descartes: 'I observe...the brain to be double, just as we have two eyes, two hands, two ears, and indeed, all the organs of our external senses double.' In 1800, the French physiologist and physician Bichat put the idea yet more strongly, emphasising the anatomical symmetry of the brain and then arguing for symmetry from first principles: 'Harmony is to the *functions* of the organs what symmetry is to their *conformation*; it supposes a perfect equality of force and action ... [T]wo parts essentially alike in their structure, cannot be different in their mode of acting.' Bichat had made the same error Ørsted would make a decade or so later, assuming that if parts *look* symmetric their functions must *be* symmetric. Having made this assumption, he went on to suggest that lack of symmetry must be the cause of improper brain function or madness: *Symmetry:Asymmetry :: Harmony:Discord :: Proper:Improper :: Sanity:Madness.*[17]

Even in 1865, after definitively showing that the left and right halves of the brain had different functions, Broca clung to Bichat's ideas:

Now there is a physiological law that holds everywhere else, without exception: it is that two organs that are equal or symmetrical have the same properties, and it would be quite strange if the law were to undergo a violent exception here.

Nevertheless, Broca knew that science is ultimately about experiment rather than theory, for he continued, 'To be sure, observation is superior to theories, and one must sometimes know how to bow before a fact, however inexplicable, however paradoxical, it might seem to us.' That word 'paradoxical' is so revealing. There can only be a paradox if facts are interpreted in terms of a preconceived, manifest truth, against which Nature somehow has been found wanting. 'Symmetry is truth; truth symmetry' seems, once more, to be the message, with all the logic of what John Stuart Mill called 'the deep slumber of a decided opinion'.[18]

The battle to save Bichat's 'law of symmetry' was not yet lost. Two variations on the theme can be found, each relying on symbolic reversal. One, mentioned by the neurologist Hughlings Jackson, suggests that the right and left hemisphere differ in development, the left hemisphere being more developed at the front of the brain (the area concerned with action), and the right hemisphere more developed at the back of the brain (the area more concerned with perception). Just as, in phrenology, the higher moral attributes were seen as being at the front and top of the brain, with the baser, more primitive instincts at the back and base, so, according to this theory, the highest achievement of man, language, is once more put in its rightful place: *Left:Right* :: *Speech:Perception* :: *Front:Back* :: *Man:Animal* :: *Morality:Instinct*. A second approach using symbolic reversal was readily taken up by Broca's contemporaries, to the effect that the left hemisphere is dominant for language in right-handers and the right hemisphere dominant for language in left-handers: *Left language:Right language* :: *Right-hander:Left-hander*. This didn't quite save Bichat, but it seemed to restore symmetry to the system. It did so at the expense of the facts, and for half a century it impeded a correct description of language in left-handers. Such is the power of theories based on ideas of symmetry – they sometimes trump even the hardest of data.[19]

If symmetry is such a deceptive theoretical companion, when *can* it be trusted? One of the few principles that can be applied was proposed in 1894 by the physicist Pierre Curie, whose work with his wife Marie on radioactivity won them both a Nobel prize. Curie's principle of symmetry states: 'When certain effects show a given asymmetry, this asymmetry must be found in the causes which have given origin to them.' In other words, if most people are right-handed, then some asymmetry must *make* them right-handed, their asymmetry not simply springing from nowhere. However, even Curie's principle must be handled with kid gloves. Think about fluctuating asymmetry. In half of us, our first incisor tooth is slightly larger on the right-hand side, and in the other half it is larger on the left. Now, if my right incisor is slightly larger, must there be an asymmetric cause that made it that way, and, if so, did that asymmetric cause itself have an asymmetric cause, and so on, right back to the beginning of the universe? Clearly, no, for it is ridiculous to suggest that the asymmetry of my teeth was determined just after the Big Bang. What happens in fluctuating asymmetry is *symmetry breaking*. The system is essentially symmetric but, on this occasion, happens to have lost its symmetry. The problem is best explained through a beast much loved of philosophers, Buridan's ass.[20]

Jean Buridan was the Rector of the Sorbonne in Paris in both 1328 and 1340, and is said to have scribbled his brief comment about the ass, on which his fame mostly rests, in the margins of his copy of the works of Aristotle. Buridan wondered what would happen to an ass placed exactly midway between two identical piles of hay. Since the piles are identical in every respect the ass would have no good reason to choose one over the other and consequently it would eventually die of starvation. The situation is entirely symmetric. Since, logically, the ass cannot eat both piles of hay at the same time, it is doomed to eat neither pile of hay.[21]

Something has to be wrong here, because surely even the most logical of asses would not starve, but what is it? The ass's situation is symmetric, but it is also unstable. It is a bit like a pencil balanced on a table. The pencil can stand there, and, while doing so, it has perfect radial symmetry. But the slightest breath of air, the merest nudge of the table, and the pencil falls over, and then shows only bilateral symmetry. The radial symmetry of the pencil has been broken and reduced to bilateral symmetry. However, imagine doing that experiment several thousand times, on each occasion watching the direction in which the pencil falls. As long as there is no draught in the room and no one keeps knocking the table from one particular side, the pencil is equally likely to fall in any direction. Draw a graph of the direction the pencil falls and the result should approximate a circle. In other words, the radial symmetry is still there but only in the set of all possible ways the pencil *can* rather than *does* fall. It is the same with Buridan's ass. The tiniest amount of sway will move the ass ever so slightly closer to one pile of hay than the other, making that pile slightly more attractive. The symmetry is then broken, and the ass has good reason for eating the pile that is that little bit closer. It could equally well have swayed to the other side. Put a thousand asses between a thousand symmetric piles of hay, and half will eat the right-hand pile, half the left-hand and none will starve. The situation retains its original symmetry but only in the set of possible outcomes – what has been called 'the extended Curie principle'. The fluctuating asymmetry of our teeth has a similar explanation: half of us have a larger incisor on the left, half on the right, and, as a population, we remain symmetric. Individual asymmetries sometimes therefore appear to arise from symmetric beginnings.[22]

Symmetry may be important as a theoretical tool, and it may seem to occur all around us, but it is not necessarily easily seen. Sometimes, it just pops out, as in the Rorschach-type ink-blot in Figure 14.1. Similarly, there is no mistaking the symmetry of the random dots shown in Figure 14.8a. Figure 14.8b, though, is another matter. It doesn't look symmetric at first glance, but a careful scrutiny shows that the inner part is identical to that in Figure 14.8a, but the addition of a few non-symmetric dots around the outside makes the underlying symmetry far less visible. Similarly, the underlying symmetry in Figure 14.8c is harder to see than in 14.8a. Here the two halves of Figure 14.8a have been pulled apart and the middle filled with random, non-symmetric dots. The symmetry of Figure 14.8d is also less apparent than in Figure 14.8a, despite the only change being to rotate the axis of symmetry forty-five degrees away from the vertical. Figure 14.8e started out the same as Figure 14.8a but here each of the dots on the right-hand side has been 'jittered' slightly so that although the basic symmetry is still present it is much harder to see. Finally Figures 14.8f and 14.8g show symmetries but they are not bilateral symmetries. In Figure 14.8f, the top half has been copied and slid

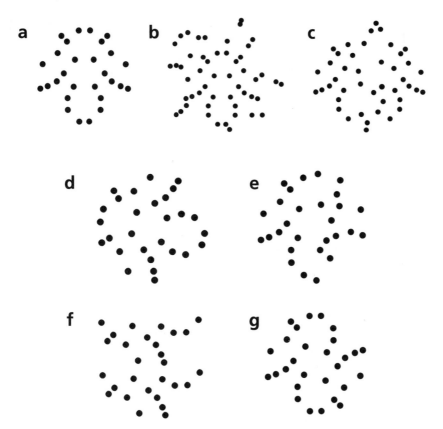

Figure 14.8 Dot figures showing different types and degrees of symmetry (see text for details).

down to the bottom half; a translation. In Figure 14.8g, the right-hand half has been rotated around the centre of the figure, so that the bottom of the left half is the top of the right half. Despite the seven variations in Figure 14.8 all being closely related to one another, the only symmetry that jumps from the page is that of Figure 14.8a.[23]

A slightly different type of stimulus for studying symmetry, developed by psychologist Bela Julesz, is shown in Figure 14.9a. One half of the image comprises randomly generated black and white squares, whereas the other half is the mirror-image reflection. The symmetry is not easy to see, as the figure has purposely been printed with a horizontal symmetry axis. Turn the book on its side and look again. Now the symmetry should jump out, particularly near the mid-line of the image. Cover the central part, as in Figure 14.9b, and again the symmetry is more difficult to see. Symmetry is a sensitive flower. Given the right conditions it blooms and cannot be ignored, but surprisingly little alteration makes it disappear from sight completely, only to be found again by laborious search.[24]

a   b

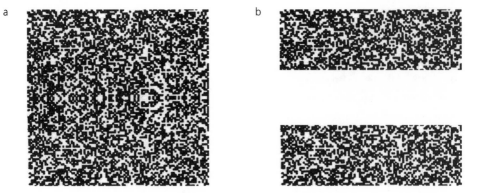

Figure 14.9 A random-dot pattern created by Julesz. In **a** the figure is seen in its entirety whereas in **b** the central portion of the figure has been removed. See text for details.

Vertical bilateral symmetry is the only symmetry that is seen easily in collections of random dots. Horizontal or oblique bilateral symmetries are more difficult to see, as also are translation, rotation and glide reflection, unless they are clearly emphasised by using geometrically regular components, as in Figures 14.2–5. Mach emphasised the primacy of vertical symmetry in a popular scientific lecture given in 1871:

> The vertical symmetry of a Gothic cathedral strikes us at once, whereas we can travel up and down the whole length of the Rhine or the Hudson without becoming aware of the symmetry between objects and their reflexions in the water.

Mach also noticed that children commonly confuse the letters 'p' and 'q' or 'b' and 'd', which have vertical symmetry, whereas confusion of 'p' and 'b' hardly ever occurs, despite having horizontal symmetry. The mechanism by which vertical symmetry is so much more salient is far from clear. Although the symmetry of the two cerebral hemispheres might seem to play a role, that cannot be the entire explanation, since symmetry is recognisable around axes that are not vertical, or when the eye looks to one side of a stimulus such as Figure 14.9a. Although the vertical axis is defined in relation to the body and to gravity, the gravitational field itself does not seem to be an important part of the equation, experiments in the zero gravity of the Mir space station still resulting in vertical symmetry being better perceived.[25]

Whatever the mechanism favouring vertical symmetry, the ultimate origin of its advantage was recognised by the philosopher Pascal, as being 'founded on the...shape of the human face. Whence it comes that we do not ask for symmetry in height or depth, but only breadth.' Because the world is everywhere full of vertical symmetries – not only the face, but also the arms and legs in people, the bodies of many other animals, and flowers and trees – it is hardly surprising that we are good at recognising vertical symmetry quickly, efficiently and automatically. Having that ability confers a biological advantage, and not just for humans. Symmetry is used by animals to find prey or avoid predators, and it is used to select a mate by birds and bees, much as humans search for the symmetric facial beauty they find so captivating (see chapter 5). Symmetry detection is so important that even quite lowly animals carry it out, and, indeed, simple neural networks develop a sense of vertical symmetry as an inevitable consequence of recognising objects.[26]

Perhaps now we have some explanation of why left and right simultaneously intrigue and yet repel us. Symmetry is a fundamental property of the vertebrates and many other animals, and is readily detected by even quite simple organisms. It is beloved of humans, who cover so many artefacts with complex symmetrical patterns. Finally, it is a powerful tool for understanding mathematics and the physical world. It is therefore hardly surprising that scientists find symmetry in so much of the biological and physical world; nor that there is a temptation to slide across and presume that the world *ought* to be symmetric. The absence of symmetry will inevitably seem conspicuous and even worrying. However, the situation is complicated because not all symmetries are equal. Brains have long been especially tuned to find vertical symmetry, so its absence is particularly obvious. However, that ancient symmetry-detection system finds itself alongside the more recent evolutionary development, resulting from the asymmetry of the human brain itself, of an ability systematically to distinguish and label left and right. A tension then results. Symmetry may be common but we can see that it is not universal. Left–right asymmetries are found everywhere in our world and are not easily explained away. It is time to try to integrate those myriad asymmetries into a complete story.[27]

THE WORLD, THE SMALL, THE GREAT

Around 670 BC, an Assyrian sculptor in the palace at Nimrud created the stylised but exquisite image represented in Figure 15.1a. Two eagle-headed, winged, magical spirits are purifying, or perhaps fertilising, a Sacred Tree using liquid from a bucket. Overall, the picture looks symmetrical, and can serve well as a model of our understanding of the physical world, symmetry seeming to be everywhere. At a glance, the figure seems identical to Figure 15.1b, in which there is perfect symmetry. However, ignoring a mass of tiny asymmetries in Figure 15.1a that merely reflect inexactness in the drawing, there is also one very big asymmetry: both figures use their right hand to hold the purifier, and their left to hold the bucket. As we notice this, so our conception of the picture, and our awareness of the subtlety of the artist, changes. Technically, it would have been easier for the artist to draw the bottom picture, and so presumably there were strong reasons – aesthetic, symbolic and representational – for the additional effort put into drawing the top picture. Not the least of those was probably that the artist knew full well that most people are right-handed, and it was reasonable to presume that eagle-headed spirits would be also. The asymmetry at the heart of the original version of the picture adds to its interest, subtlety and realism, as the eye and the mind flip back and forth between the left and right halves, which are simultaneously similar and different.[1]

The physical and biological worlds also look symmetric at first, but are found to be asymmetric on closer inspection. Whether we gain or lose by that asymmetry depends on our perspective. Cosmologists John Barrow and Joseph Silk have described in *The Left Hand of Creation* how many physicists fear the loss of 'paradise…the state of ultimate and perfect symmetry' – a perfect symmetry that may once have existed, but that was 'doomed to transience', existing only for the merest fraction of a second after the Big Bang. After that, 'Paradise was irretrievably lost', and in

a

b

Figure 15.1 a, A drawing of an Assyrian bas-relief sculpture of about 670 BC. **b,** The same drawing made completely symmetric.

apocalyptic style, 'decadence dominated the subatomic world, and the result is the varied universe of broken symmetry that now surrounds us'. However, Barrow and Silk see beyond the apparent perfection of such a pure, complete and symmetric world, realising instead the potentialities of asymmetry: 'the tiny breaches in the perfect pattern we might have expected to find are the cogs of a glittering mechanism at the centre of things, and one of the reasons our very existence is possible'.[2]

However much some theoretical physicists may prefer a world of symmetry, akin to the bland and uninteresting Figure 15.1b, the universe seems instead to have a wealth of asymmetries, akin to the far more subtle Figure 15.1a. Barrow and Silk suggest that a perfectly symmetric and regular world would also be a world without history, a world that is timeless, as is perfectly seen in that epitome of symmetry, the crystal, which, despite its profound and elegant geometries, tells us nothing about itself. When, however, a crystal has a defect, an irregularity, an imperfection in its structure, then the crystal also tells us of its past. With asymmetry comes history (and perhaps, by a further argument from symmetry, a future): 'The [imperfections] tell, as it were, the frozen story of how the crystal came about. They represent, therefore, what is interesting and specific about a chunk of crystal. The regular parts are universal and history-less, they tell us nothing.'

The icy flawlessness of a perfect crystal was also rejected by Hans Castorp in Thomas Mann's *The Magic Mountain* describing snow-flakes:

> ...there was not one like unto another; an endless inventiveness governed the development and unthinkable differentiation of one and the same basic scheme, the equilateral, equiangled hexagon. Yet each in itself – this was the uncanny, the antiorganic, the life-denying character of them all – each was absolutely symmetrical, icily regular in form. They were too regular, as substance adapted to life never was to this degree – the living principle shuddered at this perfect precision, found it deathly.[3]

The physicist Richard Feynman, in his lectures to undergraduates at the beginning of the 1960s, similarly elucidated the recently discovered asymmetries of the sub-atomic world, and put them in a wider, more human, context:

> There is a gate in Japan, a gate in Neiko, which is sometimes called by the Japanese the most beautiful gate in all Japan; it was built in a time when there was great influence from Chinese art. This gate is very elaborate, with lots of gables and beautiful carving and lots of columns and dragon heads and princes carved into the pillars, and so on. But when one looks closely he sees that in the elaborate and complex design along one of the pillars, one of the small design elements is carved upside down; otherwise the thing is completely symmetrical. If one asks why this is, the story is that it was carved upside down so that the gods will not be jealous of the perfection of man. So they purposely put an error in there, so that the gods would not be jealous and get angry with human beings.
>
> We might like to turn the idea around and think that the true explanation of the near symmetry of nature is this: that God made the laws only nearly symmetrical so that we should not be jealous of His perfection!

In his lectures Feynman also turned around the problem of symmetry in another way, saying that the problem is not explaining why the world is sometimes asymmetric, but understanding why so often it should be symmetric. Symmetry is the real surprise, and asymmetry should be our norm of expectation.[4]

Humans attach symbolic meaning to everything they touch, as Robert Hertz realised so clearly. Just as left and right have their associations, their myriad indirect references, so also, like all other binary oppositions, do the concepts of symmetry and asymmetry. The art historian Dagobert Frey summarised some of them, and Hermann Weyl repeated them in his justly famous book *Symmetry*. A table best represents them:

Table 15.1: Concepts associated with symmetry and asymmetry	
Symmetry	Asymmetry
rest	motion
binding	loosening
order	arbitrariness
law	accident
formal rigidity	life, play
constraint	freedom

One is reminded of Goethe, who said that 'the absence of symmetry seems to be evidence of the progress of evolution', or the somewhat uncomplimentary description by Goethe's disciple, Thomas Carlyle, of the literary style of Voltaire, which he compared to 'the simple artificial symmetry of a parlour chandelier'. The style had, 'not Beauty, but at best, Regularity', and compared poorly with 'the deep natural symmetry of a forest oak', Carlyle meaning by 'natural symmetry' the irregularities of life and freedom.[5]

Whatever the theoretical virtues of symmetry – and it indubitably makes the universe an easier place to explain scientifically – there also seems little doubt that for those who make their homes in the universe, much of the fun and interest arise from the asymmetry. Indeed, to take one of the pairs in Dagobert Frey's table, it is life itself that is most irrepressibly asymmetric; so much so that one has to speculate whether any life form of whatever chemistry could be entirely symmetric. The asymmetries of life are everywhere in our world, as described in the earlier chapters in this book. Based perhaps in the most primitive of our asymmetries originating in particle physics, they then appear in the chemistry of carbon and the biochemistry of amino acids, sugars, DNA and proteins, in our bodies with their asymmetric viscera, in the functional asymmetries of our brain and hands, in the cultural artefacts of writing and driving, and in right–left dual symbolic classifications. The time is therefore ripe for asking the big question of whether these myriad asymmetries are related, or whether they are all just independent, idiosyncratic examples of symmetry-breaking.[6]

Scientists have long been wary of any overarching mechanism that ties together asymmetries at all levels, from physics to biology to neuroscience to anthropology, in a way that can be represented as a chain such as this: Weak interaction → L-amino acids → Proteins → Cells → Tissues → Organs → Bodies → Brains → Cultures. Ian Stewart and Martin Golubitsky flirted with a world in which 'a weak left-handed God creates left-handed amino acids that build left-handed proteins that form a left-handed embryo that grows into a left-handed adult', but then beat a tactical retreat, saying, 'Few biologists believe that chirality in animals had anything to do with chirality in chemical molecules.' Michael Corballis and Ivan Beale also toyed with the idea, applying it to the brain itself, before suddenly declaring their agnosticism:

> Could there be a link between the asymmetry of nuclear decay, the asymmetry of living molecules, and the left–right gradient which seems to underlie the morphological asymmetries of animals and men, including cerebral asymmetries in the human brain? We have no idea.

In his Nobel address, the physicist Chen Ning Yang implicitly presumed a link between molecules and morphology when he considered 'a mirror-image man with his heart on the right side, his internal organs reversed compared to ours, and...his body molecules – for example, sugar molecules – the mirror-image of ours'. Others, though, have been far more dogmatic the other way. As we saw in chapter 6, J. B. S Haldane, Leslie Orgel and Murray Gell-Mann all argued that chance alone is responsible for our bodies being made of L- rather than D-amino acids; and, if that is true, then the first link of any chain of causality from particle physics to biology to brains will have been severed. The argument, though, is less than satisfactory, invoking chance to explain a single event. Certainly there are many physicists, nowadays, who would reject such a position.[7]

Popular writers on science tell a different story. Isaac Asimov, for instance, rejected the idea that chance was an explanation of the handedness of amino acids, because it seemed so emotionally unsatisfying:

> Of course, this postulated connection between non-conservation of parity and the asymmetry of life is, as yet, highly tentative, but I am emotionally drawn to it. I firmly believe that everything in the universe is interconnected, that knowledge is one; and it seems dramatically right to me to have a discovery concerning the non-conservation of the law of

parity, which seems so ivory-towerish and far-removed, serve to explain something so fundamental about life, about man, about you and me.

Martin Gardner, in *The Ambidextrous Universe*, was also happy to accept the possibility that the weak interaction may be responsible for our amino acids being left-handed, but then ran into what seemed an insurmountable problem with the next link along the chain: 'Our bodies have hearts on the left...No fundamental asymmetry in natural law is involved. The location of the human heart is an accident in the evolution of life on this planet.'[8]

The heart is the sticking point for many biologists, who are willing to accept that the weak interaction might mean our sugars and amino-acids are asymmetric, but cannot see how that can possibly relate to body asymmetries. Biologically, the distance is so far from molecules to organs. Hundreds or thousands of amino acids are linked together into chains which fold to make proteins, millions of copies of which form cells. Of those cells, in their myriad different types, millions and millions form tissues, which, in turn, form organs that are placed in bodies. How could a tiny asymmetry in the first part of that process, the amino acids, possibly relate to asymmetries at the level of bodies? This was the stumbling block at the seminal Ciba Symposium meeting in 1991 on Biological Asymmetry and Handedness, which was mentioned at the end of chapter 1. Cyrus Chothia, a protein chemist, presented a paper in which he talked about the asymmetries of protein molecules, which are typically richly folded and highly asymmetric. The asymmetry of a protein depends on its sequence of amino acids, but there is no overall tendency for proteins to fold in a left-handed rather than right-handed way, despite the fact that they are made of left-handed amino acids. The low-level asymmetry of the amino acids does not seem able to make the leap from amino acids to proteins. Lewis Wolpert, in summing up the Ciba Symposium, called the problem 'Chothia's gap', and, at that time, it seemed an uncrossable abyss.[9]

Impassable routes, however, have a habit of becoming passable, usually by approaching them in an entirely different way from that originally envisaged. So it may be with the chasm of Chothia's gap. On average, the handedness of proteins is indeed random, so that if the handedness of bodies depended on the aggregate handedness of large numbers of proteins, then bodies and their internal organs would also have random handedness. That is undoubtedly true, in general, but it need not be the case for the one organ that particularly concerns us in this story: the heart. The lesson of chapter 5 is that *situs* may well be the consequence of tiny amounts of fluid flowing in the nodal region of the developing embryo, wafting chemicals to one side of the body rather than the other, and causing the heart to develop on that side. The location of the heart depends, therefore, not on the handedness of huge numbers of proteins throughout the body, or even on those proteins that make up the heart itself, but on the very small number that make the cilia of the nodal region beat in a particular direction. If one could engineer it so that an organism was made entirely of L-amino acids, with the sole exception of those few that were going to make the cilia in the node, which would be built entirely from D-amino acids, then those cilia would undoubtedly turn in the opposite direction. As a result, the nodal flow would be in the opposite direction, and the heart would be on the right rather than the left. The causal chain is now very similar to that shown earlier but with the crucial difference that the L-amino acids are responsible for the handedness of just one key protein, which then determines the handedness of the nodal cilia, which at a stroke reverses the organs and the body of the entire organism. There is now a causal route straight through from the weak interaction, via the L-amino acids, to the heart being on the left.

Explaining the asymmetry of the heart had to be the real challenge for any theory linking physical and chemical asymmetries to the macroscopic asymmetries of the heart, and the cilia make that link superbly. The next step in the story – what Robert Browning called, 'how heart moves brain, and how both move hand' – is potentially much simpler, since the human brain finds itself in a body that is already massively asymmetric because of the heart. As has been seen in chapter 9, a reasonable interpretation is that the brain has become asymmetric by using a similar mechanism to that by which the heart is asymmetric – and it might even be that there are other separate cilia involved in wafting different chemicals across from one side of the organism to the other to make our brains asymmetric. Whatever the precise details, the principle point is that if there is a way of linking the heart's asymmetry to the asymmetry of L-amino acids, or even of weak interaction, then it is straightforward also to presume that the asymmetry of the brain can be determined by a similar route.[10]

Can one go any further? Beyond the asymmetries of the body and the brain lie the asymmetries of human culture. Is there any realistic sense in which these may be related to the asymmetry of the weak interaction? For some, perhaps, the answer is yes, but for others it is undoubtedly no. When we look at cars driving down the road there seems to be no way in which driving on the right or the left is a consequence of asymmetries at the level of physics or amino acids. When, though, we look at the pedals in those cars – the accelerator always on the right and the brake on the left – then there is a clear link to handedness and the asymmetry of the brain. From there, we can go straight down to the weak interaction. Part of the cultural world in which we find ourselves can be seen as a consequence of a chain of causality originating in the sub-atomic world of physics.

At this point, I should make it very clear that there is no sense in which science can be said to have *proved* every step in such a chain of causation. However, there is, it seems, a reasonable case to be made for such a picture, and it is sufficiently exciting to be worth exploring not least to see quite how far it can go. That also has the advantage that any weaknesses in the story will become more apparent as well. There have been many twists and turns in this book, and the argument has at times been complicated, seemingly taking us into many byways of the physical, biological and social world. All this may have obscured the relative simplicity of the big story. Milton recognised the difficulty when writing *Paradise Lost* and therefore prefaced each Book with an 'Argument'. Let me place the argument for this book at its end:

The argument

Most people are right-handed because they have a gene called the D *gene, and that same gene means most of us also have language in our left hemisphere. The* D *gene was the principal factor in separating humans from other apes, perhaps two to three million years ago. Language and motor control in right-handers are controlled by the left hemisphere because the* D *gene is probably a mutation of the* situs *gene, which has been responsible for humans and all other vertebrates having their heart on the left side. Vertebrates and their predecessors have had asymmetric bodies for about 550 million years. The* situs *gene causes our heart to be on the left side because, early in embryological development, cilia in the nodal region waft a current containing determinants of development in a clockwise rather than an anti-clockwise direction. The cilia beat clockwise because they are made principally of L-amino acids, rather than their mirror-image D- form. Almost all organisms on earth are made of L-amino acids; a predominance that is probably not due to pure chance, since amino acids found in meteorites from deep space show the same predominance. Early life evolved to contain only L-amino acids because they were the most abundant form, at least in the local areas of Earth where life evolved; perhaps due to them coming from meteorites. L-amino acids may also predominate because of what physicists call 'failure of conservation of parity', which is reflected in an asymmetry of the weak interaction at the sub-atomic level. The predominance of right-handedness among humans means that many artefacts in daily life on Earth and our use of symbolic terms in language and culture are also highly asymmetric, the association of 'right' with 'good' and 'left' with 'bad' being found in almost all human cultures. It is probably not an exaggeration to suggest that when we read in the Bible of God sorting the sheep to the right and the goats to the left, or when radical politicals are described as being on the left wing and conservatives on the right, these dual symbolic classifications are directly linked to the organisation of language in our brains, which is linked to our manual dexterity, which is linked to our left-sided hearts, which is linked to the clockwise beating of cilia, which, like the rest of our bodies, are composed of L-amino acids, the predominance of which reflects failure of conservation of parity in physics, which is a feature of the deepest laws of physics of which the universe is constructed.*

This argument may not be correct, but as a theory it is at least possible, and there seem to be no strong data that currently contradict it. Stating the theory as forcefully as I have done may help others either to strengthen it or, as can happen with any scientific hypothesis, to refute it utterly: as Kurt Vonnegut says in *Slaughterhouse-Five*, 'So it goes.'

When, in 1835, in a post-mortem room in a London hospital, Dr Thomas Watson contemplated the dissected body of John Reid, with its heart and viscera back to front, he could little have realised the paths that would have to be travelled to explain that seemingly simple reversal. It is a scientific route that covers almost every scale of human and physical existence. Only a poet could really find the words to describe its wonders. In 1959 the Nobel Prize-winning Greek poet Odysseus Elytis wrote a wonderful long poem called *The Axion Esti*. It has a powerful refrain:[11]

ΑΥΤΟΣ	ὁ χόσμος	ὁ μιχρός	ὁ μέγας!
AUTOS	*o cosmos*	*o micros*	*o megas!*
THIS	the world	the small	the great!

Those words are perfect for ending this book. The world (*cosmos*), the small (*micros*) and the great (*megas*) – each has its own handedness, and they are perhaps all inter-related.

NOTES AND REFERENCES

Chapter 1: Dr Watson's problem

1. ☞www☜
2. Anonymous (1882) *British Medical Journal*, ii: 1282–5; Munk, W. (1878) *The Roll of the Royal College of Physicians of London*, London: Royal College of Physicians. ☞www☜
3. Watson, T. (1836) *London Medical Gazette*, 18: 393–403. ☞www☜
4. East, T. (1957) *The Story of Heart Disease*, London: William Dawson, p. 38. ☞www☜
5. Bryan, D. (1824) *Lancet* 5: 44–6.
6. Torgersen, J. (1950) *American Journal of Human Genetics*, 2: 361–70; Cockayne, E. A. (1938) *Quarterly Journal of Medicine*, 31: 479–93. ☞www☜
7. ☞www☜
8. Nicolle, J. (1962) *Louis Pasteur: A Master of Scientific Enquiry*, London: The Scientific Book Guild, p. 25; Mason, S. F. (1989) *Chirality*, 1: 183–91. ☞www☜
9. From the notebooks of Viollet-le-Duc, reproduced in Cook (1914) *The Curves of Life*, Constable: London (reprinted in Dover Books, 1979).
10. Ravoire, J. (1933) *Le Docteur Marc Dax (de Sommières) et l'Aphasie*, Montpellier: Imprimerie Mari-Lavit, pp. 8–9; Mouret, M. A. (1959) *Chronique médicale de la ville de Sommières*, Montpellier: Ets Valette, p. 48; Dax, M. (1865) *Gazette Hebdomadaire de Médecine et de Chirurgie*, 2 (2nd Series): 259–62, (for a translation see Joynt and Benton (1964) *Neurology*, 14: 851–4); Finger, S. and Roe, D. (1999) *Brain and Language*, 69: 16–30. ☞www☜
11. Broca, P. (1861) *Bulletin de la Société Anatomique de Paris*, 2nd series, 6: 330–57 (for a partial translation see Eling, P. (1994) *Reader in the History of Aphasia: From [Franz] Gall to [Norman] Geschwind*, Amsterdam: John Benjamins;) Schiller, F. (1979) *Paul Broca: Explorer of the Brain*, Oxford: Oxford University Press; Signoret, J. L. *et al.* (1984) *Brain and Language*, 22: 303–19; Castaigne, P. *et al.* (1980) *Revue Neurologique (Paris)*, 136: 563–83. ☞www☜
12. Schiller, F. (1979) p. 192; Broca, P. (1865) *Bulletin de la Société d'Anthropologie de Paris*, 6: 377–93; Hécaen, H. and Dubois, J. (1969) *La naissance de la neuropsychologie du langage (1825–1865)*, Paris: Flammarion; Berker, E. A. *et al.* (1986) *Archives of Neurology*, 43: 1065–72. ☞www☜
13. Watson, T. (1871) *Lectures on the Principles and Practice of Physic*, vol. 1, 5th edn, London: Jon W. Parker, p. 494. ☞www☜

14. Ibid., p. 501.
15. Ibid., p. 503. ☞www☜
16. Bock, G. R. and Marsh, J. (1991) *Biological Asymmetry and Handedness (Ciba Foundation symposium 162)*, Chichester: Wiley, 1991; Etaugh, C. and Hoehn, S. (1979) *Perceptual and Motor Skills*, 48: 385–6.

Chapter 2: Death and the right hand

1. Taylor, A. J. P. (1966) *The First World War: An Illustrated History*, Harmondsworth: Penguin, p. 23; Durkheim, E. (1916) *L'annuaire de l'association des anciens élèves de l'école normale supérieure*, pp. 116–20; Needham, R. (1973) *Right and Left: Essays on Dual Symbolic Classification*, Chicago: University of Chicago Press, p. xi. ☞www☜
2. Mauss, M. (1925) *Année Sociologique*, 1 (new series): 7–29.
3. Mauss, M. (1990) *The Gift: The Form and Reason for Exchange in Archaic Societies*, London: Routledge. ☞www☜
4. Introduction by Evans-Pritchard to Herz, R. *Death and the Right Hand*, translated by R. Needham and C. Needham, Aberdeen: Cohen and West, 1960; Parkin, R. (1996) *The Dark Side of Humanity: The Work of Robert Hertz and its Legacy*, Amsterdam: Harwood. ☞www☜
5. Roodenburg, H., in Bremmer, J. and Roodenburg, H. (eds) (1991) *A Cultural History of Gesture: From Antiquity to the Present Day*, Cambridge: Polity Press, pp. 152–89; Bulwer, J. (1644) *Chirologia; or, The Natural Language of the Hand; Composed of the Speaking Motions, and Discoursing Gestures Thereof. Whereunto is Added, Chironomia; or, The Art of Manual Rhetoricke*, London: Thomas Harper, p. 109. ☞www☜
6. Hertz, R. *Death and the Right Hand*, translated by R. Needham and C. Needham. Aberdeen: Cohen and West, 1960. ☞www☜
7. Hertz, R. (1909) *Revue Philosophique*, 68: 553–80; Hertz, R. *Death and the Right Hand*; Durkheim, E. (1916) p. 118. ☞www☜
8. Kraig, B. (1978) *Journal of Indo-European Studies*, 6: 149–72. ☞www☜
9. Mallory, J. P. (1989) *In Search of the Indo-Europeans*, London: Thames and Hudson, p. 140.
10. Kraig, B. (1978) op. cit.
11. Cook, T. A. (1914) *The Curves of Life*, London: Constable (reprinted in Dover Books, 1979); Blake-Coleman, B. C. (1982) *Folklore*, 93: 151–63; Sylvester, D. (1999) *Modern Painters* 12 (3): 26–33. ☞www☜
12. Rigby, P. (1966) *Africa*, 36: 1–16; Werner, A. (1904) *Journal of the African Society*, 13: 112–16.
13. Paskauskas, R. A. (1993) *The Complete Correspondence of Sigmund Freud and Ernest Jones 1908–1939*, Cambridge, MA: Harvard University Press, p. 31; Rigby, P. *Africa*; Smith, E. W. (1952) *Journal of the Royal Anthropological Institute of Great Britain and Ireland*, 82: 13–37; Granet, M., in Needham, R. (ed.) (1973) *Right and Left: Essays on Dual Symbolic Classification*, Chicago: University of Chicago Press, pp. 43–58; Devereux, G. (1951) *Psychoanalytic Quarterly*, 20: 398–422.
14. Platt, A. (1910) *De Generatione Animalium*, Oxford: Oxford University Press, 763. 6.131; Pearsall, R. (1971) *The Worm in the Bud*, Harmondsworth: Penguin, p. 303. ☞www☜
15. Das, T. (1945) *The Purums: An Old Kuki Tribe of Manipur*, Calcutta: University of Calcutta Press; Hertz, R. *Death and the Right Hand*; Needham, R. (1973); Needham, R. (1979) *Symbolic Classification*, Santa Monica, California: Goodyear.

16. Needham, R. (1962) *Structure and Sentiment: A Test Case in Social Anthropology*, Chicago: University of Chicago Press, p. 96.

17. Wieschoff, H. A. (1938) *Journal of the American Oriental Society*, 58: 202–17; Kruyt, A. C., pp. 74–91 in Needham, R. (ed.) (1973) p. 78; Needham, R. (1967) *Africa* 37: 425–51; Beidelman, T. O., in Needham, R. (1973) pp. 128–66. ☞www☜

18. Needham, R. (1979) op. cit., p. 52. ☞www☜

19. Wilkin, S. (1852; revised version of 1836 edition), *The Works of Sir Thomas Browne*, London: Henry Bohn, Vulgar Errors, IV: iv. ☞www☜

20. Fabbro, F. (1994) *Brain and Cognition*, 24: 161–83; McManus, I. C. (1979) *Determinants of Laterality in Man*, University of Cambridge: unpublished PhD thesis, chapter 13; Sattler, J. B. (2000) *Links und Rechts in der Wahrnehmung des Menschen: Zur Geschichte der Linkshändigkeit*, Donauwörth: Auer Verlag; Fowler, A. (1971) *Milton: Paradise Lost*, London: Longman, V: 689; V: 726; VI: 79. ☞www☜

21. Psalm 118: 16; Anonymous (1916) *The Jewish Encyclopaedia*, New York: Funk and Wagnalls, X: 419; Wile, I. S. (1934) *Handedness: Right and Left*, Boston: Lothrop, Lee and Shepard, p. 218. ☞www☜

22. Chelhod, J., in Needham, R. (ed.) (1973) pp. 239–62; *The Qu'ran*, Sura 56; Walsh, J. G. and Pool, R. M. (1943) *The Journal of Southern Medicine and Surgery*, 112–25; http://islam.org/dialogue/Q325.htm. ☞www☜

23. Bousquet, G.-H. (1949) *Les grandes pratiques rituelles de l'Islam*, Paris: Presses Universitaires de France, p. 105.

24. Clark, G. (1989) *Iamblichus: On the Pythagorean Life*, Liverpool: Liverpool University Press, p. 70; Warrington, J. (1961) *Aristotle's Metaphysics*, London: Everyman, p. 65; Wender, D. T. (1973) *Hesiod and Theognis*, Harmondsworth: Penguin Books, p. 29.

25. Davidson, J. (1998) *Courtesans and Fishcakes: The Consuming Passions of Classical Athens*, London: Fontana, pp. 24, 44. ☞www☜

26. Chetwynd, T. (1993) *Dictionary for Dreamers*, London: Aquarian Press, p. 150–1; Lawrence, T. E. (1935) *The Odyssey of Homer*, London: Oxford University Press, p. 321.

27. Freud, S. ([1900] 1976) *The Interpretation of Dreams*, Harmondsworth: Penguin Books, pp. 475–503; Butler, A. J. C. (1898) *Bismarck: The Man and the Statesman; Being the Reflections and Reminiscences of Otto Prince von Bismarck*, vol. II, London: Smith, Elder and Co, p. 210; Domhoff, G. W. (1968) *Psychoanalytic Review*, 56: 587–96. ☞www☜

28. Thass-Thienemann, T. (1955) *Psychoanalytic Review*, 42: 239–61, pp. 239, 260; Masson, J. S. *The Complete Letters of Sigmund Freud to Wilhelm Fliess 1887–1904*, Cambridge, MA: Harvard University Press, 1985 pp. 292–3.

29. ☞www☜

30. Annett, M. (1972) *British Journal of Psychology*, 63: 343–58; Annett, M. (1985) *Left, Right, Hand and Brain: The Right Shift Theory*, New Jersey: Lawrence Erlbaum. ☞www☜

31. ☞www☜

32. ☞www☜

33. ☞www☜

34. Eco, U. (1984) *Semiotics and the Philosophy of Language*, London: Macmillan, pp. 131, 137. ☞www☜

35. Leach, E. (1976) *Culture and Communication: The Logic by which Symbols are Connected*, Cambridge: Cambridge University Press. Although anthropologists and

semiologists like to distinguish symbols, signals, signs and natural indices, I am here using the term 'symbols' for all of them.

36. Sperber, D. (1975) *Rethinking Symbolism*, Cambridge: Cambridge University Press, p. 113. ☞www☜
37. Ibid., pp. 21–2.
38. Ibid., p. 26. ☞www☜
39. ☞www☜

Chapter 3: On the left bank

1. Desmond, A. (1994) *Huxley: The Devil's Disciple*, London: Michael Joseph. ☞www☜
2. Huxley, T. H. (1877) *Physiography: An Introduction to the Study of Nature*, London: MacMillan; Desmond, A. (1997) *Huxley: Evolution's High Priest*, London: Michael Joseph. ☞www☜
3. ☞www☜
4. ☞www☜
5. ☞www☜
6. Lowrie, W. and Alvarez, W. (1981) *Geology*, 9: 392–7. ☞www☜
7. ☞www☜
8. Mullins, J. (1999) *New Scientist*, 25 December: 66–7. ☞www☜
9. ☞www☜
10. Cook, T. A. (1914) *The Curves of Life*, London: Constable (reprinted in Dover Books, 1979); Rybczynski, W. (2000) *One Good Turn: A Natural History of the Screwdriver and the Screw*, New York: Simon and Schuster. ☞www☜
11. Maritan, A. *et al.* (2000) *Nature*, 406: 287–90; Galloway, J. W. (1991) in Bock, G. R. and Marsh, J. (eds) *Biological Asymmetry and Handedness* (Ciba foundation symposium 162), Chichester: Wiley, pp. 16–35.
12. ☞www☜
13. ☞www☜
14. McManus, I. C. and Humphrey, N. K. (1973) *Nature*, 243: 271–2; Humphrey, N. K. and McManus, I. C. (1973) *New Scientist*, 59: 437–9. ☞www☜
15. Woodcock, T. and Robinson, J. M. (1988) *The Oxford Guide to Heraldry*, Oxford: Oxford University Press.
16. Marx, K. *Karl Marx, Frederick Engels, Collected Works, Vol. I: Karl Marx, 1835–1843*, London: Lawrence and Wishart, 1975 p. 622. ☞www☜
17. Sklar, L. (1974) *Space, Time, and Spacetime*, Berkeley, CA: University of California Press; Earman, J. (1989) *World Enough and Space-time: Absolute Versus Relational Theories of Space and Time*, Cambridge, MA: MIT Press.
18. Walford, D. and Meerbote, R. (1992) *The Cambridge Edition of the Works of Immanuel Kant: Theoretical Philosophy, 1755–1770*, Cambridge: Cambridge University Press, pp. 365–72; Van Cleve, J. and Frederick, R. E. (1991) *The Philosophy of Right and Left: Incongruent Counterparts and the Nature of Space*, Dordrecht: Kluwer, p. 30. ☞www☜
19. Van Cleve, J. and Frederick, R. E. 'Prolegomena to any future metaphysic', in Van Cleve, J. and Frederick, R. E. (1991) op. cit.
20. Wittgenstein, L. (1961) *Tractatus Logico-philosophicus* (Translated D. F. Pears and B. F. McGuinness), London: Routledge and Kegan Paul, 6.36111. ☞www☜
21. ☞www☜
22. Walford, D. and Meerbote, R. (1992) op. cit., p. 371.

23. ☞www☜

24. Frederick, R. E., in Van Cleve, J. and Frederick, R. E. (1992) op. cit., pp. 1–14; Nerlich, G. (1973) *Journal of Philosophy*, 70: 337–51. ☞www☜

25. Gardner, M. (1990) *The New Ambidextrous Universe* (1973), (revised edition), New York: W. H. Freeman; Bennett, J. (1970) *American Philosophical Quarterly* 7: 175–91. ☞www☜

26. Lanza, R. P. *et al.*, (2000) *Scientific American*, November: 67–71.

Chapter 4: *Kleiz, drept, luft, zeso, lijevi, prawy*

1. Fritsch, V. (1968) *Left and Right in Science and Life*, London: Barrie and Rockliff; Fowler, A. (1971) *Milton: Paradise Lost*, London: Longman. ☞www☜

2. Fritsch, V. (1968) op. cit., p. 54; Gleick, J. (1994) *Genius: Richard Feynman and Modern Physics*, London: Abacus, p. 331; Masson, J. M. (1985) *The Complete Letters of Sigmund Freud to Wilhelm Fliess 1887–1904*, Cambridge, MA: Harvard University Press. ☞www☜

3. Aitchison, J. (1994) *Words in the Mind*, (2nd edn), Oxford: Blackwell, p. 85. ☞www☜

4. Buck, C. D. (1949) *A Dictionary of Selected Synonyms in the Principal Indo-European Languages*, Chicago: University of Chicago Press; Gamkrelidze, T. V. and Ivanov, V. V. (1995) *Indo-European and the Indo-Europeans: A Reconstruction and Historical Analysis of a Proto-language and a Proto-culture*, Berlin: Mouton de Gruyer.

5. Crystal, D. (1987) *The Cambridge Encyclopaedia of Language*, Cambridge: Cambridge University Press; Beekes, R. S. P. (1995) *Comparative Indo-European Linguistics: An Introduction*, Amsterdam: John Benjamins. ☞www☜

6. Collinge, N. E. (1985) *The Laws of Indo-European*, Amsterdam: John Benjamins. ☞www☜

7. Beekes, R. S. P. (1995) op. cit.; Renfrew, C. (1987) *Archaeology and Language: The Puzzle of Indo-European Origins*, London: Jonathan Cape. ☞www☜

8. Beekes, R. S. P. (1995) op. cit.

9. Buck, C. D. (1949) op. cit., p. 865; Gamkrelidze, T. V. and Ivanov, V. V. (1995) p. 686; Delamarre, X. (1984) *Le vocabulaire Indo-Européen: lexique étymologique thématique*, Paris: Librairie d'Amérique et d'Orient. ☞www☜

10. Hertz, R. (1909) *Revue Philosophique* 68: 553–80; Gamkrelidze, T. V. and Ivanov, V. V. (1995); Markey, T. L. (1982) *Mankind Quarterly* 23: 183–94; Wilson, G. P. (1937) *Words* 3: 102–5. ☞www☜

11. Hamilton, H. W. and Deese, J. (1971) *Journal of Verbal Learning and Verbal Behavior* 10: 707–14; Clark, H. H. (1973), in Moore, T. E. (ed.) *Cognitive Development and the Acquisition of Language*, New York: Academic Press pp. 27–63; Markey, T. L. p. 189; Mallory, J. P. and Adams, D. Q. (1997) *Encyclopaedia of Indo-European Culture*, London: Fitzroy Dearborn, p. 349. ☞www☜

12. Laserson, M. M. (1939) *American Sociological Review* 4: 534–42; Gamkrelidze, T. V. and Ivanov, V. V. (1995) p. 687.

13. Renfrew, C. (1998), in *The Nostratic Macrofamily and Linguistic Palaeontology*, Cambridge McDonald Institute for Archaeological Research pp. vii–xxii; Ross, P. E. (1991) *Scientific American* April: 70–79; Salmons, J. C. and Joseph, B. D. (1998) (eds) *Nostratic: Sifting the Evidence*, Amsterdam: John Benjamins; Dolgopolsky, A. (1998), in *The Nostratic Macrofamily and Linguistic Palaeontology*, Cambridge: McDonald Institute for Archaeological Research. ☞www☜

14. The picture of the Queen on the right, facing left, is the correct one. The picture

of George Washington on the left, facing right is the correct one. The left-hand picture of Hale-Bopp is the correct one, with the tail pointing to the right. ☞www☜

15. Martin, M. and Jones, G. V. (1999) *Psychological Science* 10: 267–70; McKelvie, S. J. and Aikins, S. (1993) *British Journal of Psychology* 84: 355–63; Jones, G. V. and Martin, M. (1997) *British Journal of Psychology* 88: 609–19. ☞www☜

16. ☞www☜

17. Ofte, S. H. & Hugdahl, K. (2001) *Journal of Experimental and Clinical Neuropsychology* (in press); Piaget, J. (1928) *Judgement and Reasoning in the Child,* London: Kegan Paul, Trench and Trubner; Elkind, D. (1961) *Journal of Genetic Psychology* 99: 269–76; Dellatolas, G. *et al. Cortex* 34: 659–76. ☞www☜

18. Piaget, J. (1928) op. cit., pp. 109–10, 202–3. ☞www☜

19. Regal, R. (1996) *Perceptual and Motor Skills* 83: 831–42.

20. Harris, L. J. and Gitterman, S. R. (1978) *Perceptual and Motor Skills* 47: 819–23; Wolf, S. M. (1973) *Archives of Neurology* 29: 128–9; Hannay, H. J. *et al.* (1990) *Perceptual and Motor Skills* 70: 451–7; Storfer, M. D. (1995) *Perceptual and Motor Skills* 81: 491–7. ☞www☜

21. Brandt, J. and Mackavey, W. (1981) *International Journal of Neuroscience* 12: 87–94. ☞www☜

22. Sholl, M. J. and Egeth, H. E. (1981) *Memory and Cognition* 9: 339–50. ☞www☜

23. Olson, G. M. and Lanar, K. (1973) *Journal of Experimental Psychology* 100: 284–90.

24. Mach, E. (1914) *The Analysis of Sensations,* 5th edn, Chicago: Open Court. ☞www☜

25. Corballis, M. C. and Beale, I. L. (1970) *Psychological Review* 77: 451–64; Corballis, M. C. and Beale, I. L. (1971) *Scientific American* 224 (3): 96–104; Corballis, M. C. and Beale, I. L. (1976) *The Psychology of Left and Right,* Hillsdale, NJ: Lawrence Erlbaum Associates.

26. Benton, A. L. (1959) *Right–Left Discrimination and Finger Localization: Development and Pathology,* New York: Hoeber, pp. 37, 51; Storfer, M. D. (1995); Dellatolas, G. *et al.* (1998) op. cit.; Gallet, J. H. (1988) *Buffalo Law Review* 37: 739–50, p. 741.

27. Critchley, M. (1966) *Brain* 89: 183–98; Dehaene, S., Dehaene-Lambertz, G. and Cohen, L. (1998) *Trends in Neurosciences* 21: 355–61. ☞www☜

28. Mayer, E. *et al.* (1999) *Brain* 122: 1107–20.

29. Mayer *et al.* 1999; Gold, M., Adair, J. C., Jacobs, D. H. and Heilman, K. M. (1995) *Cortex* 31: 267–83; Galton, F. (1883) *Inquiries into Human Faculty and its Development,* London: Macmillan; Spalding, J. M. K. and Zangwill, O. L. (1959) *Journal of Neurology, Neurosurgery, and Psychiatry* 13: 24–9; Butterworth, B. (1999) *The Mathematical Brain,* London: Macmillan. ☞www☜

30. Turnbull, O. H. and McCarthy, R. A. (1996) *Neurocase* 2: 63–72; Milner, A. D. and Goodale, M. A. (1995) *The Visual Brain in Action,* Oxford: Oxford University Press. ☞www☜

31. Wood Jones, F. (1941) *The Principles of Anatomy as Seen in the Hand,* London: Bailliere, Tindall and Cox; Wood Jones, F. (1949) *Structure and Function as Seen in the Foot,* 2nd edn, London: Bailliere, Tindall and Cox; Quinn, H. R. and Witherell, M. S. (1998) *Scientific American* 279 (October): 50–55; www.lecb.ncifcrf.gov/~toms/LeftHanded.DNA.html; Wang, A. H.-J. *et al. Nature* (1979) 282: 680–86; Anonymous, *Nature* 15 June 2000, p. 737; Porter, C. (2000) *Nature* 406: 234. ☞www☜

Chapter 5: The heart of the dragon

1. ☞www☜
2. ☞www☜
3. James, R. R. (1994) *Henry Wellcome*, London: Hodder & Stoughton, p. 344; Vinken, P. (1999) *The Shape of the Heart*, Amsterdam: Elsevier. ☞www☜
4. Sitwell, O. (1946) *The Scarlet Tree*, London: Macmillan; Ruehm, S. G. *et al.* (2001) *Lancet* 357: 1086–91. ☞www☜
5. Oster, J. (1971) *Scandinavian Journal of Urology and Nephrology* 5: 27–32. ☞www☜
6. Haldane, J. B. S. (1985) *On Being the Right Size* (ed. John Maynard Smith), Oxford: Oxford University Press.
7. Kilner, P. J. *et al.* (2000) *Nature* 404: 759–61. ☞www☜
8. Shu, D.-G. *et al.* (1999) *Nature* 402: 42–6; Janvier, P. (1999) *Nature* 402: 21–2; Gee, H. (1996) *Before the Backbone: Views on the Origins of the Vertebrates*, London: Chapman and Hall. ☞www☜
9. Gee, H. (1996) op. cit., p. 10. ☞www☜
10. Jefferies, R. P. S. (1991) in Bock, G. R. and Marsh, J. (eds) *Biological Asymmetry and Handedness* (Ciba Foundation symposium 162), Chichester: John Wiley, p. 124. ☞www☜
11. Gee, H. (1996) op. cit., pp. 201–86; Jefferies, R. P. S. (1986) *The Ancestry of the Vertebrates*, London: British Museum (Natural History); Sutcliffe, O. E. *et al.* (2000) *Lethaia* 33: 1–12. ☞www☜
12. ☞www☜
13. Jefferies, R. P. S. (1991) op. cit., p. 116; Dawkins, R. (1997) *Climbing Mount Improbable*, London: Penguin Books, pp. 204–10. ☞www☜
14. Norman, J. R. (1934) *A Systematic Monograph of the Flatfishes (Heterosomata), Volume 1: Psettodidae, Bothidae, Pleuronectidae*, London: British Museum (Natural History); Norman, J. R. and Greenwood, P. H. (1963) *A History of Fishes*, 2nd edn, London: Ernest Benn. ☞www☜
15. ☞www☜
16. Larsen, W. J. (1998) *Essentials of Human Embryology*, New York: Churchill Livingstone; Chiang, C. *et al.* (1996) *Nature* 383: 407–13. ☞www☜
17. Driever, W. (2000) *Nature* 406: 141–2; Srivastava, D. and Olson, E. N. (2000) *Nature* 407: 221–6.
18. Ivemark, B. L. (1955) *Acta Paediatrica* 44 (Suppl 104): 1–110; Burn, J. (1991), in *Biological Asymmetry and Handedness*, pp. 282–99; Capdevila, J. *et al.* (2000) *Cell* 101: 9–21. ☞www☜
19. ☞www☜
20. ☞www☜
21. ☞www☜
22. Furlow, F. B. *et al.* (1997) *Proceedings of the Royal Society of London, Series B* 264: 823–9; Jung, R. E. *et al.* (2000) *Neuropsychiatry, Neuropsychology and Behavioural Neurology*, 13: 195–198; Yeo, R. A. *et al.* (2000) *Developmental Neuropsychology*, 17: 143–159; Møller, A. P. and Swaddle, J. P. (1997) *Asymmetry, Developmental Instability and Evolution*. Oxford: Oxford University Press, pp. 157–9.
23. Sitwell, O. (1945) *Left Hand, Right Hand!*, London: Macmillan; Mealey, L. *et al.* (1999) *Journal of Personality and Social Psychology* 76: 157–65; Thornhill, R. and Gangestad, S. W. (1999) *Trends in Cognitive Sciences* 3: 452–60; Bruce, V. and Young, A. (1998) *In the Eye of the Beholder: The Science of Face Perception*, Oxford: Oxford University Press, p. 140; Møller, A. P. *et al.* (1996) *Behav. Ecol.* 7: 247–53;

Roldan, E. R. S. *et al.* (1998) *Proceedings of the Royal Society of London, Series B* 265: 243–8; Arcese, P. (1994) *Animal Behaviour* 48: 1485–8; Møller, A. P. and Swaddle, J. P. (1997) *Asymmetry, Developmental Instability and Evolution,* Oxford: Oxford University Press; Møller, A. P. (1992) *Nature* 357: 238–40. ☞www☜

24. Wilkin, S. (1852) *The Works of Sir Thomas Browne* (revised version of 1836 edition), London: Henry Bohn, vol. 1: p. 374. ☞www☜

25. Weismann, A. (1885) *Die Kontinuität des Keimplasmas als Grundlage der Vererbung,* Jena: Gustav Fischer; Wolpert, L. *et al.* (1998) *Principles of Development,* London: Current Biology; Maienschein, J. (1994) in Gilbert, S. F. (ed.) *A Conceptual History of Modern Embryology,* Baltimore: The Johns Hopkins University Press; pp. 43–61; Willier, B. H. and Oppenheimer, J. M. (1964) *Foundations of Experimental Embryology,* Englewood Cliffs, N.J.: Prentice-Hall, pp. 2–37, 38–50; Hamburger, V. (1988) *The Heritage of Experimental Embryology: Hans Spemann and the Organizer,* New York: Oxford University Press. ☞www☜

26. Horder, T. J. and Weindling, P. J. (1986) in Horder, T. J., Witkowski, J. A. and Wylie, C. C. (eds) *A History of Embryology,* Cambridge: Cambridge University Press, p. 186; Spemann, H. and Falkenberg, H. (1919) *Wilhelm Roux' Archiv für Entwicklungsmechanik* 45: 371–422; Huxley, J. S. and de Beer, G. R. (1934) *The Elements of Experimental Embryology,* Cambridge: Cambridge University Press, pp. 75–7; Oppenheimer, J. M. (1974) *American Zoologist* 14: 867–79, pp. 871–2. ☞www☜

27. Huxley, A. (1996 [1928]) *Point Counter Point,* Illinois: Dalkey Archive Press.

28. Bedford, S. (1973) *Aldous Huxley, A Biography. Vol I: 1894–1939,* London: Chatto & Windus; Huxley, J. S. and de Beer, G. R. (1934) op. cit. ☞www☜

29. Oppenheimer, J. M. (1974); Von Kraft, A. (1999) *Laterality* 4: 209–55; Bryden, M. P. *et al.* (1996) *Laterality* 1: 1–3; Corballis, M. C. and Beale, I. L. (1976) *The Psychology of Left and Right,* Hillsdale, NJ: Lawrence Erlbaum Associates; Morgan, M. J. (1977) in Harnad, S., Doty, R. W., Jaynes, J., Goldstein, L. and Krauthamer, G (eds), *Lateralization in the Nervous System,* New York: Academic Press, pp. 173–94; Corballis, M. C. and Morgan, M. J. (1978) *Behavioral and Brain Sciences* 2: 261–9, 270–78; Wehrmaker, A. (1969) *Wilhelm Roux' Archiv für Entwicklungsmechanik* 163: 1–32. ☞www☜

30. Hummel, K. P. (1969) and Chapman, D. B. (1959) *Journal of Heredity* 50: 9–13; Layton, W. M. (1976) *Journal of Heredity* 67: 336–8; Brown, N. A. and Wolpert, L. (1990) *Development* 109: 1–9; Layton, W. M. (1978) *Birth Defects* 14: 277–93. ☞www☜

31. Kartagener, M. (1933) *Beiträge zur Klinik und Erforschung der Tuberkulose und der Lungenkrankheiten* 83: 489–501; Parraudeau, M. *et al.* (1994) *British Medical Journal* 308: 519–21; Afzelius, B. A. *et al.* (1978); *Fertility and Sterility* 29: 72–4; Sivak, B. and MacKenzie, C. L. (1989) *Brain and Cognition* 9: 109–22. ☞www☜

32. Afzelius, B. A. *et al.* (1978) op. cit.; Afzelius, B. A. (1976) *Science* 193: 317–19.

33. Afzelius, B. A. (1999) *International Journal of Developmental Biology* 43: 283–6; Waite, D. *et al.* (1978) *Lancet* ii: 132–3; Handel, M. A. and Kennedy, J. R. (1984) *Journal of Heredity* 75: 498. ☞www☜

34. Brown, N. A. *et al.* (1989) *Development* 107: 637–42; Brown, N. A. and Wolpert, L. (1990) *Development* 109: 1–9; Brown, N. A. *et al.* (1991), in *Biological Asymmetry and Handedness* pp. 182–201. ☞www☜

35. Levin, M. *et al.* (1995) *Cell* 82: 803–14; Ibid., Wolpert, L. *et al.* (1998); Goldstein, A. M. *et al.* (1998) *Developmental Genetics* 22: 278–87. ☞www☜

36. Esteban, C. R. *et al.* (1999) *Nature* 401: 243–51; Capdevila, J. *et al.* (2000) *Cell* 101:

9–21; Levin, M. *et al.* (1997) *Developmental Biology* 189: 57–67; Ryan, A. K. *et al.* (1998) *Nature* 394: 545–51; Whitman, M. and Mercola, M. (2001) *Science's STKE* http://www.stke.org/cgi/content/full/OCV_sigtrans:2001/64/re1: ☞www☜

37. Brueckner, M. *et al.* (1989) *Proceedings of the National Academy of Sciences of the USA* 86: 5035–8; Brueckner, M. *et al.* (1991) in *Biological Asymmetry and Handedness*, pp. 202–18; Supp, D. M. *et al.* (1997) *Nature* 389: 963–6. ☞www☜

38. Nonaka, S. *et al.* (1998) *Cell* 95: 829–37; Takeda, S. *et al.* (1999) *Journal of Cell Biology* 145: 825–36; Marszalek, J. R. *et al.* (1999) *Proceedings of the National Academy of Sciences of the USA* 96: 5043–8; Vogan, K. J. and Tabin, C. J. (1999) *Nature* 397: 295–8.

39. ☞www☜

40. Okada, Y. *et al.* (1999) *Molecular Cell* 4: 459–68, p. 459; Ainsworth, C. (2000) *New Scientist* 17 June: 40–45; Supp, D. M. *et al.* (2000) *Trends in Cell Biology* 10: 41–5. ☞www☜

41. Yokoyama, T. *et al.* (1993) *Science* 260: 679–82; Brown, N. A. and Lander, A. (1993) *Nature* 363: 303–4; Morgan, M. J. (1978) *Behavioral and Brain Sciences* 2: 325–31. ☞www☜

42. Okada, Y. *et al.* (1999) op. cit. ☞www☜

43. Okada, Y. *et al.* (1999) op. cit. ☞www☜

44. ☞www☜

Chapter 6: The toad, ugly and venomous

1. ☞www☜

2. ☞www☜

3. Feynman, R. P. *et al.* (1963) *The Feynman Lectures on Physics. Vol I: Mainly Mechanics, Radiation, and Heat*, Reading, MA: Addison-Wesley, p. 52–5. ☞www☜

4. Thompson, S. P. (1910) *The Life of William Thomson, Baron Kelvin of Largs*, London: Macmillan, vol. II, p. 1054; Kelvin, Lord (William Thompson) (1904) *Baltimore Lectures on Molecular Dynamics and the Wave Theory of Light*, London: Cambridge University Press, pp. 619, 637, 642. ☞www☜

5. Mendelson, E. (1976) *W. H. Auden: Collected Poems*, London: Faber and Faber, p. 345; Gardner, M. (1990) *The New Ambidextrous Universe* (revised edition), New York: W. H. Freeman, p. 124; Laska, M. *et al.* (1999) 277: R1098–R1103; Mason, S. F. (1989) *Chirality*, 1: 183–91. ☞www☜

6. ☞www☜

7. De Camp, W. H. (1989) *Chirality*, 1: 2–6; Tucker, G. T. (2000) *Lancet*, 355: 1085–7.

8. Lewis, D. L. *et al.* (1999) *Nature*, 401: 898–901; Seo, J. S. *et al.* (2000) *Nature*, 404: 982–6. ☞www☜

9. Wnendt, S. Z. and Zwingenberger, K. (1997) *Nature*, 385: 303–4; Winter, W. and Frankus, E. (1992) *Lancet*, 339: 365; Eriksson, T. *et al.* (1998) *Chirality*, 10: 223–8; Eriksson, T. *et al.* (1995) *Chirality*, 7: 44–52. ☞www☜

10. ☞www☜

11. Lamzin, V. S. *et al.* (1995) *Current Opinion in Structural Biology*, 5: 830–36; Petsko, G. A. (1992) *Science*, 256: 1403–4; Milton, R. C. D. *et al.* (1992) *Science*, 256: 1445–8. ☞www☜

12. Nicolle, J. (1962) *Louis Pasteur: A Master of Scientific Enquiry*, London: The Scientific Book Guild, pp. 33–4; Meister, A. (1965) *Biochemistry of the Amino Acids*, (2nd edn), New York: Academic Press, vol. 1, p. 113; Helfman, P. M. and Bada, J. L. (1975) *Proceedings of the National Academy of Sciences of the USA*, 72: 2891–4;

Masters, P. M. *et al.* (1977) *Nature*, 268: 71–3; Ingrosso, D. and Perna, A. F. (1998) in Jollès, P. (ed.) *D-amino Acids in Sequences of Secreted Peptides of Multicellular Organisms*, Basel: Birkhäuser Verlag pp. 119–41; Lubec, G. *et al.* (1994) *FASEB Journal*, 8: 1166–9. ☞www☜

13. Yamada, R. and Kera, Y. (1998), in Jollès, P. (ed.) op. cit., pp. 145–55; Ingrosso, D. and Perna, A. F. (1998). ☞www☜

14. Shapira, R. *et al.* (1988) *Journal of Neurochemistry*, 50: 69–74; Kaneko, I. *et al.* (1995) *Journal of Neurochemistry*, 65: 2585–93.

15. Bender, D. A. (1985) *Amino Acid Metabolism* (2nd edn), Chichester: John Wiley; Hashimoto, A. and Oka, T. (1997) *Progress in Neurobiology*, 52: 325–53; Hashimoto, A. *et al.* (1992) *FEBS Letters*, 296: 33–6; Schell, M. J. *et al.* (1997) *Proceedings of the National Academy of Sciences of the USA*, 94: 2013–18; Wolosker, H. *et al.* (1999) *Proceedings of the National Academy of Sciences of the USA*, 96: 13409–14; Schell, M. *et al.* (1997) *Journal of Neuroscience*, 17: 1604–15; Schell, M. J. *et al.* (1995) *Proceedings of the National Academy of Sciences of the USA*, 92: 3948–52. ☞www☜

16. Mor, A. *et al.* (1992) *Trends in Biochemical Science*, 17: 481–5.

17. Meister, A. (1965) op. cit., p. 115. ☞www☜

18. Meister, A. (1965) op. cit.; Erspamer, V. (1992) *International Journal of Developmental Neuroscience*, 10: 3–30; Erspamer, V. (1984) *Comparative Biochemistry and Physiology* 79C: 1–7. ☞www☜

19. Lazarus, L. H. and Attila, M. (1993) *Progress in Neurobiology*, 41: 473–507; Lazarus, L. H. *et al.* (1999) *Progress in Neurobiology*, 57: 377–420. ☞www☜

20. Lazarus, L. H. *et al.* (1999) op. cit., p. 399.

21. Volkmann, R. A. and Heck, S. D. (1998) in Jollès, P. (ed.) op. cit., pp. 87–105; Lazarus, L. H. *et al.* (1999) op. cit., p. 380; Lazarus, L. H. and Attila, M. (1993) op. cit., p. 477. ☞www☜

22. Scaloni, A. *et al.* (1998) in Jollès, P. (ed.) op. cit., pp. 3–26; Volkmann, R. A. and Heck, S. D. (1998), in Jollès, P. (ed.) op. cit., pp. 87–105; Terlau, H. *et al.* (1996) *Nature*, 381: 148–51; Jimenéz, E. C., Olivera, B. M. *et al.* (1996) *Journal of Biological Chemistry*, 271: 28002–5; Heck, S. D. *et al.* (1994) *Science*, 266: 1065–8; Heck, S. D. *et al.* (1996) *Proceedings of the National Academy of Sciences of the USA*, 93: 4036–9. ☞www☜

23. Bonner, W. (1998), in Jollès, P. (ed.) op. cit., p. 161. ☞www☜

24. Haldane, J. B. S. (1932) *The Causes of Evolution*, London: Longmans, Green and Co., p. 146; Orgel, L. E. (1973) *The Origins of Life: Molecules and Natural Selection*, London: Chapman and Hall, p. 167; Gell-Mann, M. (1995) *The Quark and the Jaguar*, London: Abacus, p. 229. ☞www☜

25. Bonner, W. (1998), in Jollès, P. (ed.) op. cit., p. 162. ☞www☜

26. Ibid., p. 184.

27. Wigner, E. P. (1967) *Symmetries and Reflections*, Bloomington: Indiana University Press; Blackett, P. M. S. (1959) *Proceedings of the Royal Society of London, Series A*, 251: 293–305; Wu, C. S. *et al.* (1957) *Physical Review*, 105: 1413–15; Gardner, M. (1990); Cloe, F. (2000) *Lucifer's Legacy: The Meaning of Asymmetry*, Oxford: Oxford University Press. ☞www☜

28. Lee, T. D. and Yang, C. N. (1956) *Physical Review*, 104: 254–8; Kurti, N. and Sutton, C. (1997) *Nature, 385: 575*. ☞www☜

29. Gardner, M. (1990); Bernstein, J. (1962) *New Yorker*, May 12: 49–104; Morrison, P. (1957) *Scientific American*, 196 (April): 45–55. ☞www☜

30. Salam, A. (1958) *Endeavour*, 17: 97–105; Mason, S. F. (1991) *Chemical Evolution: Origin of the Elements, Molecules and Living Systems*, Oxford: Clarendon Press, p.

283; Frisch, O. (1995) in Carey, J. (ed.), *The Faber Book of Science*, London: Faber and Faber, pp. 403–12. ☞www☜

31. Haldane, J. B. S. (1960) *Nature*, 185: 87; Cline, D. B. (1996), in Cline D. B. (ed.), *Physical Origin of Homochirality in Life*, Woodbury, NY, pp. 266–82. ☞www☜

32. Mason, S. F. and Tranter, G. E. (1985) *Proceedings of the Royal Society of London, Series A*, 397: 45–65. ☞www☜

33. Frank, F. C. (1953) *Biochimica et Biophysica Acta*, 11: 459–63; Kondepudi, D. K. and Nelson, G. W. (2000) *Nature*, 314: 438–41; Hegstrom, R. A. and Kondepudi, D. K. (1990) *Scientific American*, 98–105; Soai, K. *et al.* (1995) *Nature*, 378: 767–8; Shibata, T. *et al.* (1998) *Journal of the American Chemical Society*, 120: 12157–8.

34. Halpern, B., Westley, J. W., Levinthal, E. C. *et al.* (1966) *Life Science and Space Research*, 5: 239–49; MacDermott, A. J. (1996), in Cline, D. B. (ed.), p. 250.

35. Kvenvolden, K. *et al.* (1970) *Nature*, 228: 923–6; Cronin, J. R. and Pizzarello, S. (1997) *Science*, 275: 951–5; Scaloni, A. *et al.* (1998) in Jollès, P. (ed.) op. cit., pp. 3–26; Engel, M. H. and Nagy, B. (1982) *Nature*, 296: 837–40; Pizzarello, S. and Cronin, J. R. (2000) *Geochimica et Cosmochimica Acta*, 64: 329–38; Brown, P. G. *et al.* (2000) *Science*, 290: 320–25. ☞www☜

36. Curie, P. (1894) *Journal de Physique Théorique et Appliquée* (3rd Series), 3: 393–415; Bouchiat, M.-A. and Pottier, L. (1984) *Scientific American*, 250: 100–111; Norden, B. (1977) *Nature*, 266: 567–8; Cline, D. B. (1996) in Cline, D. B. (ed.) pp. 266–82; Greenberg, J. M. (1996), in Cline, D. (ed.) pp. 185–210. ☞www☜

37. Bonner, W. (1998) in Jollès, P. (ed.) op. cit.; Greenberg, J. M. (2000) *Scientific American*, 283 (December): 46–51. ☞www☜

38. Chyba, C. F. *et al.* (1990) *Science*, 249: 366–73; Holzheid, A. *et al.* (2000) *Nature*, 406: 396–9. ☞www☜

39. Madigan, M. T. and Marrs, B. L. (1997) *Scientific American*, 276 (April): 66–71; Gould, S. J. (1997) *Life's Grandeur: The Spread of Excellence from Plato to Darwin*, London: Vintage; Nagata, Y., Tanaka, K., Iida, T. *et al.* (1999) *Biochimica et Biophysica Acta*, 1435: 160–66; Stetter, K. O. (1996) in Bock, G. R. and Goode, J. A. (eds) *Evolution of Hydrothermal Systems on Earth (and Mars?)*, Chichester: John Wiley, pp. 1–18. ☞www☜

40. Barns, S. M. *et al.* (1996), in Bock, G. R. and Goode, J. A. (eds) op. cit., pp. 24–39. ☞www☜

41. MacDermott, A. J. (1996) in Cline, D. B. (ed.) pp. 241–54; MacDermott, A. J. *et al.* (1996) *Planetary and Space Science*, 44: 1441–6.

Chapter 7: The dextrous and the gauche

1. Burkhardt, F. and Smith, S. (1986) *The Correspondence of Charles Darwin, Vol. 2 1837–1843*, Cambridge: Cambridge University Press, p. 269; Bowlby, J. (1990) *Charles Darwin: A New Biography*, London: Hutchinson.

2. Bowlby, J. (1990) Barrett, P. H. *et al.* (1987) *Charles Darwin's Notebooks, 1836–1844*, Cambridge: Cambridge University Press, p. 560; Burkhardt, F. and Smith, S. (1988) *The Correspondence of Charles Darwin, Vol. 4 1847–1850*, Cambridge: Cambridge University Press, Appendix III; Burkhardt, F. and Smith, S. (1986) op. cit. ☞www☜

3. Burkhardt, F. and Smith, S. (1988) op. cit., p. 415. ☞www☜

4. Taine, H. A. (1877) *Mind*, 2: 252–9; Darwin, C. (1877) *Mind* 2: 285–94; Gruber, H. E. and Barrett, P. H. (1974) *Darwin on Man: A Psychological Study of Scientific Creativity*, London: Wildwood House, p. 224. ☞www☜

5. Burkhardt, F. and Smith, S. (1988) op. cit., pp. 413–14; Darwin, C. (1905) *The Variation of Animals and Plants under Domestication*, vol. I, (popular edn), London: John Murray, p. 545.

6. ☞www☜

7. McManus, I. C. (1996) in Beaumont, J. G., Kenealy, P. M. and Rogers, M. J. C. (eds), *The Blackwell Dictionary of Neuropsychology*, Oxford: Blackwell, pp. 367–76. ☞www☜

8. Tapley, S. M. and Bryden, M. P. (1985) *Neuropsychologia*, 23: 215–21; Van Horn, J. D. (1992) *Brain Structural Abnormality and Laterality in Schizophrenia*, University College London: unpublished PhD thesis; McManus, I. C. *et al.* (1993) *British Journal of Psychology*, 84: 517–37; McManus, I. C., in Bock, G. R. and Marsh, J. (eds) *Biological Asymmetry and Handedness* (Ciba foundation symposium 162), Chichester: Wiley, pp. 251–81. ☞www☜

9. McManus, I. C. (1991) op. cit.; McManus, I. C. and Bryden, M. P. (1992) in Rapin, I. and Segalowitz, S. J. (eds), *Handbook of Neuropsychology, Volume 6, Section 10: Child Neuropsychology (Part 1)*, Amsterdam: Elsevier, pp. 115–44; Jones, G. V. and Martin, M. (2000) *Psychological Review*, 107: 213–18. ☞www☜

10. Masson, J. M. (1985) *The Complete Letters of Sigmund Freud to Wilhelm Fliess 1887–1904* (translated by Jeffrey Mouussaieff Masson), Cambridge, MA: Harvard University Press, pp. 290–6; Marchant-Haycox, S. E. *et al.* (1991) *Cortex*, 27: 49–56; Lalumière, M. L. *et al.* (2000) *Psychological Bulletin*, 126: 575–92; Green, J. (1998) *The Cassell Dictionary of Slang*, London: Cassell; Orlebeke, J. F. *et al.* (1992) *Neuropsychology*, 6: 351–5; Zucker, K. J. *et al.* (2001) *Journal of Child Psychology and Psychiatry*, 42: 767–76. ☞www☜

11. ☞www☜

12. Peters, M. (1987) *Canadian Journal of Psychology*, 41: 91–9; Peters, M. and Servos, P. (1989) *Canadian Journal of Psychology*, 43: 341–58; Peters M. (1990) *Neuropsychologia*, 28: 279–89; McManus, I. C. *et al.* (1999) *Laterality*, 4: 173–92; Ellis, H. (1967) (first published 1940) *My Life*, London: Neville Spearman. ☞www☜

13. ☞www☜

14. Peters, M. (1988) *Psychological Bulletin*, 103: 179–92; Porac, C. and Coren, S. (1981) *Lateral Preferences and Human Behaviour*, New York: Springer Verlag; Carey, D. P. *et al.* (2001) *Journal of Sport Sciences*, in press. ☞www☜

15. Porac, C. and Coren, S. (1981) op. cit.; Noonan, M. and Axelrod, S. (1981) *Journal of Auditory Research*, 21: 263–77; Bourassa, D. C. *et al.* (1996) *Laterality*, 1: 5–34; Orton, S. T. (1925) *Archives of Neurology and Psychiatry*, 14: 581–615; Orton, S. T. (1937) *Reading, Writing and Speech Problems in Children*, London: Chapman and Hall. ☞www☜

16. Hoogmartens, M. J. and Caubergh, M. A. (1987) *Electromyography and Clinical Neurophysiology*, 27: 293–300; McManus, I. C. and Mascie-Taylor, C. G. N. (1979) *Annals of Human Biology*, 6: 527–58. ☞www☜

17. McManus, I. C. and Mascie-Taylor, C. G. N. (1979); Code, C. (1995) *Perceptual and Motor Skills* 80: 1147–54. ☞www☜

18. Gesell, A. and Ames, L. B. (1947) *Journal of Genetic Psychology*, 70: 155–76; Michel, G. F. (1983), in Young, G., Segalowitz, S. J., Corter, C. M. and Trehub, S. E. (eds) *Manual Specialisation and the Developing Brain*, New York: Academic Press, pp. 33–70; Ibid., Harris, L. J., pp. 177–247; McManus, I. C. *et al.* (1988) *British Journal of Developmental Psychology*, 6: 257–73. ☞www☜

19. Hepper, P. G. *et al.* (1991) *Neuropsychologia*, 29: 1107–11; Hepper, P. G. *et al.* (1998) *Neuropsychologia*, 36: 531–4. ☞www☜

20. Provins, K. A. (1997) *Psychological Review*, 104: 554–71; McManus, I. C. and Bryden, M. P. (1992) op. cit. ☞www☜

21. Bury, R. G. (1968) *Plato. Laws, Vol. II, Books VII–XII*, Cambridge, MA: Harvard University Press, 794e; Armstrong, G. C. (1935) *Aristotle, Vol. XVIII: Magna Moralia*, Cambridge, MA: Harvard University Press, 1194.b.32; Bell, C. (1834) *The Hand: Its Mechanism and Vital Endowments as Evincing Design*, London: John Murray, p. 142; Klar, A. J. S. (1996) *Cold Spring Harbor Symposia on Quantitative Biology*, 6: 59–65. ☞www☜

22. Bakan, P. (1971) *Nature* 229: 195; Bakan, P. *et al.* (1973) *Neuropsychologia*, 11: 363–6; Satz, P. (1972) *Cortex*, 8: 121–35. ☞www☜

23. McManus, I. C. (1979) *Determinants of Laterality in Man*, University of Cambridge: unpublished PhD thesis, chapter 3; McManus, I. C. (1981) *Psychological Medicine*, 11: 485–96. ☞www☜

24. McManus, I. C. (1980) *Neuropsychologia*, 18: 347–55; McManus, I. C. and Bryden, M. P. (1992) op. cit.; Sicotte, N. L. *et al.* (1999) *Laterality*, 4: 265–86. ☞www☜

25. Layton, W. M. (1976) *Journal of Heredity*, 67: 336–8; Cockayne, E. A. (1940) *Biometrika*, 31: 287–94; Gedda, L. *et al.* (1984) *Acta Genet Med Gamellol*, 33: 81–5.

26. McManus, I. C. (1979) op. cit.; McManus, I. C. (1984) in Rose, F. C. (ed.) *Advances in Neurology, Vol. 42: Progress in Aphasiology*, New York: Raven Press, pp. 125–38; Layton, W. M. (1976) *Journal of Heredity*, 67: 336–8. ☞www☜

27. ☞www☜

28. Van Agtmael, T. *et al.* (2001) *Laterality*, 6: 149–64; Klar, A. J. S. (1996); Crow, T. J. (1998) *Cahiers de Psychologie Cognitive/Current Psychology of Cognition*, 17 : 1079–114. ☞www☜

29. Whitman, W. (1982) *Complete Poetry and Collected Prose*, New York: Library of America, pp. 1179–80; Peters, M. and Durding, B. (1978) *Canadian Journal of Psychology*, 32: 257–61; McManus, I. C. *et al.* (1986) *Cortex* 22: 461–74.

30. McManus, I. C. *et al.* (1992) *Cortex*, 28: 373–81; Cornish, K. M. and McManus, I. C. (1996) *Journal of Autism and Developmental Disorders*, 26: 597–609. ☞www☜

31. Hécaen, H. and de Ajuriaguerra, J. (1964) *Left-handedness: Manual Superiority and Cerebral Dominance*, New York: Grune and Stratton, p. 86; McManus, I. C. and Cornish, K. M. (1997) *Laterality*, 2: 81–90. ☞www☜

32. Glick, S. D. (1983) in Myslobodsky, M. S. (ed.), *Hemisyndromes: Psychobiology, Neurology, Psychology*, New York: Academic Press, pp. 7–26. ☞www☜

33. Mach, E. (1914) *The Analysis of Sensations*, 5th edn, Chicago: Open Court, p. 112; Guldberg, F. O. (1897) *Zeitschrift für Biologie*, 35: 419–458; Ludwig, W. (1932) *Das Rechts-Links-Problem im Tierreich und beim Menschen*, Berlin: Verlag Julius Springer, pp. 324–30; Schaeffer, A. A. (1928) *Journal of Morphology*, 45: 293–398; Millar, S. (1999) *Perception*, 28: 765–80; Bracha, H. S. (1987) *Biological Psychiatry*, 22: 995–1003; Gesell, A. and Ames, L. B. (1947) *Journal of Genetic Psychology*, 70: 155–76; Gesell, A. *et al.* (1949) *Vision: Its Development in Infant and Child*, New York: Paul B. Hoeber, pp. 48–9. ☞www☜

34. Moxon, W. (1866) *The British and Foreign Medico-Chirurgical Review*, 37: 481–9; Harris, L. J. (2000) *Brain and Language*, 73: 157–60. ☞www☜

Chapter 8: The left brain, the right brain and the whole brain

1. Debré, P. (1998) *Louis Pasteur*, Baltimore: Johns Hopkins University Press; Blackmore, J. T. (1972) *Ernst Mach: His Work, Life, and Influence*, Berkeley, CA: University of California Press, p. 279. ☞www☜

2. Debré, P. (1998) op. cit., p. 209; Vallery-Radot, R. (1911) *The Life of Pasteur*, London: Constable, p. 159.

3. Mach, E. (1914) *The Analysis of Sensations*, (5th edn), Chicago: Open Court. ☞www☜

4. Anderson, M. (1976) *Practitioner*, 217: 968–74; Anderson, W. E. K. (1972) *The Journal of Sir Walter Scott*, Oxford: Clarendon Press, p. 589; Scott, W. (1891) *The Journal of Sir Walter Scott*, Edinburgh: David Douglas, p. 741; Johnson, E. (1970) *Sir Walter Scott*, London: Hamish Hamilton, p. 179; Buchan, J. (1931) *Sir Walter Scott*, London: Cassell, p. 331; Anonymous (1882) *British Medical Journal*, ii: 1282–5; Sutherland, J. (1995) *The Life of Walter Scott: A Critical Biography*, Oxford: Blackwell, p. 354; Lockhart, J. G. (1896) *The Life of Sir Walter Scott*, London: Adam and Charles Black, p. 751; Albert, M. L. (1979), in Heilman, K. M. and Valenstein, E. (eds), *Clinical Neuropsychology*, New York: Oxford University Press. ☞www☜

5. Damásio, A. R. *et al.* (1976) *Archives of Neurology*, 33: 300–301; Lecours, A. R. *et al.* (1988) *Neuropsychologia*, 26: 575–89; Coppens, P. *et al.* (1998) in Coppens, P., Lebrun, Y. and Basso, A. (eds), *Aphasia in Atypical Populations*, Mahwah, NJ: Lawrence Erlbaum, pp. 175–202; Benton, A. (1984) *Neuropsychologia*, 22: 807–11; Carlyle, T. (1874) *Wilhelm Meister's Apprenticeship and Travels (translated from Goethe)* [1824], London: Chapman and Hall, book VII, chapter 6. ☞www☜

6. Benton, A. (1984) op. cit.

7. ☞www☜

8. Anonymous (1876) *Nature*, 13: 400–406. ☞www☜

9. Osborne, J. (1834) *Dublin Journal of Medical and Chemical Science*, 4: 157–70; Christman, S. S. and Buckingham, H. W. (1991), in Code, C. (ed.), *The Characteristics of Aphasia*, Hove: Lawrence Erlbaum, pp. 111–30. ☞www☜

10. Crampton, P. (1833) *Dublin Journal of Medical and Chemical Science*, 2: 30–45, 199–211; Goldfarb, R. and Halpern, H. (1991) in Code, C. (ed.), *The Characteristics of Aphasia*, Hove: Lawrence Erlbaum, pp. 33–52. ☞www☜

11. Damasio, A. R. (1992) *New England Journal of Medicine*, 326: 531–9; Code, C. (ed.) (1991) op. cit.; Hagoort, P. *et al.* (1999) in Brown, C. M. and Hagoort, P. (eds) *The Neurocognition of Language*, Oxford: Oxford University Press, pp. 273–316; Uylings, H. B. M. *et al.*, in Brown, C. M. and Hagoort, P. (eds), pp. 319–36. ☞www☜

12. De Buck (1899) (see Wilson, S. A. K. (1908) *Brain*, 31: 164–216, p. 181).

13. Jackson, J. H. (1874) *Medical Press and Circular*, i: 19, 41, 63, (emphasis in original). ☞www☜

14. ☞www☜

15. Humphreys, G. W. and Riddoch, M. J. (1987) *To See but Not to See: A Case Study of Visual Agnosia*, London: Lawrence Erlbaum. ☞www☜

16. Sacks, O. (1985) *The man who mistook his wife for a hat*, London: Duckworth, p. 13; Poppelreuter, W. (1917) *Die psychischen Schädigungen durch Kopfschuss in Kriege*, Leipzig: Voss; Walsh, K. W. (1978) *Neuropsychology: A clinical approach*, New York: Churchill Livingstone; Humphreys, G. W. & Riddoch, M. J. (1984) *Quarterly Journal of Experimental Psychology* 36A: 385–415; Warrington, E. K. & Taylor, A. M. (1973) *Cortex* 9: 152–64. ☞www☜

17. ☞www☜

18. Ross, E. D. *et al.* (1988) *Brain and Language*, 33: 128–45; Giora, R. *et al.* (2000) *Metaphor and Symbol*, 15: 63–83; McCrone, J. (2000) *New Scientist*, 27 May: 23–6. ☞WWW☜

19. Wilson, D. H. *et al.* (1977) *Neurology*, 27: 708–15; LeDoux, J. E. *et al.* (1977) *Annals of Neurology*, 2: 417–21; Gazzaniga, M. S. *et al.* (1977) *Neurology* 27: 1144–7; Gazzaniga, M. S. (1985) *The Social Brain: Discovering the Networks of the Mind*, New York: Basic Books.

20. Hellige, J. B. (1993) *Hemispheric Asymmetry: What's Right and What's Left*, Cambridge, MA: Harvard University Press. ☞WWW☜

21. Deglin, V. L. and Kinsbourne, M. (1996) *Brain and Cognition*, 285–307. ☞WWW☜

22. Ornstein, R. (1997) *The Right Mind: Making Sense of the Hemispheres*, Harcourt Brace: San Diego, p. 106.

23. Kaplan, F. (1988) *Dickens: A Biography*, New York: William Morrow, p. 539; Forster, J. (1928) *The Life of Charles Dickens* (edited by J. W. T. Ley), London: Cecil Palmer, p. 804; Forster, J. (1872) *The Life of Charles Dickens* [three volumes, 1872–1874], London: Chapman and Hall. ☞WWW☜

24. Halligan, P. W. and Marshall, J. C. (1993) in Robertson, I. H. and Marshall, J. C. (eds), *Unilateral Neglect: Clinical and Experimental Studies*, Hove: Lawrence Erlbaum Associates, pp. 3–25; Jackson, J. H. (1876) *Royal Ophthalmological Hospital Reports*, 8: 434–44; Ellis, A. W. *et al.* (1993) in Robertson, I. H. and Marshall, J. C. (eds) op. cit., p. 233; McCarthy, R. A. and Warrington, E. K. (1990) *Cognitive Neuropsychology: A Clinical Introduction*, San Diego: Academic Press, p. 219; McManus, I. C. (2001) *Lancet*, 358: 2158–61. ☞WWW☜

25. Chandler, C. (1996) *I, Fellini*, London: Bloomsbury, p. 365; Cantagallo, A. and Della Sala, S. (1998) *Cortex*, 34: 163–89. ☞WWW☜

26. Denes, G. *et al.* (1982) *Brain*, 105: 543–52; Bowen, A. *et al.* (1999) *Stroke*, 30: 1196–202; Stone, S. P. *et al.* (1991) *Journal of Neurology, Neurosurgery, and Psychiatry*, 54: 345–50; Robertson, I. H. (1993) in Robertson, I. H. and Marshall, J. C. (eds) op. cit., pp. 257–75. ☞WWW☜

27. ☞WWW☜

28. Bisiach, E. and Luzzatti, C. (1978) *Cortex*, 14: 129–33; Azouvi, P. *et al.* (1996) *Neuropsychological Rehabilitation* 6: 133–50. ☞WWW☜

29. Marshall, J. C. and Halligan, P. W. (1993) *Nature*, 364: 193–4.

30. Marshall, J. C. and Halligan, P. W. (1993) *Journal of Neurology*, 240: 37–40. ☞WWW☜

31. Wurtz, R. H. *et al.* (1982) *Scientific American*, 246 (June): 100–107; Parkin, A. J. (1996) *Explorations in Cognitive Neuropsychology*, Hove: Psychology Press; Halligan, P. W. and Marshall, J. C. (1997) *Lancet*, 350: 139–40; Halligan, P. W. and Marshall, J. C. (2001) *NeuroImage* 14: S91–S97; Marsh, G. G. and Philwin, B. (1987) *Cortex*, 23: 149–55; Schnider, A. *et al.* (1993) *Neuropsychiatry, Neuropsychology, and Behavioral Neurology*, 6: 249–55; Vigouroux, R. A. *et al.* (1990) *Revue Neurologique*, 146: 665–70. ☞WWW☜

32. Cantagallo, A. and Della Sala, S. (1998) op. cit., p. 165; Critchley, M. (1979) *The Divine Banquet of the Brain and Other Essays*, New York: Raven Press, p. 115–120; Ghika, S. F. *et al.* (1999) *Neurology*, 52: 22–8; Stone, S. P. *et al.* (1993) *Age and Ageing*, 22: 46–52; Façon, E. *et al.* (1960) *Revue Neurologique*, 102: 61–74. ☞WWW☜

33. Gordon, C. *et al.* (1987) *British Medical Journal*, 295: 411–14; Chandler, C. (1996), p. 327; Hamdy, S. *et al.* (1997) *Lancet*, 350: 686–92.

34. Hellige, J. B. (1993) *Hemispheric Asymmetry: What's Right and What's Left*,

Cambridge, MA: Harvard University Press, p. 54; Pritchard, T. C. *et al.* (1999) *Behavioral Neuroscience*, 113: 663–71; Regard, M. and Landis, T. (1997) *Neurology*, vol. 48: 1185–90; McKinnon, M. C. and Schellenberg, E. G. (1997) *Canadian Journal of Experimental Psychology*, 51: 171–5; Plenger, P. M. *et al.* (1996) *Neuropsychologia*, 34: 1015–18; Gordon, H. W. and Bogen, J. E. (1974) *Journal of Neurology, Neurosurgery and Psychiatry*, 37: 727–38; Segalowitz, S. *et al.* (1979) *Brain and Language*, 8: 315–23; Nakada, T. *et al.* (1998) *NeuroReport*, 9: 3853–6; Robinson, G. M. and Solomon, D. J. (1974) *Journal of Experimental Psychology*, 102: 508–11; Schlaug, G. *et al.* (1995) *Science*, 267: 699–701; Sergent, J. *et al.* (1992) *Science*, 257: 106–9; Tramo, M. J. (2001) *Science*, 291: 54–6; Gordon, A. G. (1991) *British Journal of Psychiatry* 158: 715–16; Stoléru, S., Grégoire, M. C., Gérard, D. *et al.* (1999) *Archives of Sexual Behaviour* 28: 1–21. ☞www☜

35. Nicholls, M. E. R. and Bradshaw, J. L. M. J. B. (1999) *Neuropsychologia*, 37: 307–14; Bowers, D. and Heilman, K. M. (1980) *Neuropsychologia*, 18: 491–8; Jewell, G. and McCourt, M. E. (2000) *Neuropsychologia*, vol. 38: 93–110; Turnbull, O. H. and McGeorge, P. (1998) *Brain and Cognition*, vol. 37: 31–3; Adair, H. and Bartley, S. H. (1958) *Perceptual and Motor Skills*, 8: 135–41; Nelson, T. M. and MacDonald, G. A. (1971) *Perceptual and Motor Skills*, 33: 983–6; Jaynes, J. (1976) *The Origin of Consciousness in the Breakdown of the Bicameral Mind*. Harmondsworth: Penguin Books, p. 120. ☞www☜

36. Kimura, D. (1981) *Canadian Journal of Psychology*, 15: 166–71; Bryden, M. P. (1962) *Canadian Journal of Psychology*, 16: 291–9; Bryden, M. P. (1982) *Laterality: Functional Asymmetry in the Intact Brain*, New York: Academic Press; Hugdahl, K. E. (1988) *Handbook of Dichotic Listening: Theory, Methods and Research*, Chichester, England: John Wiley & Sons. ☞www☜

37. Harris, L. J. (1999) *Schizophrenia Bulletin*, 25: 11–39; p. 11; Hirsch, E. D. Jr (1987) *Cultural Literacy: What Every American Needs to Know*, Boston, MA: Houghton Mifflin.

38. Hécaen, H. and Piercy, M. (1956) *Journal of Neurology, Neurosurgery, and Psychiatry*, 19: 194–201; Jackson, J. H. (1866) *Medical Times and Gazette*, 2: 210. ☞www☜

39. James, W. (1890) *The Principles of Psychology*, London: Macmillan, vol. I: p. 39; McManus, I. C. (1979) *Determinants of Laterality in Man*, University of Cambridge: unpublished PhD thesis, 1979 Fig 6.1; Bramwell, B. (1898) *Brain*, 21: 343–73; Bramwell, B. (1899) *Lancet*, i: 1473–9; Harris, L. J. (1991) *Brain and Language* 40: 1–50, p. 27. ☞www☜

40. Luria, A. R. (1970) *Traumatic Aphasia: Its Syndromes, Psychology and Treatment*, The Hague: Mouton, (originally published in Russian in 1947); McManus, I. C. (1983) *Journal of Communication Disorders*, 16: 315–44. ☞www☜

41. Harris, L. J. (1991) pp. 31–3; McManus, I. C. (1985) *Handedness, Language Dominance and Aphasia: a Genetic Model*, Monograph Supplement No. 8. to *Psychological Medicine*.

42. Bryden, M. P. *et al.* (1983) *Brain and Language*, 20: 249–62; Walters, J. (1882) *British Medical Journal*, ii: 914–15.

43. McManus, I. C. (1979) op. cit.; McManus, I. C. (1985) op. cit. ☞www☜

44. ☞www☜

45. Gardner, W. J. (1941) *Archives of Neurology and Psychiary*, 46: 1035–8; Harris, L. J. and Snyder, P. J. (1997) *Brain Language*, 56: 377–96; Wada, J. A. (1997) *Brain and Cognition*, 33: 7–10. ☞www☜

46. ☞www☜

47. Wada, J. A. (1997) op. cit., pp. 11–13; Wada, J. and Rasmussen, T. (1960) *Journal of Neurosurgery*, 17: 266–82.

48. ☞www☜
49. Knecht, S. *et al.* (2000) *Brain*, 123: 74–81; Knecht, S. *et al.* (2000) *Brain*, 123: 2512–18. ☞www☜

Chapter 9: Ehud, son of Gera

1. *New English Bible*, Judges 3: 20a, 21–22a; 20: 14–16; Anonymous (1971) *Encyclopaedia Judaica*, Jerusalem: Keter, 8: 583; Cook, T. A. (1914) *The Curves of Life*, London: Constable (reprinted in Dover Books, 1979); p. 243. ☞www☜
2. Scott-Kilvert, I. (1960) *The Rise and Fall of Athens: Nine Greek Lives by Plutarch*, Harmondsworth: Penguin, p. 294. ☞www☜
3. Ogle, W. (1871) *Transactions of the Royal Medical and Chirurgical Society of London* 54: 279–301; Harris, L. J. (1990), in Coren, S. (ed.) *Left-handedness: Behavioral Implications and Anomalies*, Amsterdam: North-Holland pp. 195–258; Beeley, A. L. (1918) *An Experimental Study in Left-handedness*, Chicago: University of Chicago Press. ☞www☜
4. Gilbert, A. N. and Wysocki, C. J. (1992) *Neuropsychologia* 30: 601–8. ☞www☜
5. Singh, M. and Bryden, M. P. (1994) *International Journal of Neuroscience* 74: 33–43; Ida, Y. and Bryden, M. P. (1996) *Canadian Journal of Experimental Psychology*, 50: 234–9; De Agostini, M. *et al.* (1997) *Brain and Cognition*, 35: 151–67; Amir, T. and McManus, I. C. (in preparation).
6. McManus, I. C. *et al.* (1996) *Laterality*, 1: 257–68; Bulman-Fleming, M. B. (1998) *Brain and Cognition*, 36: 99–103.
7. ☞www☜
8. Bryden, M. P. *et al.* (1997) *Laterality*, 2: 317–36; De Agostini, M. *et al.* (1997) op. cit. ☞www☜
9. Chamberlain, H. D. (1928) *Journal of Heredity*, 19: 557–9; McManus, I. C. and Bryden, M. P. (1992), in Rapin, I. and Segalowitz, S. J. (eds) *Handbook of Neuropsychology, Volume 6, Section 10: Child Neuropsychology (Part 1)*, Amsterdam: Elsevier, pp. 115–44. ☞www☜
10. McManus, I. C. and Bryden, M. P. (1992) op. cit. ☞www☜
11. Coren, S. and Porac, C. (1997) *Science*, 198: 631–2. ☞www☜
12. Painter, K. S. (1977) *The Mildenhall Treasure: Roman Silver from East Anglia*, London: British Museum. ☞www☜
13. Spindler, K. (1996) in Spindler, K., Wilfing, H., Rastbichler-Zissernig, E., zur Nedden, D. and Nothdurfter, H. (eds) *Human Mummies: A Global Survey of their Status and the Techniques of Conservation*, Wien: Springer-Verlag, pp. 249–63; Egg, M. (1992), in Höpfel, F., Platzer, W. and Spindler, K. (eds) *Der Mann im Eis: Band 1*, Innsbruck: University of Innsbruck, pp. 254–72. ☞www☜
14. Cahen, D. *et al.* (1979) *Current Anthropology*, 20: 661–83, (pp. 667–8). ☞www☜
15. Pitts, M. and Roberts, M. (1997) *Fairweather Eden: Life in Britain Half a Million Years Ago as Revealed by the Excavations at Boxgrove*, London: Century, pp. 174–6 and plates 7 and 28; Roberts, M. B. and Parfitt, S. A. (1999) *Boxgrove: A Middle Pleistocene Hominid Site at Eartham Quarry, Boxgrove, West Sussex*, London: English Heritage, pp. 332–9. ☞www☜
16. Steele, J. (2000) *Laterality*, 5: 193–220; Bahn, P. G. (1989) *Nature* 337: 693; Walker, A. and Leakey, R., in Walker, A. and Leakey, R. (eds) *The Nariokotome* Homo erectus *Skeleton*, Berlin: Springer-Verlag, pp. 95–160.
17. Ambrose, S. H. (2001) *Science* 291: 1748–53; Toth, N. (1985) *Journal of Human Evolution*, 14: 607–14; Schick, K. D. and Toth, N. (1993) *Making Silent Stones Speak:*

Human Evolution and the Dawn of Technology, London: Weidenfeld and Nicolson. ☞www☜

18. Cole, J. (1955) *Journal of Comparative and Physiological Psychology*, 48: 137–40; Fabre-Thrope, M. *et al.* (1993) *Cortex* 29: 15–24. ☞www☜

19. Byrne, R. W. and Byrne, J. M. (1991) *Cortex*, 27: 521–46; Byrne, R. W. and Byrne, J. M. (1993) *American Journal of Primatology*, 31: 241–61; Byrne, R. W. (1995) *Natural History*, 10: 13–15; McGrew, W. C. and Marchant, L. F. (1992) *Current Anthropology*, 33: 114–19; McGrew, W. C. *et al.* (1999) *Laterality*, 4: 79–87; Marchant, L. F. and McGrew, W. C. (1996) *Journal of Human Evolution*, 30: 427–43; Hopkins, W. D. (1995) *Journal of Comparative Psychology*, 109: 291–7; Wysocki, C. J. and Gilbert, A. N. (1989), in Murphy, C., Cain, W. S. and Hegsted, D. M. (eds) *Nutrition and the Chemical Senses in Aging: Recent Advances and Current Research Needs*, New York: New York Academy of Sciences, pp. 12–28. ☞www☜

20. Berman, D. S. *et al.* (2000) *Science*, 290: 969–72.

21. Collins, R. L. (1968) *Journal of Heredity*, 59: 9–12; Collins, R. L. (1969) *Journal of Heredity*, 60: 117–19; Collins, R. L. (1975) *Science*, 187: 181–4; Signore, P. *et al.* (1991) *Behavior Genetics*, 21: 421–9; Clapham, P. J., Leimkuhler, E., Gray, B. K. and Mattila, D. K. (1995) *Animal Behaviour* 50: 73–82.

22. Humphrey, N. (1998) *Laterality*, 3: 289; Harris, L. J. (1999) *Laterality*, 3: 291–4; Harris, L. J. (1989) *Canadian Journal of Psychology*, 43: 369–76; Snyder, P. J. *et al.* (1996) *Brain and Cognition*, 32: 208–11; Rogers, L. J. and Workman, L. (1993) *Animal Behaviour*, 45: 409–11; Vallortigara, G. *et al.* (1999) *Cognitive Brain Research*, 7: 307–20; Mascetti, G. G. *et al.* (1999) *Cognitive Brain Research*, 7: 451–63; Bisazza, A. *et al.* (1997) *Laterality*, 2: 49–64; Vallortigara, G. *et al.* (1998) *NeuroReport*, 9: 3341–4; Bisazza, A. *et al.* (1997) *Behavioural Brain Research*, 89: 237–42; Facchin, L. *et al.* (1999) *Behavioural Brain Research*, 103: 229–34; Bisazza, A. *et al.* (1999) *Animal Behaviour*, 57: 1145–9; Vallortigara, G. *et al.* (1999) *Brain Research Reviews*, 30: 164–75; Babcock, L. E. and Robison, R. A. (1989) *Nature*, 337: 695–6; Babcock, L. E. (1993) *Journal of Paleontology*, 67: 217–29.

23. Peckham, M. (1959) *The Origin of Species by Charles Darwin: A Variorum Text*, Philadelphia: University of Pennsylvania Press, p. 250–3; Norman, J. R. (1934) *A Systematic Monograph of the Flatfishes (Heterosomata) Volume 1: Psettodidae, Bothidae, Pleuronectidae*, London: British Museum (Natural History); Norman, J. R. and Greenwood, P. H. (1963) *A History of Fishes*, (2nd edn), London: Ernest Benn; Policansky, D. (1982) *Scientific American*, 246 (May): 96–102; Darling, K. F. *et al.* (2000) *Nature*, 405: 43–7. ☞www☜

24. Neville, A. C. (1976) *Animal Asymmetry*, London: Edward Arnold; Boycott, A. E. *et al.* (1931) *Philosophical Transactions of the Royal Society of London, Series B*, 219: 51–131; DeForest, M. J. (1981) *Trends in Neurosciences* October: 245–8; Cunningham, C. W. *et al.* (1992) *Nature*, 355: 539–42; Bisazza, A. *et al.* (1997) *Behavioural Brain Research*, 89: 237–42; Rust, J., Stumpner, A. and Gottwald, J. (1999) *Nature*, 399; Bisazza, A. *et al.* (1999) op. cit.

25. Smith, A. (1986 [1776]) *The Wealth of Nations, Books I–III*, Harmondsworth: Penguin, pp. 112–13; McGrew, W. C. and Marchant, L. F. (1999) *Primates*, 40: 509–13. ☞www☜

26. Peck, A. L. (1937) *Aristotle: Parts of Animals*, London: Heinemann, p. 687.a. ☞www☜

27. Feuillerat, A. (1922) *Sir Philip Sidney: The Countesse of Pembrokes Arcadia*, Cambridge: Cambridge University Press, pp. 132–7; Rowe, K. (1999) *Dead Hands: Fictions of Agency, Renaissance to Modern*, Stanford, CA: Stanford University Press;

Schultz, A. H. (1969) *The Life of Primates*, London: Weidenfeld & Nicolson. ☞www☜

28. Napier, J. R. (1956) *Journal of Bone and Joint Surgery*, 38B: 902–13; MacKenzie, C. L. and Iberall, T. (1994) *The Grasping Hand*, Amsterdam: North-Holland, pp. 15–36; Napier, J. R. (1962) *Scientific American*, 207 (December): 56–62; Marzke, M. W. (1983) *Journal of Human Evolution*, 12: 197–211. ☞www☜

29. Kelley, J. (1992) in Jones, S., Martin, R. and Pilbeam, D. (eds) *The Cambridge Encyclopaedia of Human Evolution*, Cambridge: Cambridge University Press; pp. 223–30; Trinkhaus, E. (1992), in Jones, S., Martin, R. and Pilbeam, D. (eds), pp. 346–9; Marzke, M. W. (1983) *Journal of Human Evolution*, 12: 197–211; McHenry, H. M. (1983) *American Journal of Physical Anthropology*, 62: 187–98; Marzke, M. W. *et al.* (1992) *American Journal of Physical Anthropology*, 89: 283–98; Johanson, D. C. and Edey, M. A. (1981) *Lucy: The Beginnings of Humankind*, London: Granada, pp. 348–9; Marzke, M. W. *et al.* (1988) *American Journal of Physical Anthropology*, 77: 519–28; Wilson, F. R. (1998) *The Hand: How Its Use Shapes the Brain, Language, and Human Culture*, New York: Vintage. ☞www☜

30. Napier, J. (1962) *Nature*, 196: 409–11; Napier, J. R. (1980) *Hands*, London: George Allen & Unwin, pp. 100–104; Marzke, M. W. *et al.* (1992) *American Journal of Physical Anthropology*, 89: 283–98; Musgrave, J. H. (1971) *Nature*, 233: 538–41; Wood Jones, F. (1941) *The Principles of Anatomy as Seen in the Hand*, London: Bailliere, Tindall and Cox, p. 301. ☞www☜

31. Wood Jones, F. (1941) op. cit.

32. Springer, S. and Deutsch, G. (1981) *Left Brain, Right Brain*, San Francisco: W. H. Freeman, p. 186.

33. Nicholls, M. E. R. (1993) *Cerebral Asymmetries for Temporal Resolution*, (unpublished PhD thesis); Nicholls, M. E. R. (1994) *Quarterly Journal of Experimental Psychology*, 47A: 291–310; Nicholls, M. (1994) *Neuropsychologia*, 32: 209–20; Nicholls, M. E. R. (1996) *Laterality*, 1: 97–137; Guylee, M. J. *et al.* (2000) *Brain and Cognition* 43: 234–8; Elias, L. J. (1999) *Cerebral Asymmetries in Processing Language and Time*, University of Waterloo: (unpublished PhD thesis); Brown, S. and Nicholls, M. E. (1997) *Perception and Psychophysics*, 59: 442–7. ☞www☜

34. Elias, L. J. *et al.* (1999) *Neuropsychologia*, 37: 1243–9; Guylee, M. J. *et al.* (2000) op. cit.; Hagoort, P. *et al.* (1999) in Brown, C. M. and Hagoort, P. (eds) *The Neurocognition of Language*, Oxford: Oxford University Press, pp. 273–316.

35. Calvin, W. H. (1982) *Ethology and Sociobiology* 3: 115–24; Calvin, W. H. (1983) *Journal of Theoretical Biology*, 104: 121–35; Calvin, W. H. (1983) *The Throwing Madonna: Essays on the Brain*, New York: McGraw-Hill; Knüsel, C. J. (1992) *Human Evolution*, 7: 1–7; Hore, J., Watts, S. and Tweed, D. (1996) *Journal of Neurophysiology*, 75: 1013–25; Hore, J. *et al.* (1996) *Journal of Neurophysiology* 76: 3693–704. ☞www☜

36. Maynard Smith, J. and Szathmáry, E. (1995) *The Major Transitions in Evolution*, Oxford: W. H. Freeman.

37. Lynch, M. and Conery, J. S. (2000) *Science*, 290: 1151–5.

38. Jacobs, B. *et al.* (1993) *Journal of Comparative Neurology*, 327: 97–111; Galuske, R. A. W. *et al.* (2000) *Science* 289: 1946–9; Buxhoeveden, D. and Casanova, M. (2000) *Laterality*, 5: 315–30; McManus, I. C. (1991) in Bock, G. R. and Marsh, J. (eds) *Biological asymmetry and handedness* (Ciba foundation symposium 162) Chichester: Wiley, pp. 251–81; McManus, I. C., in Corballis, M. C. and Lea, S. E. G. (eds) *The Descent of Mind: Psychological Perspectives on Hominid Evolution*, Oxford: Oxford University Press, pp. 194–217.

39. ☞www☜
40. ☞www☜
41. ☞www☜
42. ☞www☜
43. Kauffman, S. A. (1993) *The Origins of Order: Self-organization and Selection in Evolution*, New York: Oxford University Press; Kauffman, S. (1996) *At Home in the Universe*, London: Penguin; Miller, G. (2000) *The Mating Mind: How Sexual Choice Shaped the Evolution of Human Nature*, London: William Heinemann, pp. 392–406. ☞www☜
44. Eglinton, E. and Annett, M. (1994) *Perceptual and Motor Skills*, 79: 1611–16; Records, M. A. *et al.* (1977) 2: 271–82; McManus, I. C. *et al.* (1992) *Cortex*, 28: 373–81; Cornish, K. M. and McManus, I. C. (1996) *Journal of Autism and Developmental Disorders*, (1992) 26: 597–609; Crow, T. J. (1990) *Schizophrenia Bulletin*, 16: 433–43; Shenton, M. E. *et al.* (1992) *New England Journal of Medicine*, 327: 604–12.
45. ☞www☜

Chapter 10: Three men went to mow

1. LeQuesne, A. L. (1982) *Carlyle*. Oxford: Oxford University Press, p. 57; Litchfield, H. E. (1915) *Emma Darwin: A Century of Family Letters, 1792–1896*, London: John Murray, vol. II, p. 52; Froude, J. A. (1885) *Thomas Carlyle: A History of His Life in London, 1834–1881*, London: Longman, Green & Co, vol. II, p. 406. ☞www☜
2. Kaplan, F. (1983) *Thomas Carlyle: A Biography*, Cambridge: Cambridge University Press; Wilson, D. A., and MacArthur, D. W. (1934) *Carlyle in Old Age (1865–1881)*, London: Kegan Paul, Trench and Trubner; Froude, J. A. (1885). ☞www☜
3. Wilson, D. A. and MacArthur, D. W. (1934) op. cit., p. 250; Heffer, S. (1995) *Moral Desperado: A Life of Thomas Carlyle*, London: Weidenfeld & Nicolson, p. 369. ☞www☜
4. ☞www☜
5. ☞www☜
6. ☞www☜
7. ☞www☜
8. Walker, C. B. F. (1990) in Hooker, J. T. (ed.), *Reading the Past*; London: British Museum, pp. 15–74; Healey, J. F. (1990) *The Early Alphabet*, London: British Museum Publications.
9. Jeffrey, L. H. (1990) *The Local Scripts of Archaic Greece* (revised edition), Oxford: Clarendon Press. ☞www☜
10. ☞www☜
11. Gaur, A. (1984) *A History of Writing*, London: British Library, p. 55.
12. www.vis.colostate.edu/~traevoli/; (www.braille.org/papers/lorimer/chap3.html). ☞www☜
13. Healey, J. F. (1990) p. 61; Coulmas, F. (1996) *The Blackwell Encyclopaedia of Writing Systems*, Oxford: Blackwell; Diringer, D. and Regensburger, R. (1968) *The Alphabet: A Key to the History of Mankind*, (3rd edn), London: Hutchinson; Daniels, P. T. and Bright, W. (1996) *The World's Writing Systems*, New York: Oxford University Press. ☞www☜
14. De Selincourt, A. and Burn, A. R. (1972) *Herodotus: The Histories*, (revised edn), Harmondsworth: Penguin Books, p. 143; Skoyles, J. R. (1988) in De Kerckhove, D. and Lumsden, C. J. (eds), *The Alphabet and the Brain: the Lateralization of Writing*,

Berlin: Springer-Verlag, pp. 363–80; Walker, C. B. F. (1990) op. cit.; Englund, R. K. (1996), in Daniels, P. T. and Bright, W. (eds) *The World's Writing Systems*, New York: Oxford University Press, pp. 160–4; Powell, M. A. (1981) *Visible Language*, 15: 419–40. ☞www☜

15. Hewes, G. (1949) *Human Biology*, 21: 233–45; Gould, S. J. (1989) *Wonderful Life: The Burgess Shale and the Nature of History*, New York: W. W. Norton. ☞www☜

16. De Luna, F. A. (1993) *Beaver*, 73 (4): 17–21, p. 19; Goethe, J. W. (1970), *Italian Journey*, translated by W. H. Auden and Elizabeth Mayer, London: Penguin, p. 448; Darnton, R. (1999) *New York Review of Books*, 14th January. ☞www☜

17. Buchanan, M. (2000) *New Scientist*, 15 July: 28–31; Sebald, W. G. (1998) *The Rings of Saturn*, London: The Harvill Press, p. 18; Nabokov, V. (1995) *Lolita*, London: Penguin, p. 306. ☞www☜

18. Miller, G. (2000) *The Mating Mind: How Sexual Choice Shaped the Evolution of Human Nature*, London: William Heinemann, pp. 315–16.

19. Feldman, D. (1987) *Why Do Clocks Run Clockwise? and Other Imponderables*, New York: Harper and Row, p. 238; Kincaid, P. (1986) *The Rule of the Road: An International Guide to History and Practice*, New York: Greenwood Press; www.travel-library.com/ general/driving/drive_which_side.html; www.mmailbase.ac.uk/lists/int-boundaries/1999-09/0000.html. ☞www☜

20. De Luna, F. A. (1993) op. cit.; Gould, G. M. (1908) *Righthandedness and Lefthandedness*, Philadelphia: J. B. Lippincott, p. 75; www.littletechshoppe.com/ns1625/techdt06.html. ☞www☜

21. Kincaid, P. (1986) op. cit., pp. 14–17, 125; Hamer, M. (1986) *New Scientist*, 25 December: 16–18; www.travel-library.com/ general/driving/drive_which_side.html; www.last-word.com. ☞www☜

22. Kincaid, P. (1986) op. cit. ☞www☜

23. Honour, H. (1995) *New York Review of Books*, 42: 56–61. ☞www☜

24. www.czbrats.com/ Articles/left.htm

25. ☞www☜

26. ☞www☜

27. ☞www☜

28. ☞www☜

29. The *Guardian*, weekend supplement, 23 March, 1996, p. 3. ☞www☜

30. Leeming, J. J. (1969) *Road Accidents: Prevent or Punish?*, London: Cassell. ☞www☜

31. Kincaid, P. (1986) op. cit.; Mestel, R. (1998) *New Scientist*, 28 March: 38–9; Hamer, M. (1986) *New Scientist* 25 December: 16–18. ☞www☜

32. Turnbull, O. H. *et al.* (1995) *Journal of Genetic Psychology*, vol. 156: 17–21.

33. Froude, J. A. (1885) op. cit., vol. II, pp. 407–8; Pye-Smith, P. H. (1871) *Guy's Hospital Reports*, 16: 141–6. ☞www☜

34. Orwell, G. (1970) *The Collected Essays, Journalism and Letters of George Orwell: Volume IV: In Front of Your Nose, 1945–1950*, Harmondsworth: Penguin, p. 62.

35. McLean, J. M. and Churczak, F. M. (1982) *New England Journal of Medicine* 307: 1278–9; Cooke, A. and Mullins, J. (1994) *New Scientist*, 12 March: 40–42; Estberg, L. *et al.* (1996) *Journal of the American Veterinary Association*, 208: 92–6. ☞www☜

36. Wood, C. J. and Aggleton, J. P. (1989) *British Journal of Psychology*, 80: 227–40; Bisiacchi, P. S. *et al.* (1985) *Perceptual and Motor Skills*, 61: 507–13; Gordon, N. and McKinlay, I. (1980) *Helping Clumsy Children*, Edinburgh: Churchill Livingstone; Gubbay, S. S. (1975) *The Clumsy Child: A Study of Developmental Apraxis and*

Agnosic Ataxia, London: W. B. Saunders; Grouios, G. *et al.* (2000) *Perceptual and Motor Skills*, 90: 1273–82. ☞www☜

37. Wood, C. J. and Aggleton, J. P. (1989); Raymond, M. *et al.* (1996) *Proceedings of the Royal Society of London, Series B*, 263: 1627–33; 1996; Anonymous, *Independent on Sunday*, 18 April: 24 1993; Bisiacchi, P. S. *et al.* (1985) *Perceptual and Motor Skills*, 61: 507–13; Azemar, G. (1993) *Escrime Internationale*, 7: 15–19. ☞www☜

38. Raymond, M. *et al.* (1996) op. cit. ☞www☜

39. Withington, E. T. (1927) *Hippocrates*, London: Heinemann, p. 63; Spencer, W. G. (1938) *Celsus: De Medicina*, London: William Heinemann, vol. III, p. 297.

40. Brodie, B. C. (1862) *Psychological Inquiries* (4th edn), London: Longman, Green, Longman, Roberts and Green; Schott, J. and Puttick, M. (1995) *British Medical Journal*, 310: 739; Dudley, H. (1995) *The left-hander*, No 21: 8. ☞www☜

41. Chaplin, C. (1964) *My Autobiography*, London: Bodley Head, p. 131; W. J. W. *The Guardian*, 3 March: 35, 1992; Aggleton, J. P. *et al.* (1994) *Psychology of Music* 22: 148–56; Oldfield, R. C. (1969) *British Journal of Psychology*, 60: 91–9. ☞www☜

42. Safire, W. (1978) *Safire's Political Dictionary*, New York: Random House; Parkin, R. (1996) *The Dark Side of Humanity: The Work of Robert Hertz and its Legacy*, Amsterdam: Harwood, p. 69; Carlyle, T. (1871) *The French Revolution: A History [1837]*, London: Chapman and Hall, vol. I, VI: II, p. 192. ☞www☜

43. Safire, W. (1978) op. cit.; Eysenck, H. J. (1954) *The Psychology of Politics*, London: Routledge and Kegan Paul. ☞www☜

44. Bobbio, N. (1996) *Left and Right: The Significance of a Political Distinction*, Cambridge: Polity Press, pp. 56–7. ☞www☜

45. Ibid., p. 2; Domhoff, G. W. (1968) *Psychoanalytic Review*, 56: 587–96, p. 594; Wieschoff, H. A. (1938) *Journal of the American Oriental Society*, 58: 202–17; Brimnes, N. (1999) *Constructing the Colonial Encounter: Right and Left Hand Castes in Early Colonial South India*, Richmond, Surrey: Curzon Press, pp. 26–30; Nicholson, S. (1926) *Journal of the Royal Anthropological Institute of Great Britain and Ireland*, 56: 91–103. ☞www☜

46. Tucker, D. M. (1985) *Lateral Dialectics*, unpublished manuscript; Gur, R. C. *et al.* (1976) *Journal of Abnormal Psychology*, 85: 122–4; Gur, R. E. and Gur, R. C. (1975) *Journal of Consulting and Clinical Psychology*, 43: 416–20. ☞www☜

47. Mair, P. (2000) *Guardian*, 2000. ☞www☜

48. Boucher, J. and Osgood, C. E. (1969) *Journal of Verbal Learning and Verbal Behaviour* 8: 1–8. ☞www☜

49. Winder, B. C. (2000) *Calibration, Misleading Questions and Medical Knowledge*, (PhD thesis), London: University of London.

50. Tuohy, A. P. and Stradling, S. G. (1987) *British Journal of Psychology*, 78: 457–64; Berlyne, D. E. (1971) *Aesthetics and Psychobiology*, New York: Appleton-Century-Crofts. ☞www☜

51. Heffer, S. (1995) op. cit., pp. 19–25. ☞www☜

Chapter 11: Keggie-hander

1. Ziegler, P. (1999) *Osbert Sitwell: A Biography*, London: Pimlico. ☞www☜

2. Sitwell, O. (1945) *Left Hand, Right Hand!*, London: Macmillan; Anonymous (1945) *Times Literary Supplement*, 7 April: Orwell, S. and Angus, I. (1970). *The Collected Essays, Journalism and Letters of George Orwell: Vol. II, In Front of Your Nose, 1945–1950*, Harmondsworth: Penguin, p. 505. ☞www☜

3. Watson, F. (1950) *Dawson of Penn*, London: Chatto and Windus; Sitwell, O. (1946) *The Scarlet Tree*, London: Macmillan, pp. 239, 244. ☞www☜

4. Ireland, W. W. (1880) *Brain*, 3: 207–14.

5. Comment of Kathleen Stacey in the Vestry House Museum exhibition. 'A sinister way of life? The story of left-handedness', 13 August–16 November 1996; Wordsworth, H. *The Left-hander*, (1995) 21: 8–8; Comment of Neil Houghton about his grandmother, Doris Rayner, in the Vestry House Museum exhibition, 'A sinister way of life? The story of left-handedness', 13 August–16 November 1996.

6. www.faqs.org/faqs/left-faq/; www.scican.net/~ptjones/left.html; www.indiana.edu/~primate/lspeak.html. ☞www☜

7. Kidd, D. (1906) *Savage Childhood: A Study of Kafir Children*, London: Adam and Charles Black; Wallace, W. J. (1948) *Primitive Man*, 21: 19–38, (p. 28); Hertz, R. (1960) *Death and the Right Hand*, Aberdeen: Cohen and West, p. 92; Barsley, M. (1970) *Left-handed Man in a Right-handed World*, London: Pitman, p. 35; Smith, E. W. (1952) *Journal of the Royal Anthropological Institute of Great Britain and Ireland*, 82: 13–37, (p. 22); Wile, I. S. (1934) *Handedness: Right and Left*, Boston: Lethrop, Lee and Shepard, pp. 40, 334; Kruyt, A. C. (1973) in Needham, R. (ed.), *Right and Left: Essays on Dual Symbolic Classification*, Chicago: University of Chicago Press pp. 74–91; Bulwer, J. (1644) *Chirologia: or, the Natural Language of the Hand; Composed of the Speaking Motions, and Discoursing Gestures Thereof. Whereunto is Added, Chironomia; or, The Art of Manual Rhetoricke*, London: Thomas Harper, p. 126; Lyttelton, O. (1962) *The Memoirs of Lord Chandos*, London: Bodley Head, p. 230.

8. Wile, I. S. (1934) *Handedness: Right and Left*, Boston: Lothrop, Lee and Shepard, p. 40; Rigby, P. (1966) *Africa*, 36: 1–16, p. 266; Parkin, R. (1996) *The Dark Side of Humanity: The Work of Robert Hertz and its Legacy*, Amsterdam: Harwood, p. 71; Coel, M. (1981) *Chief Left Hand: Southern Arapaho*, Norman, Oklahoma: University of Oklahoma Press.

9. Goffman, E. (1963) *Stigma: Notes on the Management of Spoiled Identity*, Englewood Cliffs, NJ: Prentice-Hall.

10. Page, R. M. (1984) *Stigma*, London: Routledge and Kegan Paul, chapter 1.

11. Hare, D. (1991) *Writing Left Handed*, London: Faber and Faber; Milton, J. (1953) *Complete Prose Works of John Milton, Volume I, 1624–1642*, New Haven: Yale University Press, p. 808; Shelby, R. A. *et al.* (2001) *Science*, 292: 77–9, Wilson, G. P. (1937) *Words*, 3: 102–5; Kelley, E. (1992) *The Metaphysical Basis of Language: A Study in Cross-cultural Linguistics, or, The Left-handed Hummingbird*, Lewiston, NY: Edwin Miller. ☞www☜

12. Anonymous (1958) *The Complete Letters of Vincent Van Gogh*, London: Thames and Hudson, pp. 364, 372. ☞www☜

13. Fincher, J. (1977) *Sinister People: The Looking-glass World of the Left-hander*, New York: G. P. Putnam's Sons., p. 30. ☞www☜

14. McGuire, W. J. and McGuire, C. V. (1980) *Perceptual and Motor Skills*, pp. 3–7. ☞www☜

15. Orton, H. *et al.* (1978) *The Linguistic Atlas of England*, London: Croom Helm; Upton, C. and Widdowson, J. D. A. (1996) *An Atlas of English Dialects*, Oxford: Oxford University Press. ☞www☜

16. ☞www☜

17. Trudgill, P. (1990) *The Dialects of England*, Oxford: Blackwell, p. 103; Runciman, D. (1997) *Guardian*, 3 October, Sport section: 16. ☞www☜

18. Mather, J. Y. and Speitel, H. H. (1975) *The Linguistic Atlas of Scotland: Scots Section, Volume 1*, London: Croom Helm. ☞www☜

19. Jennie Cook, comment in the Vestry House Museum exhibition, 'A sinister way of life? The story of left-handedness', 13 August–16 November 1996.

20. ☞www☜

21. Carlyle, T. (1984) *Sartor Resartus*, London: Dent, p. 30; Coren, S. (1992) *The Left-hander Syndrome: the Causes and Consequences of Left-handedness*, London: John Murray; Chwast, S. (1985) *The Left-handed Designer*, Paris: Booth-Clibborn; Petroski, H. (1994) *The Evolution of Useful Things*, New York: Vintage Books, pp. 243, 249–50. ☞www☜

22. *Hansard*, 22 July 1998, columns 1085–93. ☞www☜

23. Sadler, N. (1997) in Pearce, S. M. (ed.) *Experiencing Material Culture in the Western World*, London: Leicester University Press, pp. 140–53.

24. Anonymous (1986) *The Economist*, 20 December: 25–8; Anonymous (1962) *The Economist*, 105 (Dec 22): 1177–9; Dalrymple, T. (1995) *So Little Done: The Testament of a Serial Killer*, London: Andrew Deutsch, p. 93. ☞www☜

25. Rawls, J. (1972) *A Theory of Justice*, Oxford: Oxford University Press. ☞www☜

26. Editorial, *Guardian*, 1 January 2000: 1–19; Braverman, A. M. (2001) *Fertility and Sterility*, 59: 1216–20; ds.dial.pipex.com/town/plaza/gb54/web50p99.htm. ☞www☜

Chapter 12: Vulgar errors

1. Wilkin, S. (1852) *The Works of Sir Thomas Browne*, (revised version of 1836 edition), London: Henry Bohn; Bennett, J. (1962) *Sir Thomas Browne*, Cambridge: Cambridge University Press; Blackmore, S. (1999) *The Meme Machine*, Oxford: Oxford University Press. ☞www☜

2. Sebald, W. G. (1998) *The Rings of Saturn*, London: The Harvill Press, p. 19; Bennett, J. (1962), p. 5. ☞www☜

3. Bennett, J. (1962) op. cit., p. 20; Needham, J. (1934) *A History of Embryology*, Cambridge: Cambridge University Press, vol. 1, p. 112; Needham, J. (1931) *Chemical Embryology*, Cambridge: Cambridge University Press, vol. 1, p. 137. ☞www☜

4. Wilkin, S. (1852) op. cit., vol. I, p. xvii. ☞www☜

5. ☞www☜

6. Dawkins, R. (1997) *Climbing Mount Improbable*, London: Penguin Books, pp. 208–9. ☞www☜

7. Molière, *Le Médecin malgré lui*, II, 4. ☞www☜

8. Edinburgh International Science Festival (1993) *How Much Do Scots Know About Their Own Bodies? A Survey on Public Awareness of the Body*, News Release, 1 February, Edinburgh. ☞www☜

9. Ackermann, A. S. E. (1950) *Popular Fallacies: A Book of Common Errors: Explained and Corrected with Copious References to Authorities*, (4th edn), London: Old Westminster Press, pp. 47–8. ☞www☜

10. Burnett, A. C. and Thompson, D. G. (1986) *Medical Education*, 20: 424–31.

11. Henderson, A. H. (1996) in Weatherall, D. J., Ledingham, J. G. G. and Warrell, D. A. (eds), *Oxford Textbook of Medicine* (3rd edn) Oxford: Oxford University Press, pp. 2165–9; Swanton, R. H. (1996) in *Oxford Textbook of Medicine*, (3rd edn), pp. 2321–31. ☞www☜

12. Coffin, W. S. (1999) *The Heart is a Little to the Left*, University Press of New England. ☞www☜

13. Lauterbach, C. E. (1925) *Genetics*, 10: 525–68; Newman, H. H. *et al.* (1966) *Twins: A Study of Heredity and Environment*, Chicago: The University of Chicago Press; Townsend, G. C. *et al.* (1986) *Acta Genetica Medica et Gemellogia*, 35: 179–92; Newman, H. H. in *The Physiology of Twinning*, Chicago: The University of Chicago Press, 1111 pp. 164–89 (pp. 185–189); Larsen, W. J. (1998) *Essentials of Human Embryology*, New York: Churchill Livingstone, p. 326; Sicotti, N. L. *et al.* (1999) *Laterality*, 4: 265–86. ☞www☜

14. Winckelmann, J. J. (1968 [1764]) *History of Ancient Art* (translated by A. Gode), New York: Book V, VI; McManus, I. C. (1976) *Nature* 259: 426–26; Peck, A. L. (1953) *Aristotle: Generation of Animals*, Cambridge, MA: Loeb, 717.a.34, 788.a.10; Nasmyth, D. G. *et al.* (1991) *British Medical Journal* 302: 93–4. ☞www☜

15. Editorial (2001) *Guardian*, 5 February; Halpern, D. F. and Coren, S. (1988) *Nature*, 333: 213; Halpern, D. F. and Coren, S. (1991) *New England Journal of Medicine*, 324: 998; Rothman, K. J. (1991) *New England Journal of Medicine*, 325: 1041. ☞www☜

16. Harris, P. (2001) *The Observer*, 4 February; Geschwind, N. and Behan, P. (1982) *Proceedings of the National Academy of Sciences of the USA*, 79: 5097–100.

17. Geschwind, N. and Galaburda, A. M. (1985) *Archives of Neurology*, 42: 428–59, 521–52, 634–54; Geschwind, N. and Galaburda, A. M. (1987) *Cerebral Lateralization: Biological Mechanisms, Associations, and Pathology*, Cambridge, MA: MIT Press; McManus, I. C. and Bryden, M. P. (1991) *Psychological Bulletin*, 110: 237–53.

18. Bryden, M. P. *et al.* (1994) *Brain and Cognition*, 26: 103–67. ☞www☜

19. Gould, S. J. (1994) *New York Review of Books*, 41, 20 October: 24–32; Gooch, S. (1977) *The Neanderthal Question*, London: Wildwood House; Holloway, R. and Coste-Lareymondie, M. C. d. l. (1982) *American Journal of Physical Anthropology*, 58: 101–10; Wong, K. (2000) *Scientific American* April: 79–87. ☞www☜

20. Stevenson, R. L. (1979 [1886]) *The Strange Case of Dr Jekyll and Mr Hyde and Other Stories*, Harmondsworth: Penguin, p. 82.

21. Pedersen, D. L. (1994) *Cameral Analysis: A Method of Treating the Psychoneuroses Using Hypnosis*, London: Routledge; Schiffer, F. (1998) *Of Two Minds: The Revolutionary Science of Dual-brain Psychology*, New York: Free Press; McManus, I. C. (1998) *Nature*, 396: 132.

22. Wolfe, T. (1998) *A Man in Full*, London: Jonathan Cape, p. 38; McManus, I. C. (1985) *Current Psychological Research and Reviews*, 4: 195–203; Tapley, S. M. and Bryden, M. P. (1983) *Neuropsychologia*, 21: 129–38; Levy, J. and Reid, M. (1978) *Journal of Experimental Psychology*, 107: 119–44; Peters, M. and McGrory, J. (1987) *Brain and Language*, 32: 253–64; Meulenbroek, R. G. J. and Van Galen, G. P. (1989) *Journal of Human Movement Studies*, 16: 239–54.

23. Clarke, G. (1993) *Left-handed Children: The Teacher's Guide*, London: Left-Handers Club, p. 3; Perelle, I. B. and Ehrman, L. (1982) *Experientia*, 38: 1257–8; Storfer, M. D. (1995) *Perceptual and Motor Skills*, 81: 491–7; Hardyck, C., Petrinovich, L. F. and Goldman, R. D. (1976) *Cortex* 12: 266–79; McManus, I. C. and Mascie-Taylor, C. G. N. (1983) *Journal of Biosocial Science*, 151: 289–306; McManus, I. C. *et al.* (1993) *British Journal of Psychology*, 84: 517–37. ☞www☜

24. Cassy, J. (2000) *Guardian*, 31 August: 22; O'Boyle, M. W. and Benbow, C. P. (1990) in Coren, S. (ed.) *Left-handedness: Behavioral Implications and Anomalies*, Amsterdam: North-Holland, pp. 343–72; Peterson, J. M. and Lansky, L. M. (1974)

Perceptual and Motor Skills, 38: 547–50; Gotestam, K. O. (1990) *Perceptual and Motor Skills*, 70: 1323–7; Wood, C. J. and Aggleton, J. P. (1991) *Canadian Journal of Psychology*, 45: 395–404. ☞www☜

25. Beaumont, G. *et al.* (1984) *Cognitive Neuropsychology*, 1: 191–212; Wells, C. G. (1989) *Right Brain Sex*, New York: Avon Books; Edwards, B. (1989) *Drawing on the Right Side of the Brain: A Course in Enhancing Creativity and Artistic Confidence* (revised edition) New York: Perigee Books; Springer, S. and Deutsch, G. (1981) *Left Brain, Right Brain*, San Francisco: W. H. Freeman, pp. 186–7; www.csicop.org/si/9903/ten-percent-mth.html.

26. Paul, D. (1990) *Living Left-handed*, London: Bloomsbury, p. 14; Langford, S. (1984) *The Left-handed Book*, London: Granada Publishing, p. 100; Fincher, J. (1977) *Sinister People: The Looking-glass World of the Left-hander*, New York: G. P. Putnam's Sons, p. 27; Lindsay, R. (1980) *The Left-handed Book*, New York: Franklin Watts, p. 5; Clarke, G. (1993) *Left-handed Children: The Teacher's Guide*, London: Left-Handers Club, p. 3; Paul, D. (1993) *Left-handed Helpline: An Essential Guide for Teachers, Teacher Trainers and Parents of Left-handed Children*, Manchester: Dextral Books; Bernadac, M.-L. (1991) *Faces of Picasso*, Paris: Editions de la Réunion des Musées Nationaux. ☞www☜

27. Edwards, B. (1989) *Drawing on the Right Side of the Brain: A Course in Enhancing Creativity and Artistic Confidence* (revised edition), New York: Perigee Books, p. 39. ☞www☜

28. De Kay, J. T. (1994) *The Left-hander's Handbook* (incorporating) *The Left-handed Book*, 1966; *The Natural Superiority of the Left-hander*, 1979; *The World's Greatest Left-handers*, 1985, and *Left-handed Kids*, (1989), New York: Quality Paperback Book Club, pp. 2, 55; Barsley, M. (1970) *Left-handed Man in a Right-handed World*, London: Pitman; Spitz, B. (1989) *Dylan: A Biography*, New York: W. W. Norton; Gross, M. and Alexander, R. (1978) *Bob Dylan: An Illustrated History*, London: Elm Tree Books. ☞www☜

29. Lindsay, R. (1980) op. cit. ☞www☜

30. Bigelow, J. (1887) *The Complete Works of Benjamin Franklin*, New York: G. P. Putnam's Sons, vol. IV, and vol VI, pp. 242–4; Lindsay, R. (1980) op. cit. ☞www☜

31. Rutledge, L. W. and Donley, R. (1992) *The Left-hander's Guide to Life: A Witty and Informative Tour of the World According to Southpaws*, New York: Penguin, p. 14.

32. Research Unit (1974) *Journal of the Royal College of General Practitioners*, 24: 437–9; Shaw, D. and McManus, I. C. (1993) *British Journal of Psychology*, 84: 545–51.

33. Wilson, D. (1891) *The Right Hand: Left-handedness*, London: Macmillan, pp. 66–7; Paul, D. (1990) *Living Left-handed*, London: Bloomsbury, p. 23; Harris, L. J. (2000) *Brain and Language*, 73: 132–88, p. 145; Anonymous (1916) *The Jewish Encyclopaedia*, New York: Funk and Wagnalls, X: 420.

34. Feldman, D. (1987) *Why Do Clocks Run Clockwise? and Other Imponderables*, New York: Harper and Row. ☞www☜

35. Landes, D. S. (2000) *Revolution in Time: Clocks and the Making of the Modern World*, London: Viking. ☞www☜

36. Pope-Hennessy, J. (1969) *Paolo Uccello: Complete Edition*, (2nd edn), London: Phaidon, pp. 144–5. ☞www☜

37. Bo, C. and Mandel, G. (1978) *L'opera completa del Botticelli*, Milan: Rizzoli Editore, Plate XXV; Simoni, A. (1965) *Orologi Italiani dal cinquecento all' ottocento*, Milan: Antonio Vallardi Editore. ☞www☜

38. Arthur, W. B. (1990) *Scientific American* 262 (February): 80–85; Arthur, W. B. (1989) *Economic Journal*, 99: 116–31; Arthur, W. B. (1988) in Anderson, P. W., Arrow, K. J. and Pines, D. (eds), *The Economy as a Complex Evolving System*, New York: Addison-Wesley, pp. 9–31; David, P. A. (1985) *American Economic Review (Papers and Proceedings)*, 75: 332–7; Bonabeau, E. *et al.* (2000) *Nature* 406: 39–42. ☞www☜

39. Thorndike, L. (1941) *Speculum*, 16: 242–3. ☞www☜

40. Price, C. M. and Gilden, D. L. (2000) *Journal of Experimental Psychology Human Perception and Performance*, 26: 18–30. ☞www☜

41. *Tlön, Uqbar, Orbis Tertius*, in Borges, J.-L. (2000) *Labyrinths*, Harmondsworth: Penguin. ☞www☜

42. Lee, D. (1965) *Plato: Timaeus and Critias*, Harmondsworth: Penguin, p. 63; Latham, R. E. and Godwin, J. (1994) *Lucretius: On the Nature of the Universe*, Harmondsworth: Penguin, p. 102–3.

43. Eco, U. (1984) *Semiotics and the Philosophy of Language*, London: Macmillan, p. 205. ☞www☜

44. Gregory, R. (1998) *Mirrors in Mind*, London: Penguin Books.

45. Gleick, J. (1994) *Genius: Richard Feynman and Modern Physics*, London: Abacus, p. 331. ☞www☜

46. ☞www☜

47. Lee, D. (1965) op. cit., pp. 62–3.

48. There is an odd number of reflections, three, so the image will appear reversed. ☞www☜

Chapter 13: The handedness of Muppets

1. Elias, L. J. (1998) *Laterality*, 3: 193–208.

2. Sadler, N. (1997) in Pearce, S. M. (ed.), *Experiencing Material Culture in the Western World*, London: Leicester University Press, pp. 140–53. ☞www☜

3. Graham, A. (1996) *Radio Times*, 8 June: 29–30.

4. Peters, M. (1997) *Laterality*, 2: 3–6; Cary, E. (1927) *Dio's Roman History*, London: William Heinemann, vol. IX, pp. 109–10, book LXIII. ☞www☜

5. Davidson, J. (2000) *Times Literary Supplement*, 12 May: 11–12; Utley, R. M. (2000) *Billy the Kid: A Short and Violent Life*, London: Tauris Parke Paperbacks. ☞www☜

6. Marsh, E. W. (1998) *James Cameron's* Titanic, London: Boxtree, pp. 52–3. ☞www☜

7. Maur, K. V. (1989) *Salvador Dali 1904–1989*, Stuttgart: Gerd Hatje, Plate D6.

8. Carey, P. (1997) *Jack Maggs*, St Lucia, Queensland: University of Queensland Press, p. 89. ☞www☜

9. Critchley, M. (1928) *Mirror-writing*, London: Kegan Paul, Trench and Trubner, p. 9; Paul, D. (1990) *Living left-handed*, London: Bloomsbury, p. 155; Beeley, A. L. (1918) *An Experimental Study in Left-handedness*, Chicago: University of Chicago Press.

10. Critchley, M. (1928) op. cit., p. 49.

11. Schott, G. D. (1999) *Lancet*, 354: 2158–61.

12. Critchley, M. (1928) op. cit., p. 13; Posèq, A. W. G. (1997) *Konsthistorisk tidskrift*, 66: 37–50; Clayton M. (1996) *Leonardo da Vinci: A Curious Vision*, London: Merrell Holberton, pp. 97–9. ☞www☜

13. Capener, N. (1952) *Lancet*, i: 813–14. ☞www☜

14. McManus, I. C. (1999) *Laterality* 4: 193–6.

15. Gardner, M. (1990) *The New Ambidextrous Universe: Revised Edition*, New York: W. H. Freeman, pp. 47–51.

16. *New Scientist*, 11 November 2000, p. 104 and 25 November, p. 59.

17. Seife, C. *New Scientist*, 21 March, 1998, p. 10.

18. Hori, M. (1993) *Science*, 260: 216–19.

19. Anonymous (2000) *New Scientist*, 8 January: 12; Nagarajan, R. (2000) PhD thesis, Exeter: University of Exeter.

20. ☞www☜

21. Palmer, S. (1995) *Jack the Ripper: A Reference Guide*, Lanham, Md: Scarecrow Press, p. 1.

22. Barsley, M. (1970) *Left-handed Man in a Right-handed World*, London: Pitman, pp. 89–90; Coren, S. (1992) *The Left-hander Syndrome: The Causes and Consequences of Left-handedness*, London: John Murray, p. 249; Morris, D. (1997) *Behind the Oval Office: Getting Re-elected Against All Odds*, New York: Renaissance. ☞www☜

23. Butterfield, L. H. and Rice, H. C. Jr (1948) *William and Mary Quarterly*, (3rd series) 5: 26–33; Bullock, H. P. (1945) *My Head and My Heart: A Little History of Thomas Jefferson and Maria Cosway*, New York: G. P. Putnam's Sons, p. 27.

24. Weinstein, E. A. (1970) *Journal of American History*, 57: 324–51; Weinstein, E. A. (1981) *Woodrow Wilson: A Medical and Psychological Biography*, Princeton, NJ: Princeton University Press. ☞www☜

25. Hardell, L. *et al.* (1999) *International Journal of Oncology*, 15: 113–16; Rothman, K. J. (2000) *Lancet*, 356: 1837–40; Hardell, L. and Hansson Mild, K. (2001) *Lancet*, 357: 960–61; Rothman, K. J. (2001) *Lancet*, 357: 961. ☞www☜

26. Ecclesiastes 10: 2; Critchley, M. and Critchley, E. A. (1978) *Dyslexia Defined*, London: Heinemann, p. 94; Anonymous (1971) *Encyclopaedia Judaica*, Jerusalem: Keter, vol. 14: 178–9. ☞www☜

27. Wölfflin, H. (1941) in *Gedanken zur Kunstgeschichte*, Basel: pp. 82; Keller, R. (1942) *Ciba Symposia*, 3: 1139–42; Blount, P. *et al.* (1975) *Perception*, 4: 385–9. ☞www☜

28. Wolff, W. (1933) *Character and Personality*, 2: 168–76; Gilbert, C. and Bakan, P. (1973) *Neuropsychologia*, 11: 355–62. ☞www☜

29. Sun, J. and Perona, P. (1998) *Nature Neuroscience*, 1: 183–4. ☞www☜

30. Salk, L. (1966) *Canadian Psychiatric Association Journal* 11: S295–S305; Salk, L. (1973) *Scientific American*, 228 (May): 24–9; Reissland, N. (2000) *British Journal of Developmental Psychology*, 18: 179–86. ☞www☜

31. Shorr, D. C. (1954) *The Christ-child in Devotional Images in Italy During the XIV Century*, New York: Glückstadt Press; McManus, I. C. (1979) *Determinants of Laterality in Man*, University of Cambridge: unpublished PhD thesis, chapter 13.

32. Weiskrantz, L. *et al.* (1971) *Nature*, 230: 598–9; Smith, J. L. and Cahusac, P. M. B. (2001) *Laterality*, 6: 233–8. ☞www☜

33. Gillett, H. G. du P. (1983) *The Chiropodist*, 38: 162–77.

34. Anonymous (2000) *Metromnia: News from the National Physical Laboratory*, Issue 7, Summer. ☞www☜

35. Melville, H. (1972 [1851]) *Moby Dick; or, The Whale*, Harmondsworth: Penguin, chapter 32, p. 23, Thompson, D. W. (1971) *On Growth and Form* (abridged edition, edited by J. T. Bonner), Cambridge: Cambridge University Press, pp. 216–20; Neville, A. C. (1976) *Animal Asymmetry*, London: Edward Arnold, pp. 48–9. ☞www☜

36. ☞www☜

37. Hamilton, E. and Cairns, H. (1961) *The Collected Dialogues of Plato*, Princeton, NJ: Princeton University Press, Laws, 795.c; Watson, J. B. (1924) *Behaviorism*, New

York: People's Institute Publishing; Harris, L. J. (1983) in Young, G., Segalowtiz, S. J., Corter, C. M. and Trehub, S. E. (eds), *Manual Specialization and the Developing Brain*, New York: Academic Press, pp. 177–247 (p. 217); Jackson, J. (1905) *Ambidexterity: or Two-handedness and Two-brainedness*, London: Kegan, Paul, Trench, Trübner & Co.; Gould, G. M. (1908) *Righthandedness and Lefthandedness*, Philadelphia: J. B. Lippincott, p. 20.

38. Anderson, J. (1977) *Toward the Liberation of the Left Hand*, Pittsburgh: University of Pittsburgh Press, 1977.

39. Barsley, M. (1970) *Left-handed Man in a Right-handed World*, London: Pitman, p. 89; Hay, I. (1939 [1907]) *'Pip': A Romance of Youth*, Harmondsworth: Penguin. ☞www☜

40. Carlyle, T. (1874) *Wilhelm Meister's Apprenticeship and Travels* (translated from Goethe), 1824, London: Chapman and Hall, III: p. 163. ☞www☜

41. ☞www☜

42. Barsley, M. (1970) op. cit., p. 5. ☞www☜

Chapter 14: Man is all symmetrie

1. Tobin, J. (1991) *George Herbert: The Complete English Poems*, Harmondsworth: Penguin; Van Fraassen, B. C. (1989) *Laws and Symmetry*, Oxford: Clarendon Press, p. 239. ☞www☜

2. Bochner, S. (1973) in Wiener, P. P. (ed.) *Dictioinary of the History of Ideas*, New York: Charles Scribner's Sons, pp. 345–53; Fowler, H. N. (1962) *Plato: Philebus*, London: Heinemann, p. 65a; Winterson, J. (1990 [1987]) *Oranges Are Not the Only Fruit*, London: Vintage, p. 58. ☞www☜

3. Pervin, L. A. (1993) *Personality: Theory and Research*, New York: John Wiley, p. 117; Lilenfeld, S. O. *et al.* (2001) *Scientific American* 284 (May): 73–9; Roe, A. (1900) in Sherman, M. H. (ed.) *A Rorschach Reader*, New York: International Universities Press, pp. 121–36. ☞www☜

4. Stewart, I. and Golubitsky, M. (1992) *Fearful Symmetry: Is God a Geometer?*, Oxford: Blackwell; Van Fraassen, B. C. (1989) *Laws and Symmetry*, Oxford: Clarendon Press, p. 262–3; Coxeter, H. S. M., in *Symmetry and Function of Biological Systems at the Macromolecular Level*, Engström, A. and Strandberg, B. (eds), Stockholm: Almqvist and Wiksell, pp. 29–33. ☞www☜

5. Balmelle, C. *et al.* (1985) *Le décor géometrique de la mosaique Romaine*, Paris: Picard, pp. 177, 255, 277, 335; Washburn, D. K. and Crowe, D. W. (1988) *Symmetries of Culture: Theory and Practice of Plane Pattern Analysis*, Seattle: University of Washington Press, Fig. 5.78, p. 171; Jablan, S. V. (1995) *Theory of Symmetry and Ornament*, Belgrade: Matematicki Institut; Stewart, I. and Golubitsky, M. (1992) op. cit., pp. 238–9; Weyl, H. (1952) *Symmetry*, Princeton, NJ: Princeton University Press, p. 103; Gombrich, E. H. (1979) *The Sense of Order*, London: Phaidon Press. ☞www☜

6. Altmann, S. L. (1992) *Icons and Symmetries*, Oxford: Clarendon Press, p. 13. ☞www☜

7. Ibid., p. 33.

8. Yang, C. N. (1962) *Elementary Particles: A Short History of Some Discoveries in Atomic Physics*, Princeton, NJ: Princeton University Press, pp. 61–2. ☞www☜

9. Altmann, S. L. (1992) op. cit., p. 35. ☞www☜

10. Radicati di Brozolo, L. (1991) in Froggatt, C. D. and Nielsen, H. B. (eds), *Origin of Symmetries*, Singapore: World Scientific, pp. 523–35; Fraser, G. (2000) *Antimatter:*

The Ultimate Mirror, Cambridge: Cambridge University Press; Feynman, R. P., Leighton, R. B. and Sands, M. (1963) The Feynman Lectures on Physics. Vol I: Mainly Mechanics, Radiation, and Heat, Reading, MA: Addison-Wesley. ☞www☜

11. Van Fraassen, B. C. (1989) Laws and Symmetry, Oxford: Clarendon Press, p. 242; Feynman, R. P. (1986) 'Surely You're Joking, Mr. Feynman!', London: Unwin, p. 247; Feynman, R. P. et al. (1963) op. cit., pp. 52–1, 52–12; Feynman, R. P. (1999) Six Not-so-easy Pieces: Einstein's Relativity, Symmetry, and Space-time, London: Penguin. ☞www☜

12. Corballis, M. C. and Beale, I. L. (1976) The Psychology of Left and Right, Hillsdale, NJ: Lawrence Erlbaum Associates, p. 197. ☞www☜

13. Salam, A. (1958) Endeavour, 17: 97–105, p. 105; Van Fraassen, B. C. (1989) op. cit., pp. 260, 262; Rosen, J. (1975) Symmetry Discovered: Concepts and Applications in Nature and Science, Cambridge: Cambridge University Press, pp. 120–22. ☞www☜

14. Johnson, G. (1997) Fire in the Mind: Science, Faith, and the Search for Order, London: Penguin, pp. 314–15; Feynman, R. P. et al. (1963) pp. 52–4, 52–11; Bouchiat, M.-A. and Pottier, L. (1984) Scientific American, 250: 100–111; Adair, R. K. (1988) Scientific American, 258 (February): 30–36; Quinn, H. R. and Witherell, M. S. (1998) Scientific American, 279 (October): 50–55; Close, F. (2000) Lucifer's Legacy: The Meaning of Asymmetry, Oxford: Oxford University Press, p. 228; Kane, G. (2000) Supersymmetry: Unveiling the Ultimate Laws of Nature, Cambridge, MA: Perseus; Nabokov, V. (1974 [947]) Bend Sinister, Harmondsworth: Penguin, pp. 144–5. ☞www☜

15. Poe, E. A. (1841) Never Bet the Devil Your Head: A Tale with a Moral, in Mabbott, T. O. (ed.) (1978) Collected Works of Edgar Allan Poe: Tales and Sketches 1831–1842, Cambridge, MA: Belknap Press, p. 619.

16. Kögl, F. and Erxleben, H. (1939) Nature, 144: 111; Miller, J. A. (1950) Cancer Research, 10: 65–72; Clemmit, M. (1991) The Scientist, August 19th: 1. ☞www☜

17. Harris, L. J. (1991) Brain and Language, 40: 1–50; Harrington, A. (1987) Medicine, Mind, and the Double Brain; a Study in Nineteenth-Century Thought, Princeton, NJ: Princeton University Press, pp. 16–17.

18. Broca, P. (1865) Bulletin de la Société d'Anthropologie de Paris, 6: 377–93; Harris, L. J. (1991) p. 37.

19. Jackson, J. H. (1874) Medical Press and Circular, i: 19, 41, 63; Harrington, A. (1987) op. cit., pp. 223–4. ☞www☜

20. Curie, P. (1894) Journal de Physique Théorique et Appliquée, 3rd Series, 3: 393–415; Altmann, S. L. (1992) p. 28; Van Fraassen, B. C. (1989) op. cit., p. 240. ☞www☜

21. Zupko, J. (1998) in Craig, E. (ed.), Routledge Encyclopaedia of Philosophy, London: Routledge. ☞www☜

22. Stewart, I. and Golubitsky, M. (1992) op. cit., p. 58. ☞www☜

23. Wenderoth, P. (1995) Spatial Vision, 9: 57–77. ☞www☜

24. Julesz, B. (1969) in Reichardt, W. (ed), Processing of Optical Data by Organisms and by Machines, New York: Academic Press, pp. 580–88.

25. Wagemans, J. (1995) Spatial Vision, 9: 9–32; Mach, E. (1910) Popular Scientific Lectures (4th edn), Chicago: Open Court, pp. 89–106; Tyler, C. W. (1995) Spatial Vision 8: 383–91; Ibid., 9: 1–7; Herbert, A. M. and Humphrey, G. K. (1996) Perception, 25: 463–80; Corballis, M. C. and Beale, I. L. (1976) pp. 74–88; Leone, G. et al. (1995) Spatial Vision, 9: 127–37.

26. Pascal, Pensées, section 28.

27. Heilbronner, E. and Dunitz, J. (1992) *Reflections on Symmetry: in Chemistry ... and Elsewhere*, Weinheim: VCH.

Chapter 15: The world, the small, the great

1. Weyl, H. (1952) *Symmetry*, Princeton, NJ: Princeton University Press, Fig. 4; Swindler, M. H. (1929) *Ancient Painting: From the Earliest Times to the Period of Christian Art*, New Haven: Yale University Press. ☞www☜

2. Barrow, J. D. and Silk, J. (1993) *The Left Hand of Creation*, (revised edition), New York: Oxford University Press, pp. xxiii–xxiv.

3. Weisskopf, V. F. (1969) in Engström, A. and Strandberg, B. (eds), *Symmetry and Function of Biological Systems at the Macromolecular Level*, Stockholm: Almqvist and Wiksell, pp. 35–9 (p. 38); Weyl, H. (1952) op. cit., pp. 64–5.

4. Feynman, R. P. *et al.* (1963) *The Feynman Lectures on Physics. Vol I: Mainly Mechanics, radiation, and heat*, Reading, MA: Addison-Wesley, p. 52–12. ☞www☜

5. Frey, D. (1949) *Studium Generale* 2: 268–78; Weyl, H. (1952) op. cit.; Riese, W. (1949) *Bulletin of the History of Medicine*, 23: 546–53, (p. 552); p. 50. ☞www☜

6. Shelstan, A. (1971) *Thomas Carlyle: Selected Writings*, London: Penguin. ☞www☜

7. Stewart, I. and Golubitsky, M. (1972) *Fearful Symmetry: Is God a Geometer?*, Oxford: Blackwell, pp. 178, 182; Corballis, M. C. and Beale, I. L. (1976) *The Psychology of Left and Right*, Hillsdale, NJ: Lawrence Erlbaum Associates, p. 197; Bernstein, J. (1962) *New Yorker*, 12 May: 49–104, (p. 72).

8. Asimov, I. (1976) *The Left Hand of the Electron*, London: Panther, p. 76; Gardner, M. (1990) *The New Ambidextrous Universe: Revised Edition* New York: W. H. Freeman, p. 192.

9. Chothia, C. (1991), in Bock, G. R. and Marsh, J. (eds), *Biological Asymmetry and Handedness* (Ciba foundation symposium 162), Chichester: Wiley, pp. 36–57; Wolpert, L. (1991) in *Biological Asymmetry and Handedness*, see above.

10. Altick, R. D. (1971) *Robert Browning: The Ring and the Book*, Harmondsworth: Penguin, Book I, line 828. ☞www☜

11. Elytis, O. (1997) *The Collected Poems of Odysseus Elytis*, translated by J. Carson and N. Sarris, Baltimore MD: Johns Hopkins University Press. ☞www☜

PICTURE AND TEXT CREDITS

The author and publisher have either sought or are grateful to the copywright holders for permission to reproduce the following illustrations and quotations. Weidenfeld & Nicolson apologise for any unintentional omissions and, if informed of any such cases, would be pleased to update future editions.

Figure 1.2 Sir Thomas Watson: reprinted with permission of the Wellcome Museum, London.

Figure 1.3 Louise Pasteur (1852): reprinted with permission of the Institut Pasteur, Paris.

Figure 1.4 b Reprinted with permission of the Institut Pasteur, Paris.

Figure 1.5 From the architectural notebooks of Viollet-le-Duc, reproduced from Cook, T. A. (1914) *The Curves of Life*, London: Constable.

Figure 1.6 From the architectural notebooks of Viollet-le-Duc, reproduced from Cook, T. A. (1914) *The Curves of Life*, London: Constable.

Figure 1.7 Paul Broca.

Figure 1.8 Dr Juster: Institut de Parasitologie, École Pratique, Paris.

Figure 1.9 Dr Juster: Institut de Parasitologie, École Pratique, Paris.

Figure 2.1 Robert Hertz: Institute of Social Anthropology, University of Oxford.

Figure 2.2 Adapted from Kraig, B. (1978) *Journal of Indo-European Studies*, 6: 149–172.

Figure 2.4 Adapted from Das, T. (1945) *The Purums: An Old Kuki Tribe of Manipur*, Calcutta: University of Calcutta Press.

Figure 2.5 Adapted from Sattler, J. B. (2000) *Links und Rechts in der Warnehmung des Menschen: Zur Geschichte der Linkshändigkeit*, Donauwörth: Auer Verlag.

Figure 3.1 Reproduced from Huxley, J. (1970) *Memories*: Allen and Unwin.

Figure 3.3 Ötzi necklace: reproduced with permission of the Römisch-Germanischez Zentralmuseum, Mainz. Photo: Christin Beeck.

Figure 3.4 Reproduced with permission from Stead, I. and Rigby, V. *Iron Age Antiques from Champagne in the British Museum: The Morel Collection*, London: British Museum Press.

Figure 3.5 Reproduced from Cook, T. A. (1914) *The Curves of Life*, London: Constable.

Figure 3.6 Reproduced from Cook, T. A. (1914) *The Curves of Life*, London: Constable.

Figure 4.3 Hale-Bopp comet photo: reproduced with kind permission of Howard C. Taylor.

Figure 4.4 Reproduced with permission from Ofte, S. H. and Hugadl, K. (2001) *Journal of Experimental and Clinical Neuropsychology* (in press).

Figure 4.5 Reproduced with permission of Taylor & Francis Ltd from Brandt, J. and Mackavey, W. (1981) *Journal of Neuroscience* 12: 87–94.

Figure 4.6 Reproduced with permission of Taylor & Francis Ltd from Brandt, J. and Mackavey, W. (1981) *Journal of Neuroscience* 12: 87–94.

Figure 4.7 Adapted from Spalding, J. M. K. and Zangwill, O. L. (1959) *Journal of Neurology, Neurosurgery and Psychiatry* 13: 24–9.

Figure 4.8 Reproduced with permission of Oxford University Press from Turnbull, H. and McCarthy, R. A. (1996) *Neurocase* 2: 63–72.

Figure 4.9 Reproduced with permission of Oxford University Press from Turnbull, H. and McCarthy, R. A. (1996) *Neurocase* 2: 63–72.

Figure 4.10 Reproduced from Wood Jones, F. (1919) *The Principles of Anatomy as Seen in the Hand*, London: Balliere, Tindall and Cox.

Figure 4.11 Reproduced with permission from Tischbein, J. H. W. *Goethe in der römischen Campagna*, Frankfurt: Städelsches Kunstinstitut. Copyright Ursula Edelmann.

Figure 5.1 Wellcome Library, London.

Figure 5.2 Reprinted with permission from Ruehm, S. G. *et al.* (2001) *Lancet* 357: 1086–1091.

Figure 5.3 Reproduced with permission from Bock, G. R. and Marsh, J. (eds) *Biological Asymmetry and Handedness* (Ciba Foundation symposium 162), Chichester: John Wiley, p. 99.

Figure 5.4 Reproduced with permission from Jefferies, R. P. S. (1986) *The Ancestry of Vertebrates*, London: British Museum (Natural History). Copyright (2001) The Natural History Museum, London.

Figure 5.5 Reproduced with permission of Taylor and Francis Ltd from Sutcliffe, O. E., *et al.* (2000) *Lethaia* 33: 1–12, p. 5.

Figure 5.6 Reproduced with permission from Jefferies, R. P. S. (1986) *The Ancestry of Vertebrates*, London: British Museum (Natural History). Copyright (2001) The Natural History Museum, London.

Figure 5.7 Redrawn from Gee, H. (1996) *Before the Backbone: Views on the Origins of Vertebrates*, London: Chapman and Hall, p. 217.

Figure 5.8 Reproduced from Norman, J. R (1931) *A History of Fishes*, London: Ernest Benn.

Figure 5.9 Reprinted with permission from Capdevila, J., *et al.* (2000) *Cell* 93: 9–21. Copyright (2000) Elsevier Science.

Figure 5.10 Reproduced with kind permission of Gillian Rhodes, University of Western Australia from Bruce, V. and Young, A. (1998) *In the Eye of the Beholder: The Science of Face Perception*, Oxford: Oxford University Press. Copyright G. Rhodes.

Figure 5.11 Reproduced with permission from Kessel, R. G. (1974).

Figure 5.12 Reprinted with permission from Hyatt, B. A. and Yost, H. J. (1998) *Cell*, 93: 37–46. Copyright (1998) Elsevier Science.

Figure 5.13 Reproduced with permission from Gilbert, S. G. (2000) *Developmental Biology*, 6th edn. Copyright Sinauer Associates.

Figure 5.14 Hans Spemann.

Figure 5.14 a, Reproduced from Spemann, H. and Falkenberg. H. (1919) *Wilhelm Roux' Archiv fur Entwicklungsmechanick* 45: 371–422.

Figure 5.14 b, Reproduced from Spemann, H. and Falkenberg. H. (1919) *Wilhelm Roux' Archiv fur Entwicklungsmechanick* 45: 371–422.

Figure 5.14 c, Reproduced from Spemann, H. and Falkenberg. H. (1919) *Wilhelm Roux' Archiv fur Entwicklungsmechanick* 45: 371–422.

Figure 5.15 d, Reproduced with permission of Cambridge University Press from Huxley, J. S. and de Beer, G. R. (1934) *The Elements of Experimental Biology*.

Figure 5.16 Reproduced with permission of the University of Chicago Press from Pennarun, G., *et al.* (1999) *American Journal of Human Genetics* 65: 1508–19.

Figure 5.17 Photos courtesy of Professor Nigel Brown, St. George's Hospital Medical School, London.

Figure 5.18 Reprinted with permission from Levin, M., *et al.* (1995) *Cell* 93: 803–814. Copyright (1995) Elsevier Science.

Figure 5.19 Reproduced with permission from Vogan, K. J. and Tabin, C. J. (1999) *Nature* 397: 295–8.

Figure 5.20 Adapted from Okada, Y., *et al.* (1999) *Molecular Cell* 4: 459–68.

Figure 6.2 Reproduced with permission of the California Institute of Technology from Feynman, R. P., *et al.* (1963) The *Feynman Lectures on Physics. Vol. I: Mainly Mechanics, Radiation and Heat*, Reading, MA: Addison-Wesley.

Figure 6.3 Reproduced with permission of the Royal Swedish Academy of Sciences.

Figure 7.1 Darwin and William: reproduced with permission of English Heritage.

Figure 7.2 Based with permission on unpublished data of Nigel Sadler.

Figure 7.3 Based on Tapley, S. M. and Bryden, M. P. (1985) *Neuropsychologia* 23: 215–21.

Figure 7.6 Reproduced from Schaeffer, A. A. (1928) *Journal of Morphology* 45: 293–398.

Figure 8.1 a Pasteur (1892): reprinted with permission of the Instut Pasteur, Paris.

Figure 8.1 b Mach (circa 1913): reprinted with permission of Peter Indefrey from Hagoort, P., *et al.* (1999) *The Neurocognition of Language*, Oxford: Oxford University Press, pp. 273–316.

Figure 8.6 Reprinted with permission of the publishers from Riddoch, M. J . and Humphreys, G. W. (1995) *Birmingham Object Recognition Battery*, Hove: Lawrence Erlbaum, p. 150.

Figure 8.7 Reprinted from Gazzaniga, M. S. (1985) *The Social Brain: Discovering the Networks of the Mind*, New York: Basic Books.

Figure 8.8 Reprinted with permission of Masson Italia and Professor Della Sala from Cantagallo, A. and Della Sala, S. (1998) *Cortex*, 34: 163–189.

Figure 8.10 Reprinted with permission from Marshall, J. C. and Halligan, P. W. (1993) *Nature*, 364: 193–194.

Figure 8.11 Reproduced with permission of Springer-Verlag from Marshall, J. C. and Halligan, P. W. (1993) *Journal of Neurology*, 240: 37–40.

Figure 8.12 Reprinted with permission of Masson Italia and Professor Della Sala from Cantagallo, A. and Della Sala, S. (1998) *Cortex*, 34: 163–189.

Figure 8.13 Reprinted with permission of Elsevier Science from Halligan, P. W. and Marshall, J. C. (1997) *Lancet* 350: 139–140.

Figure 8.15 Adapted from Jaynes, J. (1976) *The Origin of Consciousness in the Breakdown of the Bicameral Mind*, Harmondsworth: Penguin Books.

Figure 9.1 Based on data from Gilbert, A. N. and Wysocki, C. J. (1992) *Neuropsychologia*, 30: 601–608.

Figure 9.2 Based on data from Coren, S. and Porac, C. (1977) *Science* 198: 631–632.

Figure 9.3 Reproduced with permission from Painter, K. S. (1977) *The Mildenhall Treasure: Roman Silver from East Anglia*, London: British Museum. Copyright The British Museum.

Figure 9.4 Reprinted with permission of Masson Italia from Byrne, R. W. and Byrne, J. M. (1991) *Cortex*, 27: 521–546,

Figure 9.5 Reproduced with permission from Schultz, A. H. (1969) *The Life of Primates*, London: Weidenfeld and Nicolson.

Figure 9.6 Reproduced with permission of the American Physiological Society from Hore, J., *et al.* (1996) *Journal of Neurophysiology*, 76: 3693–3704.

Figure 10.1 Thomas Carlyle (1857): Hulton Archive.

Figure 10.5 Based on data from Kincaid, P. (1986) *The Rule of the Road: An International Guide to History and Practice*, New York: Greenwood Press.

Figure 10.6 Based on data from Kincaid, P. (1986) *The Rule of the Road: An International Guide to History and Practice*, New York: Greenwood Press.

Figure 11.1 Sir Osbert Sitwell: reproduced with kind permission of Sir Reresby Sitwell.

Figure 11.2 Vincent van Gogh *The Potato Eaters*: reproduced with permission of the Vincent van Gogh Museum (Vincent van Gogh Foundation), Amsterdam.

Figure 11.3 Based on data from Orton, H., *et al.* (1978) *The Linguistic Atlas of England*, London: Croom Helm.

Figure 12.1 Reprinted with permission of Blackwell Science from Burnett, A. C. and Thompson, D. G. (1986) *Medical Education*, 20: 424–31.

Figure 12.1 Sir Thomas Browne.

Figure 12.2 Adapted from Levy, J. and Reid, M. (1976) *Science*, 194: 337–9.

Figure 12.4 Copyright the President and Fellows of Queens' College Cambridge. By permission.

Figure 13.1 Billy the Kid: reproduced by permission of the A & S Upham Collection.

Figure 13.2 Reproduced by permission. Leonardo, RL 12278, RL 12682. Copyright (2001) The Royal Collection, Her Majesty Queen Elizabeth II.

Figure 13.3 Rembrandt *The Three Trees*: Teylers Museum, Holland.

Figure 14.4 Reproduced from Balmelle, C., *et al.* (1985) *Le Décor Géometrique de la Mosaïque Romaine*, Paris: Picard.

Figure 14.5 Detail of Raffia cloth from Kuba in Zaire: copyright the Africa-Museum Tervuren, Belgium.

Figure 14.6 Reproduced from Polya, G. (1924) *Zeitschrift für Kristallographie*, 60: 278–88.

Figure 14.9 Reproduced with permission from Julesz, B. (1969) in Reichardt, W. (ed) *Processing of Optical Data by Organisms and by Machines*, New York: Academic Press.

Figure 15.1 a, Reproduced from Swindler, M. H. (1929) *Ancient Painting: From the Earliest Times to the Period of Christian Art*, New Haven: Yale University Press.

All original artwork © I. C. McManus.

Extract from Sacks, O. *The Man Who Mistook his Wife for a Hat*: reproduced by permission of the Wylie Agency. Copyright (1985) Oliver Sacks.

Extract from Auden, W. H. *Horace Canonicae*: reproduced with permission of Faber and Faber.

Table, chapter 10: reproduced with permission from Mair, P. (21 February 2000) *Guardian*, p. 15. Copyright *Guardian*.

Extract from Dalrymple, T. (1995) *So Little Done: the Testament of a Serial Killer*, London: Andre Deutsch, pp. 92–4: reprinted with permission of Andre Deutsch Limited.

Extract from Anderson, J. (1997) *Toward the Liberation of the Left Hand*: reprinted by permission of the University of Pittsburgh Press. Copyright (1977) Jack Anderson.

Extract from *Oranges are not the Only Fruit*: reprinted by permission of International Creative Management, Inc. Copyright (2000) Jeanette Winterson.

Extract from Feynman, R. P., *et al.* (1963) *The Feynman Lectures on Physics, Vol. I: Mainly Mechanics, Radiation and Heat*, Reading, MA: Addison-Wesley: reprinted with permission of the California Institute of Technology.

Extract from Carson, J. and Sarris, N. (1997) *The Collected Poems of Odysseus Elytis*, Baltimore MD: Johns Hopkins University Press. Copyright Johns Hopkins University Press.

INDEX